The Calculus PRIMER

William L. Schaaf

Dover Publications, Inc.
Mineola, New York

Bibliographical Note

This Dover edition, first published in 2011, is an unabridged republication of *The Calculus,* which was originally published in 1963 by Doubleday and Company, Inc., New York. The main text of this book begins on page 13; no pages have been omitted.

Library of Congress Cataloging-in-Publication Data

Schaaf, William Leonard, 1898–
    The calculus primer / William L. Schaaf
        p. cm.
    Originally published: The calculus. Garden City, N.Y. : Doubleday, 1963.
    Includes bibliographical references and index.
    ISBN-13: 978-0-486-48579-9 (alk. paper)
    ISBN-10: 0-486-48579-X (alk. paper)
    1. Calculus. I. Schaaf, William Leonard, 1898– Calculus. II. Title.

QA303.2.S37 2011
515—dc23

20110171575

Manufactured in the United States by Courier Corporation
48579X02
www.doverpublications.com

PREFACE

It is our express purpose in this book to present a comprehensive yet concise introductory course in the Differential and Integral Calculus, covering all the topics usually included in a first course.

We have endeavored to achieve clarity of exposition, and have introduced many carefully worked out model examples. A generous number of diagrams supplements the development throughout, and in addition there is a special section of curves for reference toward the end of the book.

The development of the subject is direct and straightforward, without the distraction of unimportant details. Consequently, the demands of mathematical rigor, though fully recognized, have been less stressed in this presentation.

Since the needs of the beginner have constantly been kept in mind, the emphasis is placed upon fundamental concepts: functions, derivatives, differentiation of algebraic functions, differentiation of transcendental functions, partial differentiation, indeterminate forms, general and special methods of integration, the definite integral, partial integration, and other fundamentals. Ample exercises have been furnished after each important topic to enable the student to test his grasp of the subject matter before he proceeds to the next topic. Thus, the book is appropriate not only for classroom use, but also notably for review purposes and for home study.

The deliberate choice of an informal style will, it is hoped, hold the reader's attention and encourage him to master the Calculus, perhaps the most basic area of mathematics in the world today, whether applied to the physical sciences, the social studies, or the behavioral sciences.

W. L. S.

CONTENTS

Preface   5

Introduction   13

One   **Functions, Rates, and Limits**   18

Variables and Functions, 18; Average and Instantaneous Rates, 27; The Limit Concept, 32; Some Special Limits, 42.

Two   **The Derivative of a Function**   47

Increment Notation, 47; The Meaning of the Derivative, 50; Differentiation: Finding the Derivative, 54.

Three   **Differentiation of Algebraic Functions**   60

The Derivative of a Constant, a Variable, and a Sum, 60; Derivative of the Power Function, 64; Derivative of Products and Quotients, 67; Differentiation of Implicit Functions, 73.

Four   **Using the Derivative**   77

The Derivative as a Tool, 77; Instantaneous Rates of Change, 79; Distance, Velocity, and Acceleration, 81; Maxima and Minima, 87.

Five   **Differentiation of Transcendental Functions**   99

Derivatives of Logarithmic Functions, 99; Derivatives of Exponential Functions, 106; Derivatives of Trigonometric Functions, 112; Derivatives of the Inverse Trigonometric Functions, 116.

Six   **Further Applications of the Derivative**   126

Slopes, Tangents, and Normals, 126; Points of Inflection and Curve Tracing, 133; Parametric Equations, 139; Rectilinear and Circular Motion, 143; Related Time Rates, 148.

Seven   **Differentials**   155

Increments and Infinitesimals, 155; Using Differentials, 160; Summary of Differential Notation, 160; Approximate Calculations, 163.

Eight   **Curvature**   169

Length of Arc, 169; Meaning of Curvature, 173; Circle of Curvature, 179; The Evolute, 187.

Nine   **Indeterminate Forms**   191

Theorem of Mean Value, 191; Evaluation of Indeterminate Forms, 194.

Ten   **Partial Differentiation**   206

Partial Derivatives, 206; The Total Derivative, 211; Significance of
Partial and Total Derivatives, 214; Singular Points of a Curve, 217.

Eleven   **Expansion of Functions**   222

Infinite Series and Sigma Notation, 222; Tests for Convergence and
Divergence, 227; Power Series, 235; Expansion of Functions, 238;
The Value of π; Euler's Formulas, 247.

Twelve   **General Methods of Integration**   252

Integration as the Inverse of Differentiation, 252; Fundamental
Principles of Integration, 255; Standard Elementary Integral Forms,
264.

Thirteen   **Special Methods of Integration**   275

Integration by Parts, 275; Trigonometric Integrals, 279; Integration
by Substitution; Change of Variable, 283; Tables of Integrals, 288;
Integration of Rational Fractions, 290.

Fourteen   **The Definite Integral**   299

Integration Between Limits, 299; Area Under a Curve, 303; The
Definite Integral and Its Limits, 307; Derived Curves and Integral
Curves, 313.

Fifteen   **Integration as a Process of Summation**   318

A Basic Principle, 318; Areas of Plane Curves, 322; Length of a
Curve, 329; Solids of Revolution, 332.

Sixteen   **Successive and Partial Integration;**
            Approximate Integration   338

Multiple Integrals, 338; Areas and Volumes, 344; Approximate Integration, 350; Practical Applications, 360.

Tables   371

Integrals, 371; Values of e^x and e^{-x}, 379; Common Logarithms, 380.

Reference Material from Other Branches of Mathematics   383

Algebra, 383; Geometry, 387; Trigonometry, 388; Plane Analytic
Geometry, 392; Curves for Reference, 394; Solid Analytic Geometry, 402; Quadric Surfaces, 402; Greek Alphabet, 404.

Answers to Problems   405

Index   433

INTRODUCTION

The creation of Analytic Geometry by Descartes in the early part of the seventeenth century was a milestone of tremendous significance. Indeed, it was the beginning of modern mathematics in the broad perspective of history. ("Modern" mathematics in the sense of contemporary mathematics did not commence until about 1900, with the advent of functional analysis, abstract spaces, set theory, and symbolic logic.) This great step of recognizing the relation between the numbers of algebra and the entities of geometry once having been taken, it is perhaps not surprising that further advances were soon to follow, culminating in the invention of the Calculus by Isaac Newton and by Gottfried Leibniz. This was a classic illustration of nearly simultaneous, but presumably independent creation. In this instance the invention was attended with unfortunate and, at times, bitter dispute. Newton had been using his new method of fluxions since 1674 or even earlier, but he did not publish a direct statement of his method until 1687. Leibniz, using a different approach and a different symbolism, published his first statement in 1684. Each accused the other of plagiarism, yet from the vantage point of historical perspective there can be little doubt that each must be credited with independent creation.

What is perhaps more important is the fact that the ground had been prepared for this new advance by many predecessors over a long period of time. There are traces of the methods of the Calculus discernible even among the Greeks. Thus Democritus (c.400 B.C.) suspected the relation that exists between the volumes of cones and cylinders, although he was unable to prove it. But the influence of Zeno's famous paradoxes about infinitely small quantities discouraged the further use of infini-

tesimal quantities from the study of geometry. Shortly thereafter
another forerunner of the Calculus came into prominence, the method
of exhaustions. This was first used in connection with the circle, and
consisted essentially of doubling and redoubling the number of sides of
a regular inscribed polygon on the assumption that as the process of
doubling is continued indefinitely, the difference in area between the
circle and the polygon would finally be "exhausted." The method was
later extended and refined by Eudoxus (c.360 B.C.). The nearest approach to the process of summation, however, was made by Archimedes
(c.225 B.C.) in connection with his study of the area under a parabola
and certain other curves, as well as the volumes of certain well-known
solids.

No further step toward the Calculus was taken until during or after
the Middle Ages. The story is a long and interesting one, and only a
few of the high spots can be touched upon here. The astronomer
Kepler (c.1610) developed certain crude methods of integration, in
connection with problems of gauging, by which the volume of a solid
was regarded as composed of many infinitely small cones or infinitely
thin disks. His contemporary, the Jesuit priest Bonaventura Cavalieri, made considerable use of the so-called method of indivisibles, in
accordance with which he regarded a surface as the smallest element of
a solid, a line as the smallest element of a surface, and a point as the
smallest element of a line. Although the method lacked logical rigor
and was essentially intuitive in nature, the results obtained by its application were valid. Another contemporary, Roberval, using a similar
method, but employing the device of an infinite number of infinitely
narrow rectangular strips, succeeded in solving many problems of finding
lengths of curves as well as areas and volumes.

About 1636 the French mathematician Fermat extended the methods
of his predecessors, attacking also the problem of finding maximum and
minimum values of curves, focusing attention on the determination of
tangents to a curve. In the course of his work, Fermat enunciated the
general principle for determining a tangent, substantially equivalent to
the modern method of setting the derivative of a function equal to zero.
Thus Fermat may also, in a very real sense, be regarded as a co-inventor
of the Calculus.

Two other fertile minds contributed to the pregnant atmosphere of
this eventful period—John Wallis and Isaac Barrow, both British
mathematicians. Wallis' *Arithmetica Infinitorum*, which appeared in
1655, showed how to apply methods of summation to the determination
of the areas of triangles, the lengths of spirals, and the volumes of

paraboloids and other solids. Wallis freely acknowledged his indebtedness to Torricelli, Cavalieri, Christopher Wren, and others. While Wallis was devoting his efforts to problems of integration, Barrow (1663) drew attention to the problem of tangents and methods of differentiation. He recognized the fact that integration is the inverse of differentiation.

Newton at various times employed three different methods of procedure. At first he made some use of infinitely small quantities, but soon recognized that this was not a sound basis on which to operate. His second method, the method of fluxions, was a unique contribution. According to the concept of fluxions, a curve was considered as being generated by a moving point. The change in the position of a point in an "infinitely short" time was called its "momentum"; this momentum divided by the infinitely short time was the "fluxion." If the "flowing quantity" was x, its fluxion was denoted by $\dot{x}$. In our notation, if x is the function $f(t)$ of the time t, Newton's "$\dot{x}$" is our $\dfrac{"dx"}{dt}$; his $\ddot{x}$ is our $\dfrac{d^2x}{dt^2}$; etc. Nevertheless, Newton eventually was dissatisfied with his own device of fluxions, and toward the latter part of his career he attempted to refine the method by a theory of limits and the notion of continued motion, or continuity.

Whereas Newton was impelled toward the Calculus by his burning interest in experimental and applied science, Leibniz was drawn to the Calculus via a purely intellectual and philosophical route. In brief, Leibniz was seeking a universal system of symbolic reasoning. Just as the philosopher-mathematician Descartes had reduced all geometry to a universal method or system, so the philosopher-mathematician Leibniz hoped to reduce all reasoning of whatever sort to a "universal characteristic," or as we might say today, to a symbolic logic.

Unfortunately for the development of mathematics in England for the next hundred years, the notation introduced by Newton was not as convenient or as significant as the symbolism used by Leibniz; stubborn adherence to the geometric methods and fluxional notation of Newton retarded further advances by British mathematicians for nearly two or three generations. Eventually, however, the method and notation of Leibniz became universal, but not until they had been popularized on the continent by such writers as L'HOSPITAL, D'ALEMBERT, and the BERNOULLIS. Thereafter continued advances were made with breathtaking strides, heralded by the extension and ever-wider applications made by such brilliant minds as those of EULER, LAMBERT, LAGRANGE, LAPLACE, and LEGENDRE.

In retrospect, then, modern mathematics may be said to have begun at the start of the seventeenth century and advanced in five major directions: (1) the Analytic Geometry of Descartes and Fermat; (2) the Calculus of Newton and Leibniz; (3) the probability theory of Fermat and Pascal; (4) the higher arithmetic of Fermat; and (5) the mechanics of Galileo and Newton. In a very real sense we may say that modern physical science and technology are the direct product of the experimental method espoused by Galileo combined with the Calculus created by Newton and Leibniz.

The beginning of the nineteenth century saw more refinement and elaboration of the methods of the Calculus, as well as the infusion of greater rigor into the basis of analysis. Outstanding contributions in this connection include the work of Cauchy, Gauss, Jacobi, Abel, and Dirichlet. By the end of the century, modern analysis, a mighty superstructure resting securely upon solid foundations, had been essentially completed—a triumph of modern thought. However, the triumph was short lived. The dawn of the twentieth century witnessed the recurring struggle to understand the infinite, together with renewed desire to create an even more rigorous and unassailable foundation for all mathematics, and for analysis in particular. The magnificent efforts of Cantor, Kronecker, Dedekind, Weierstrass, Brouwer, Hilbert, Frege, Bertrand Russell, A. N. Whitehead, and others constitute an epic saga in the history of mathematics. The final chapter has not yet been written. In the deceptive nearness of history in the making, Gödel's theorem of 1931 is disconcerting, to say the least. The theorem shows that it is impossible to prove the consistency of a system S of what is essentially modern mathematics by the methods of proof used within the system S.

The Calculus is a powerful but subtle mathematical tool. It is not child's play. There is an honest difference of opinion as to whether the Calculus should be preceded by a more or less thorough course of study of analytic geometry. It is generally conceded that anything like a complete systematic course in analytics is not a necessary prerequisite. Yet a certain knowledge of fundamentals is indispensable.

In exploring the Calculus, the beginner will face a rather novel experience—that of dealing with limiting processes. He should not be discouraged if at first (or even for some time thereafter) the processes concerning limits should seem somewhat vague and elusive. To be sure, the terms "infinity" and "limit" may be troublesome for a while, and can easily be misunderstood. These two concepts—that of the infinite and that of a limit—are subtle ideas which may well be consistently and meaningfully used before (or instead of) being formally defined. He should not allow himself to think about "infinitely small quantities," but to think rather of definite intervals, however large or small, and to regard a limit as an approximation, as a property of the neighborhood in which the limit lies. These ideas cannot be grasped quickly; one must work with them for some time. The reward of such patience will be most gratifying.

Functions, Rates, and Limits

CHAPTER ONE

VARIABLES AND FUNCTIONS

1—1. The Calculus. Emerson once said that it didn't matter much where a man was, so long as you knew the direction in which he was moving. In somewhat the same way, the significance of a graph or curve often lies not so much in what height a point on the curve has reached, as it does in how fast its height is changing, and whether it is increasing or decreasing.

These observations suggest, in a crude way to be sure, the keynote of the Calculus. In the study of Analytic Geometry our concern is with the interrelations between equations and loci; we examined the properties of "static" curves, and studied the relations between the variables at any given time. In the Calculus, on the other hand, we shall be concerned chiefly with change rather than with "frozen figures." Speaking rather broadly, the Calculus enables us to understand the nature of changes, their magnitude and their rate; we focus the searchlight on the anatomy of change, revealing the behavior of related variables.

1—2. Variables and Constants. Consider the statement $y = 2x + 5$. This is an open sentence; that is, it is neither true nor false. Clearly we can substitute many numbers for the letter x: for example, when x is assigned the number 6, the number that then corresponds to y is 17; when x is -4, y then becomes -3; etc. When symbols are used in this way, we call them variables. Usually letters in the latter part of the alphabet, such as s, t, u, v, w, x, y, and z, and also the Greek letters ρ, θ, and ω are used for this purpose. A *variable*, then, is a symbol used to represent any one of a given set of numbers.

A *constant*, on the other hand, is a symbol denoting one particular number of a set. The constant may be represented by a specific numeral, such as the "2" and the "5" in the sentence $y = 2x + 5$; or the constant may be represented by a "general numeral," i.e., a letter, such as the "m" and the "b" in the sentence $y = mx + b$. In any event, the "value" or meaning of a constant is regarded as *not changing throughout a given discussion*. Constants are usually represented by letters in the early part of the alphabet, as, for example, a, b, c, h, k, m, n, p, or by Greek letters such as α, β, γ, ϵ, λ, and μ.

An absolute constant is a constant whose value never changes; examples are 8, -100, $\frac{3}{4}$, $\sqrt{5}$, log 2, π, e. An *arbitrary constant* is one whose value changes from one problem to another, but does not change during the course of any given discussion, as in the equations $x^2 + y^2 = r^2$, $y = mx + b$, $y = a \sin bx$, and $y = \log_a x$, where a, b, m, and r are constants. The *absolute value* of a constant or a variable is its *arithmetical* value, irrespective of its algebraic sign, and is designated as $|a|$. Thus, $|-5| = 5$; $|3| = 3$; $|k| = \sqrt{k^2}$.

1—3. Meaning of Function. If two variables x and y are so related that, whenever a value is assigned to x, there is automatically assigned, by some rule or correspondence, a value to y, we call y a *single-valued function* of x. Some relations admit of assigning two (or more) values to y whenever we assign a value to x. For example, consider the relation $y^2 = 2x$; when $x = 2$, $y = +2$, and also -2; when $x = 12\frac{1}{2}$, $y = +5$ as well as -5; when $x = 50$, $y = \pm 10$; when $x = 10$, $y = \pm 2\sqrt{5}$; etc. Such a relation is sometimes called a *multiple-valued function*, or more properly, simply a *relation* between variables. In any event, the variable x, to which values are assigned at will, is known as the *independent* variable; the variable y, whose corresponding values depend upon the value chosen for x, is known as the *dependent* variable. The permissible values that may be assigned to x constitute the *domain of the function*, or relation; the set of corresponding values taken on by y constitutes the *range of the function*, or relation.

It should be noted, in connection with functions, that corresponding values of the variables may be regarded as *ordered pairs*. Any value of the independent variable, say x_1, is written first, and the value of the dependent variable which corresponds to it, say y_1, is written second. Thus we have an ordered pair of numbers: (x_1, y_1).

A function, then, consists of two things:

(1) A collection of numbers, called the domain of definition of the function.

(2) A rule, verbal or symbolic, that assigns to each number in the domain of definition one and only one number.

The range of values of a function is the collection of all the numbers which the rule of the function assigns to the numbers in the domain of definition of the function. Note that in a function, when the first element of an ordered pair is known, there is no ambiguity as to what the second element is, since only one ordered pair in a given function has the specified first element. The fact that different first elements may have the same second element is immaterial. In other words, a function is a special kind of relation in which the first element of every ordered pair has a unique second element.

These ideas will become clearer from the following. In the relation $y = x^2 - 4$, we say that "y is a function of x," since to any x selected

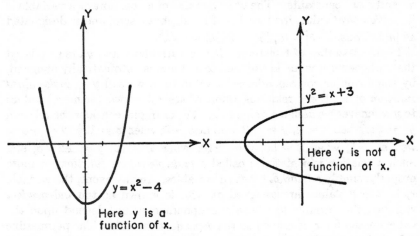

there corresponds one and only one y. (The fact that two different x's may yield the same y-value doesn't matter.) On the other hand, in the relation $y^2 = x + 3$, we do not say that y is a function of x, since to any x selected there correspond *two* y-values.

1—4. Explicit and Implicit Functions. Let us look at these two relations a little more closely. The relation $y = x^2 - 4$ has been "solved" for one variable in terms of the other. We generally refer to this kind of a relation as an *explicit function*. The relation of y to x is explicitly stated; for every x there is a unique y. Other examples of explicit functions are:

(1) $y = 3 \sin x$; y is an explicit function of x.
(2) $v = t^2 + 5t - 3$; v is an explicit function of t.
(3) $x = \pm\sqrt{a^2 - y^2}$; x has an explicit relation to y.
(4) $z = a + x \cos^2 y$; z has an explicit relation to x and y.

Statements (1) and (2) are examples of functions; (3) and (4) are examples of relations (or multiple-valued "functions"). In all four statements the relation has been solved explicitly for the dependent variable.

Now let us look at $y^2 = x + 3$ again. As it stands, it has not been solved for either variable. If we solve for y, we get

$$y = \pm \sqrt{x + 3},$$

where y is not a function of x, but it is stated as having an explicit relation to x. If we solve it for x, we get

$$x = y^2 - 3,$$

and we see that x is a function of y. Thus:

written $x = y^2 - 3$, x is an *explicit* function of y;
written $y^2 = x + 3$, x is an *implicit* function of y.

In other words, an equation involving x and y may not express y explicitly in terms of x, as for example:

(1) $xy = 10$
(2) $y^2 = x$
(3) $x^2 + y^2 = 9$
(4) $x^2 - xy + y^2 = 4$

However, each of these equations defines y in the sense that for any given value of x, there correspond one or more values of y. Such equations are examples of *implicit* functions, or implicit relations; equation (1) is an implicit function, and the remaining three equations are implicit relations (or multiple-valued "functions").

The graph of a function consists of all points on the graph of the equation for which the x-coordinate is in the domain of the function.

EXERCISE 1—1

1. If $v = 3t^2 - 4$, and the domain of t is the set of natural numbers, find the first four corresponding values of v.

2. If the set of all positive odd integers is denoted by $2x + 1$, what is the domain of the variable x?

3. What is the set of integers which leave a remainder of 3 when divided by 4? If this set of integers is represented by y, where $y = 4x + 3$, what is the domain of the variable x? What is the range of y?

4. Any three-digit number may be represented by $100h + 10t + u$. What is the domain of the variable u? of t? of h?

5. The relation $A = P(1 + ni)$ gives the amount A of a given principal P at various annual rates of interest i for various numbers of years n.

    (a) What is the domain of the variable i if the interest rate is taken by steps of $\frac{1}{4}\%$ and may not exceed 4%?

    (b) What is the domain of P if no interest is allowed on fractional parts of a dollar?

    (c) What is the domain of n if interest is computed by months? by days?

6. Find two or three values for the function $y = x^2 - 3x + 5$ when the domain of x is (a) the set of natural numbers; (b) the set of positive rational numbers; (c) the set of positive irrationals.

7. Which of these graphs represent functions of x, where the domain of x is represented along the X-axis?

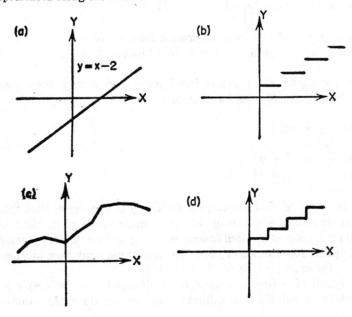

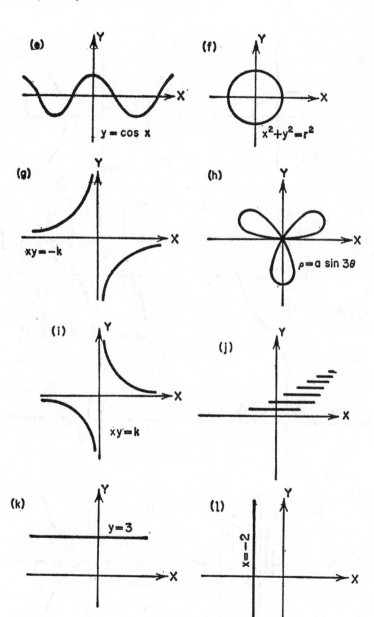

(e) $y = \cos x$

(f) $x^2 + y^2 = r^2$

(g) $xy = -k$

(h) $\rho = a \sin 3\theta$

(i) $xy = k$

(j)

(k) $y = 3$

(l) $x = -2$

(m)
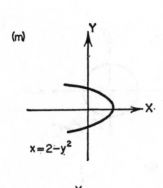
$x = 2 - y^2$

(n)
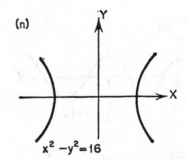
$x^2 - y^2 = 16$

(o)
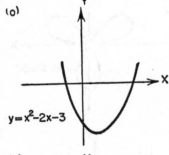
$y = x^2 - 2x - 3$

(p)
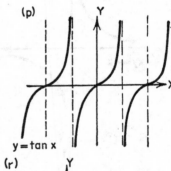
$y = \tan x$

(q)
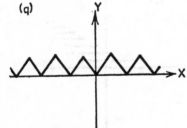

(r)

(s)
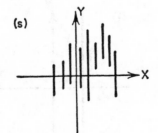

(t)
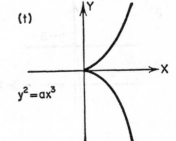
$y^2 = ax^3$

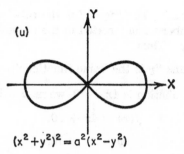

(u)

$$(x^2+y^2)^2 = a^2(x^2-y^2)$$

1—5. Functional Notation. There are many kinds of functions in mathematics. The functions $y = 4x + 5$ and $y = ax + b$ are *polynomial functions* of the first degree in x. As they are of the first degree in x, they are called *linear* functions. The "5" and the "b" are called the *absolute term* in their respective equations since neither involves a variable, that is, they are constants. Or again, the function $y = ax^2 + bx + c$, where a, b, and c are constants and $a \neq 0$, is a polynomial function of the second degree in x. In general, a polynomial function in x may be written in the form

$$y = a_0x^n + a_1x^{n-1} + a_2x^{n-2} + \cdots + a_{n-1}x + a_n,$$

or $\quad y = x^n + p_1x^{n-1} + p_2x^{n-2} + \cdots + p_{n-1}x + p_n,$

where the a's and the p's are constants, and a_n and p_n are the absolute terms.

The product of two polynomials in x is always another polynomial in x. Thus if u and v are polynomials in x, then uv is a polynomial in x; also, if u is a polynomial in x and n is a positive integer, then u^n is also a polynomial in x. However, the ratio $\dfrac{u}{v}$ of two polynomials is, in general, *not* a polynomial in x. Instead, it is called a *rational function* of x. Here the word "rational" derives its meaning from the word *ratio*, just as a rational number is defined as the ratio of two integers.

Since a functional relation is essentially a collection of ordered pairs of numbers which associate each element x of its domain with exactly one element y of its range, we often use the symbols $f(x)$ or $g(x)$ to designate the second element of the ordered pair whose first element is x. The symbol "$f(x)$" is read "f of x"; it means the "value of the dependent variable when any of a given set of values has been assigned to x". Thus the "f" symbolizes the particular rule which makes the association between the variables. The reader is warned not to think of "$f(x)$" as meaning "*f*" times "*x*." The "*f*( )" is the rule indicating the association; a specific number when inserted in the parenthesis yields a corresponding value for *y*. Thus

"*y* = *f*(*x*)" means "*y* is the value functionally associated with *x*."

For example, if the function is $4x + 10$, we may write

$$f(x) = 4x + 10,$$

or $$y = 4x + 10,$$

or $$y = f(x).$$

For this particular function, we note that

if $$f(x) = 4x + 10,$$

then $$f(1) = 4(1) + 10 = 14$$

$$f(-3) = 4(-3) + 10 = -2$$

$$f(\tfrac{3}{4}) = 4(\tfrac{3}{4}) + 10 = 13$$

$$f(0) = 4(0) + 10 = 10$$

Or again:

if $$F(x) = x^3 - 4x^2 + 2x - 8$$

$$F(3) = (3)^3 - 4(3)^2 + 2(3) - 8 = -11$$

$$F(-1) = (-1)^3 - 4(-1)^2 + 2(-1) - 8 = -15$$

$$F(0) = (0)^3 - 4(0)^2 + 2(0) - 8 = -8$$

$$F(10) = (10)^3 - 4(10)^2 + 2(10) - 8 = 612$$

1—6. Continuity of a Function. If in the equation $y = f(x)$, y is defined for every possible value assigned to x, then as x varies continuously from a to b, if y varies continuously from $f(a)$ to $f(b)$, we say that the function is *continuous* for all values of x. If, however, there is some value assigned to x such that there is no corresponding value of y, that is, for which y is not defined, then the function is said to be *discontinuous* at that point, or for that value of x. For example, in the function $xy = k$, or $y = \dfrac{k}{x}$, y is not defined when $x = 0$, since $\dfrac{k}{0}$ is an impossible operation; hence the function is discontinuous at $x = 0$. Again, in the function $x^2 - y^2 = 36$, or $y = \pm\sqrt{x^2 - 36}$, y is not defined for absolute values of $x < 6$, that is, when $-6 < x < +6$; hence the function is discontinuous in the interval from $x = -6$ to $x = +6$.

Intuitively, a continuous function may be said to have an "unbroken" graph. If there is a break of some kind in the graph, the function is discontinuous. This crude "definition" is, of course, merely a very rough description. A more precise definition of the continuity of a function will be given later.

EXERCISE 1—2

1. Write, in function notation, the general rational integral function of the fifth degree in x.

2. Write the general form of the polynomial of the ninth degree which contains only the odd powers of the variable.

3. If $f(x) = ax^2 - bx + c$, write $f(m); f(5)$.

4. If $F(x) = 3x^3 + 4x^2 + 2x - 6$, what is the value of $F(-1)$? of $F(0)$?

5. If $f(x) = 2x^2 + 5x - 3$, what function does $f(x + h)$ represent? Write this function as a general polynomial in x; what is the absolute term of this latter polynomial?

6. Given $f(x) = 3^{x-2}$; find $f(0); f(1); f(2); f(-1)$.

7. Given $\phi(x) = x^3 - 4x^2 + 8x - 8$; find $\phi(x + 1); \phi(x - h)$.

8. Given $Q(x) = \dfrac{x^2}{x+1}$; find $Q(x - 1) - Q(x)$.

9. Given $\phi(x) = ax^2 + bx + c$; find $\phi(x + h) - \phi(h)$.

10. Given $F(z) = \log \dfrac{z}{z - m}$; find $F(u) - F(v)$.

11. Given the function $f(x) = x^2 + 4x - 2$; using the set of real numbers as the domain of definition, draw the graph of $f(x)$, and state the range of $f(x)$, or y.

12. Given the domain of definition for each of the following functions as the set $\{0, 1, 2, 3, 4, 5, 6\}$, draw the graph, and in each case state the range of the function:

   (a) $f(x) = 3x + 2$

   (b) $g(x) = 2x^2$

   (c) $F(x) = 3x$

AVERAGE AND INSTANTANEOUS RATES

1—7. Idea of a Rate of Change. In a function, as the independent variable varies or changes by taking on one successive value after

another, the value of the dependent variable also changes. But we must
not confuse the value of the variable at any time with the rate at which it
is changing at that time. Thus, as time goes on, the *value* of the public
debt may be very great, but the *rate at which it is changing* may be
comparatively small. Or, in a chemical reaction, the temperature may
be relatively low, but it may be increasing very rapidly. We shall now
examine the nature of a rate of change.

1—8. Constant Rate of Change. Let us consider the rate of change
of a simple linear function, namely, the amount of simple interest earned
by a principal of \$500 at 6% per year. The graph of this function,
$I = (500)\left(\dfrac{6}{100}\right)t$, or $I = 30t$, is, of course, a straight line, and therefore
has a constant slope. A moment's reflection will show that if the curve

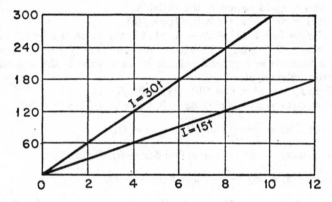

representing a function is a straight line, it rises by a fixed amount in
each horizontal unit interval; in this case, the constant increase is \$30
per year. Thus the function I, where t is the independent variable, is
increasing at a constant rate. If the interest rate were smaller, say 3%,
the function $I = 15t$ also increases at a constant, though smaller rate;
the curve is again a straight line, but not as steep. Thus the rate of
change is shown by the slope, or steepness, of the curve, not by its
height at any point.

The important concept to be grasped here is that *the rate of change is
the amount of change in the function (or dependent variable) per unit change
in the independent variable.* Similarly, if the rate of change of a function
is constant, the curve of the function must be a straight line, since it
rises (increase in ordinate) by the same amount in each horizontal unit
(increase in abscissa).

1—9. Graphic Determination of Average and Instantaneous Rates.
It so happens, however, that most quantities change at a varying rate.
For example, a train may cover a total distance of 150 miles in 3 hours,
or at an average rate of 50 miles per hour. Yet it may well have been
moving at rates greater or less than 50 miles per hour at any particular
instant during those three hours, or even for the duration of some
interval within that period of three hours.

It is therefore necessary to distinguish between an *average rate of
change during an interval,* on the one hand, and an *instantaneous rate at
some particular instant,* on the other hand.

These rate of change concepts may be illustrated by the following
examples. In our first illustration, the curve shows the variation of the

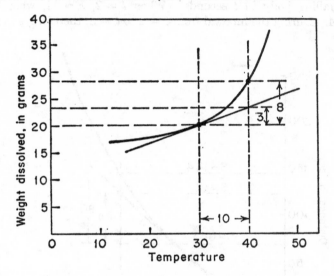

amount of a chemical dissolved as the temperature changes. Thus,
when the temperature increases from 30° to 40°, the amount dissolved
increases from 20 gm. to 28 gm.; or, a total increase of 8 gm. in an
interval of 10°. Thus the *average rate* of change equals 8 ÷ 10, or .8 gram
per degree, during this 10-degree interval. Had we selected a larger or
a smaller interval, or an equal interval elsewhere along the curve, the
average rate of change would, of course, have been different. But now
consider this question: how fast was the amount dissolved increasing
at the instant when the temperature was 30°? To find the answer, we
determine *how steep* the curve is at the point where $T = 30°$; that is,
the tangent to the curve at this point measures the steepness of the

curve at that point, and so indicates the rate at which it *would continue*
to increase *if the rate of change were to become constant at that instant.*
Naturally, drawing a tangent to a curve at a given point "free-hand,"
or by inspection, even with the aid of a ruler, is at best only an approximation. In this case, a tangent to the curve at $T = 30°$, when extended,
meets the ordinate drawn through the abscissa $T = 40°$ at the point
where the amount is 23 gm. Hence, a *uniform increase* of 3 gm. in an
interval of 10°, or a constant rate of change of .3 of a gram per degree,
which is the same as the instantaneous rate at the particular instant
$T = 30°$, by virtue of the significance of the tangent at that point.

Let us take one further illustration. Consider the function $s = \frac{1}{2}gt^2$,
where $g = 32$ ft. per sec. per sec., and s gives the distance covered by a
freely falling body in t seconds. When $t = 2$, $s = 64$; when $t = 3$,
$s = 144$. During the interval from $t = 2$ to $t = 3$, therefore, s increases

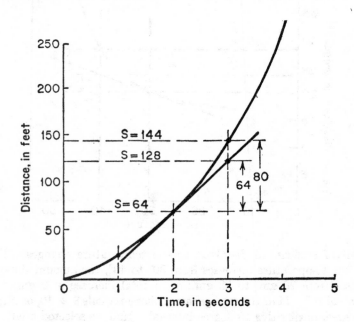

from 64 ft. to 144 ft.; it is thus changing at an average rate of 80 ft.
per sec. The tangent drawn at the point where $t = 2$ is found to cut
the ordinate through the abscissa $t = 3$ at the point where $s = 128$;
hence the instantaneous rate at $t = 2$ is found to be $128 - 64 = 64$ ft.
per sec.

1—10. Instantaneous Rate and Very Small Intervals. To appreciate more fully the relationship of an instantaneous rate to an average rate, let us study one more illustration; for these concepts, although here presented intuitively, nevertheless lie at the very foundation of the calculus. Consider the function $y = x^2$. Let us examine the instantaneous rate of change in the dependent variable y at the point where $x = 5$. We shall begin with the average rate in the interval from $x = 2$ to $x = 5$, and find the average rate as the interval taken becomes smaller

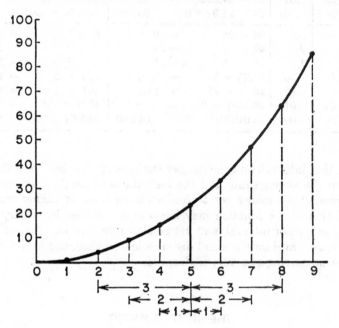

and smaller. The upper half of the table shows how the average rate of change of y approaches 9.999 as the interval is taken smaller and smaller. The lower half of the table shows how the average rate of change varies as we approach the point $x = 5$ from the right, beginning with the interval from $x = 5$ to $x = 8$, and taking the interval smaller and smaller; the average rate of change of y now approaches 10.001. Thus the rate of change of y at the instant that $x = 5$ lies between 9.999 and 10.001; we shall soon learn how to compute the instantaneous rate exactly, without resorting to cumbersome graphic methods.

In short, the average rate of change for a very small interval is seen to be very nearly the same as the instantaneous rate at any instant

Interval Taken	*Change in x*	*Change in y*			*Average Rate of Change in y*		
2 to 5	3.	25 − 4		= 21.	21 ÷ 3	=	7.00
3 to 5	2.	25 − 9		= 16.	16 ÷ 2	=	8.00
4 to 5	1.	25 − 16		= 9.	9 ÷ 1	=	9.00
4.5 to 5	.50	25 − 20.25		= 4.75	4.75 ÷ .5	=	9.50
4.9 to 5	.10	25 − 24.01		= .99	.99 ÷ .1	=	9.90
4.99 to 5	.01	25 − 24.9001		= .0999	.0999 ÷ .01	=	9.99
4.999 to 5	.001	25 − 24.990001	=	.009999	.009999 ÷ .001	=	9.999
5 to 8	3.	64 − 25		= 39	39 ÷ 3	=	13.00
5 to 7	2.	49 − 25		= 24	24 ÷ 2	=	12.00
5 to 6	1.	36 − 25		= 11	11 ÷ 1	=	11.00
5 to 5.5	.50	30.25 − 25		= 5.25	5.25 ÷ .5	=	10.50
5 to 5.1	.10	26.01 − 25		= 1.01	1.01 ÷ .1	=	10.10
5 to 5.01	.01	25.1001 − 25		= .1001	.1001 ÷ .01	=	10.01
5 to 5.001	.001	25.010001 − 25	=	.010001	.010001 ÷ .001	=	10.001

during that interval, and the smaller the interval, the less the difference between the average rate and the instantaneous rate. Once more we must remind the reader not to confuse the *amount* of change with the *rate* of change. A function may increase or decrease by a very small *amount* in a short interval, and yet be changing very *rapidly*—just as a bullet may travel only a small distance in one thousandth of a second and yet be moving at a very high speed.

THE LIMIT CONCEPT

1—11. The Idea of a Limit Is Familiar. The reader may not realize that he is already familiar with the notion of a limiting value from his geometry. For example, he will recall that the circumference of a circle may be regarded as the *limit* of the perimeters of a series of inscribed and circumscribed regular polygons as the number of sides becomes indefinitely greater. This simply means that as the number of sides becomes infinitely greater, the *difference* between the length of the perimeter and the length of the circle becomes smaller and smaller; it does *not* mean that eventually the length of the circle *equals* the perimeter of any particular polygon, or that a polygon can ever *coincide* with a circle.

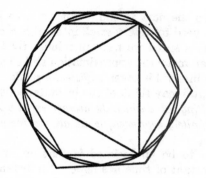

Again, from his algebra, the reader will recall that the sum of an infinite number of terms of a geometric series, when $r < 1$, also approaches a limit, namely, the value $\dfrac{a}{1-r}$. For example, in the series $1, \frac{1}{2}, \frac{1}{4}, \frac{1}{8}, \frac{1}{16}, \ldots$ to infinity, it will be recalled that as more and more terms are taken, the sum in each instance becomes greater. It appears,

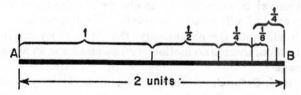

also, that as the number of terms increases, the sum, while getting larger, is increasing less rapidly; in other words, there appears to be an upper *limit* to the sum, beyond which we can never go, no matter how many terms are taken—even an infinite number. In the case of the present series, that limit equals 2, as may be seen from the geometric representation of the terms of the above series by means of the segments on a line 2 units long, formed by successively bisecting the right-hand halves. At least intuitively it is clear that even when we have added all the terms that can possibly be thought or imagined—the series never ends, however—the sum could never exceed 2. As long as we add a *finite* number of terms, the sum will always be a little less than 2; but we can make the sum *as close to 2 as we wish*, merely by taking a sufficiently large number of terms.

1—12. Sharpening Our Ideas. When we think of the speed of a moving object at a particular *instant*, exactly what do we mean? We certainly do not mean the average speed for the *next* hour, or half hour, or minute, or even the next second. But we have just seen (§1—10) that the average speed for a *very small* interval is a *close approximation* of the instantaneous speed we have in mind. By taking the interval still smaller, we can make the approximation still closer; in fact, as close as we wish, if the interval is taken *sufficiently small*.

In other words, we may think of the instantaneous speed at a certain instant as *the limiting value which the average speed would approach if the interval were indefinitely shortened, while always including the instant in question.*

The distinction to be appreciated is that an interval of time has *extent*, while an instant of time has none. No distance, however small, can be covered "during an instant"; indeed, the phrase is meaningless, since an instant has no duration. Thus we cannot say an instantaneous rate is "the distance covered during the instant divided by the duration of the instant." Furthermore, it is equally meaningless to refer to the "speed during the shortest possible interval of time"; this is merely an average rate once more, since any "possible *interval*," however small, has *some* size or extent. Moreover, "*shortest* interval" can mean nothing, since any interval, however small, can always be subdivided into millions or billions, etc., of still smaller intervals. Finally, we must not allow ourselves to slip into the phraseology, the "rate at an instant." This begs the question: there is no motion *at an instant* for even an exceedingly small interval.

1—13. Limit Defined. Is there, then, no precise way of defining an instantaneous rate? There is, by using the idea of a *limiting value*, already referred to in §1—11. We shall now define the term *limit*.

A variable v is said to approach a constant l as a limit when the successive values of v are such that the numerical value of the difference v − l ultimately becomes and remains less than any previously specified value, no matter how small.

As the reader comes to understand more fully the meaning of a limit, he will appreciate the fact that *every word and phrase* in the above definition is significant and essential. Indeed, an even more precise definition of the limiting value of a function will be presented later. It should be carefully noted that the question of whether a variable ever *reaches* its limit or not has nothing to do with the question of its *approaching* a limit. The crux of the matter is that the difference between *v* and *l* shall, sometime, as *v* varies, become less than, and *thereafter remain less than*, some specified number, however small.

1—14. Limiting Value of a Function. When dealing with limiting values, we use the notation $v \to l$, which is read "v approaches l as a limit." Consider the function $f(x) = x^2 + 3x + 1$.

x	$f(x)$
0	1
1	5
2	11
3	19
4	29
5	41
6	55

When $x = 2, f(x) = 11$; when $x = 4, f(x) = 29$. As we take the value of x close to 3, the value of $f(x)$ will be close to $3^2 + 3(3) + 1$, or 19. If we take values of x successively nearer to 3, so that $|x - 3|$ becomes less and less, the quantity $|f(x) - 19|$ will also become less and less. Now let us select any number, as small as we please, say ϵ. It is then possible to choose $|x - 3|$ so as to make $|f(x) - 19|$ less than ϵ. We may thus say that

$$\lim_{x \to 3} (x^2 + 3x + 1) = 19,$$

which is read "the limiting value of the function $x^2 + 3x + 1$ as x approaches 3 is 19." The limit in this illustration is exactly 19.

The limit of a function is the exact value toward which the function *approaches* as x comes sufficiently near to some fixed value. The value of the function *may or may not reach* the limit toward which it approaches. In general, the limit of $f(x)$ as x approaches some constant a is equal to some other constant b; that is,

$$\lim_{x \to a} f(x) = b,$$

always provided, of course, that the value of $f(x)$ can be made to differ from b by as little as we please whenever x is taken sufficiently near to a.

1—15. Laws of Limits. We shall set forth, without proof, certain basic principles concerning variables and limits. For a further discussion and proof of these theorems, the reader is referred to more advanced treatises on the calculus. For the purposes of an introductory study of the subject such as this, no harm will be done by foregoing mathematical rigor to some extent; in fact, there is much to be gained by so doing.

Let us consider that u, v, and w represent *functions* of an independent variable x. Let us also suppose that

$$\lim_{x \to k} u = A, \qquad \lim_{x \to k} v = B, \qquad \lim_{x \to k} w = C.$$

Remember that a function is a dependent variable; thus, if $u = f(x)$, then u is a variable. Then the following laws can be shown to hold.

I. *The limit of the sum, or difference, of two variables is equal to the sum, or difference, of their respective limits.*

Thus $\qquad \lim\limits_{x \to k} (u + v - w) = A + B - C.$

II. *The limit of the product of a constant and a variable equals the product of the constant and the limit of the variable.*

Thus $\qquad \lim\limits_{x \to k} (cu) = cA.$

III. *The limit of the product of a finite number of variables is equal to the product of their respective limits.*

Thus $\qquad \lim\limits_{x \to k} (uvw) = ABC.$

IV. *The limit of the quotient of two variables is equal to the quotient of their respective limits, provided that the limit of the denominator is not zero.*

Thus $\qquad \lim\limits_{x \to k} \left(\dfrac{u}{v}\right) = \dfrac{A}{B}, \quad B \neq 0.$

EXAMPLE 1.

$$\lim_{x \to 0} (3x^2 + 5x - 8) = \lim_{x \to 0} 3x^2 + \lim_{x \to 0} 5x - \lim_{x \to 0} 8$$
$$= 0 + 0 - 8 = -8.$$

EXAMPLE 2.

$$\lim_{x \to 3} (x^3 - 2x + 5) = \lim_{x \to 3} x^3 - \lim_{x \to 3} 2x + \lim_{x \to 3} 5$$
$$= 27 - 6 + 5 = 26.$$

EXAMPLE 3.

$$\lim_{x \to \pi} x \cos x = \lim_{x \to \pi} x \cdot \lim_{x \to \pi} \cos x = \pi(-1) = -\pi.$$

EXAMPLE 4.

$$\lim_{x \to 4} (x + 3)(x - 2) = \lim_{x \to 4} (x + 3) \cdot \lim_{x \to 4} (x - 2) = (7)(2) = 14.$$

EXAMPLE 5.

$$\lim_{x \to 2} \left(\frac{x^2 - 8}{x + 1}\right) = \frac{\lim\limits_{x \to 2} (x^2 - 8)}{\lim\limits_{x \to 2} (x + 1)} = \frac{-4}{3} = -4/3.$$

1—16. Infinity and Zero. We shall have occasion to make frequent use of another basic idea in connection with limits. If the numerical value of a variable x ultimately *becomes and remains* greater than any specified number, no matter how great, we say that x *becomes infinite*. This does *not* mean that infinity (∞) is a number, or any fixed quantity. The symbol ∞ does not denote a constant. We do not say "x approaches infinity"; we say instead, "x becomes infinite." This means that x increases without limit; we write

$$x \to \infty.$$

If the values of x are positive only, we say that x becomes positively infinite, or $x \to +\infty$; if the values of x are negative only, we say that x becomes negatively infinite, or $x \to -\infty$. The symbol ∞ is thus not a symbol for any number, quantity, or value; used in conjunction with other symbols, it is part of an abbreviation for the limiting results of a process of change.

Understood in this way, we may employ the symbol of infinity to abbreviate two very important limits, namely:

$$\lim_{x \to 0} \frac{1}{x} = \infty, \tag{1}$$

and
$$\lim_{x \to \infty} \frac{1}{x} = 0. \tag{2}$$

These may be interpreted, informally, as follows. In equation [1], as the denominator x becomes smaller and smaller, the value of the quotient $\frac{1}{x}$ becomes greater and greater. Thus:

$$\frac{1}{2} = .5; \quad \frac{1}{1} = 1.0; \quad \frac{1}{0.1} = 10; \quad \frac{1}{.01} = 100;$$

$$\frac{1}{.001} = 1000; \quad \frac{1}{.0001} = 10,000; \quad \text{etc.}$$

The quotient $\frac{1}{x}$ can literally be made as great as we wish; that is, no matter how great a number may previously be designated, the value of $\frac{1}{x}$ can be made greater than this number simply by selecting a sufficiently small value of x. But the indicated quotient $\frac{1}{0}$ has no "value"; there is no number which results from dividing 1 by 0. That is what is meant by saying that "division by zero is impossible."

Similarly, in equation [2], as the denominator x becomes greater and greater, the value of the quotient $\dfrac{1}{x}$ becomes smaller and smaller. Thus:

$$\frac{1}{.001} = 1000; \qquad \frac{1}{.01} = 100; \qquad \frac{1}{0.1} = 10; \qquad \frac{1}{10} = 0.1;$$

$$\frac{1}{100} = .01; \qquad \frac{1}{1000} = .001; \quad \text{etc.}$$

The quotient $\dfrac{1}{x}$ can literally be made as small as we wish, or as close to zero as we like, simply by taking a value for x which is sufficiently great.

Other important and frequently used limits are given below for reference; in the symbolical equations [3] to [14], it should be noted that $a > 0$, and $c \neq 0$.

$$\lim_{x \to 0} \frac{c}{x} = \infty, \quad \text{or} \quad \frac{c}{0} = \infty. \tag{3}$$

$$\lim_{x \to \infty} cx = \infty, \quad \text{or} \quad c \cdot \infty = \infty. \tag{4}$$

$$\lim_{x \to \infty} \frac{x}{c} = \infty, \quad \text{or} \quad \frac{\infty}{c} = \infty. \tag{5}$$

$$\lim_{x \to \infty} \frac{c}{x} = 0, \quad \text{or} \quad \frac{c}{\infty} = 0. \tag{6}$$

$$\lim_{x \to -\infty} a^x = \infty, \quad \text{when } a < 1. \tag{7}$$

$$\lim_{x \to +\infty} a^x = 0, \quad \text{when } a < 1. \tag{8}$$

$$\lim_{x \to -\infty} a^x = 0, \quad \text{when } a > 1. \tag{9}$$

$$\lim_{x \to +\infty} a^x = \infty, \quad \text{when } a > 1. \tag{10}$$

$$\lim_{x \to 0} \log_a x = +\infty, \quad \text{when } a < 1. \tag{11}$$

$$\lim_{x \to +\infty} \log_a x = -\infty, \quad \text{when } a < 1. \tag{12}$$

$$\lim_{x \to 0} \log_a x = -\infty, \quad \text{when } a > 1. \tag{13}$$

$$\lim_{x \to +\infty} \log_a x = +\infty, \quad \text{when } a > 1. \tag{14}$$

EXAMPLE 1. $\lim\limits_{x \to \infty} \dfrac{4}{x} = 0$

EXAMPLE 2. $\lim\limits_{x \to 0} \dfrac{2}{x^2} = +\infty$

EXAMPLE 3. $\lim\limits_{x \to 1} \dfrac{-3}{(x-1)^2} = -\infty$

EXAMPLE 4. $\lim\limits_{x \to -\infty} (2x + 1) = -\infty$

EXAMPLE 5. $\lim\limits_{x \to \infty} 2^x = +\infty$

EXAMPLE 6. $\lim\limits_{x \to \infty} \log x = +\infty$

EXAMPLE 7. $\lim\limits_{x \to -1} (3x^2 + 4x - 2) = 3 - 4 - 2 = -3$

EXAMPLE 8. $\lim\limits_{x \to \infty} (x^2 + x - 2) = \lim\limits_{x \to \infty} [x(x + 1) - 2]$
$$= \infty(\infty + 1) - 2 = \infty$$

EXAMPLE 9. $\lim\limits_{x \to -\infty} 2x^2 = 2(-\infty)(-\infty) = \infty$

EXAMPLE 10. Find $\lim\limits_{x \to 0} \left(\dfrac{x^3 - 2x^2}{3x^4 + 8x^2} \right)$

Solution. Substituting 0 for x yields $\dfrac{0-0}{0-0} = \dfrac{0}{0}$, which is indeterminate. Hence, divide both numerator and denominator by x^2, the lowest power of x in the function:

$$\lim\limits_{x \to 0} \left(\dfrac{x - 2}{3x^2 + 8} \right) = \dfrac{0 - 2}{0 + 8} = -\dfrac{1}{4}.$$

EXAMPLE 11. Find $\lim\limits_{x \to \infty} \left(\dfrac{x^3 + 3x^2 - x - 4}{2x^3 - x^2 + 6x} \right)$.

Solution.   Divide both numerator and denominator by x^3, the highest power of x in the function.

$$\lim_{x \to \infty} \left(\frac{1 + \dfrac{3}{x} - \dfrac{1}{x^2} - \dfrac{4}{x^3}}{2 - \dfrac{1}{x} + \dfrac{6}{x^2}} \right) = \frac{1 + 0 - 0 - 0}{2 - 0 + 0} = \frac{1}{2}.$$

EXAMPLE 12.   Find $\lim\limits_{x \to \pi/2} \left(\dfrac{1}{\cos^2 x} \right)$.

Solution. $\cos \dfrac{\pi}{2} = 0$

Hence $\lim\limits_{x \to \pi/2} \left(\dfrac{1}{\cos x} \right) = \dfrac{1}{0} = \infty.$

EXAMPLE 13.   Find $\lim\limits_{x \to 3} \left(\dfrac{x^2 - 9}{x - 3} \right)$.

Solution.   Substituting $x = 3$ yields $\dfrac{9 - 9}{3 - 3} = \dfrac{0}{0}$, which is indeterminate.

By factoring:

$$\lim_{x \to 3} \left(\frac{x^2 - 9}{x - 3} \right) = \lim_{x \to 3} \left[\frac{(x + 3)(x - 3)}{x - 3} \right] = \lim_{x \to 3} (x + 3) = 6.$$

1—17. A More Precise Definition of a Limit. Suppose that a toolmaker were asked to fashion a piece of work to a specified "ideal" dimension D, with an allowable leeway or tolerance limit of $\pm.002$ inch. Call the measured dimension of the finished piece D' and the difference $D' - D = E$. (In this discussion we need not consider the error of measurement, since its exclusion does not affect the argument.)   By using appropriate tools and skill, he can make the value of E as close to zero as he wishes. He can therefore make E lie between $-.002$ and $+.002$ simply by making E sufficiently close to zero. How "close to zero" should this be? He can presumably find a positive number δ, which may be very small, such that if the work is properly executed, he will have

$$-\delta < E < +\delta \quad (E \neq 0).$$

We are now in a position to define the limiting value of a function more rigorously. Consider the function $F(v)$, which is defined on the domain

$$v_1 < v < k \quad \text{and} \quad k < v < v_2.$$

We then say that a number L is the limit of a function $F(v)$ as v approaches the number k, or

$$\lim_{v \to k} F(v) = L,$$

if for every number $\epsilon > 0$, however small, there is another number $\delta > 0$ such that, when v is in the domain of definition of F,

$$0 < (v - k) < \delta,$$

then
$$|F(v) - L| < \epsilon.$$

Note that when we say "$F(v)$ approaches L," we are also saying that $F(v) - L$ approaches zero. Note also that it does not matter whether the function $F(v)$ is defined at $v = k$ or not.

EXAMPLE. For $F(v) = v^2$, $k = 4$,

$$\lim_{v \to 4} F(v) = 16.$$

To find a "tolerance" limit, $|\epsilon| > 0$, no matter how small, we ask: How close to $k = 4$ must v become in order to make values of $F(v)$, or v^2, lie between $(16 + \epsilon)$ and $(16 - \epsilon)$? In short, we wish to find a number δ such that

$$|v^2 - 16| < \epsilon \quad \text{when } 0 < |v - 4| < \delta.$$

Since $v^2 - 16 = (v + 4)(v - 4)$, it is clear that we can make the factor $(v - 4)$ small by taking v close to 4; if we do this, the other factor, $(v + 4)$, becomes approximately equal to $4 + 4$, or 8, when v is close to 4. Thus the latter factor, $(v + 4)$, can be made less than 9, for example, if we say let $|v - 4| < 1$. If we do let $|v - 4| < 1$, then v must lie between 3 and 5, and hence $|v + 4| < 9$. We therefore have:

$$|v^2 - 16| = |(v + 4)(v - 4)| < 9|v - 4|.$$

We can then put a further limitation upon v so that

$$9|v - 4| < \epsilon, \quad \text{or} \quad |v - 4| < \frac{\epsilon}{9}.$$

So, if we finally take δ equal to $\dfrac{\epsilon}{9}$ or 1, whichever is smaller, we know that

$$|v^2 - 16| < \epsilon,$$

provided that $0 < |v - 4| < \delta$. The diagram will clarify these ideas. Geometrically, $F(v) = v^2$ lies between $(16 + \epsilon)$ and $(16 - \epsilon)$ when v lies between $(4 + \delta)$ and $(4 - \delta)$; $\lim\limits_{v \to 4} F(v) = 16$.

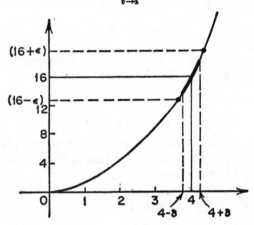

SOME SPECIAL LIMITS

1—18.   The Limit of $(1 + x)^{1/x}$.   One of the most useful limits in the calculus is the

$$\lim_{x \to 0} (1 + x)^{1/x} = 2.718 \ldots = e. \qquad [15]$$

Consider the function

$$y = (1 + x)^{1/x}.$$

x	y	x	y
10	1.001	−.5	4.000
2	1.732	−.1	2.868
1	2.000	−.01	2.732
.5	2.250	−.001	2.720
.1	2.594		
.01	2.705		
.001	2.717		

From graphic considerations, therefore, it appears that as $x \to 0$, the limit of y, or of $(1 + x)^{1/x}$, lies between 2.717 and 2.720. By using

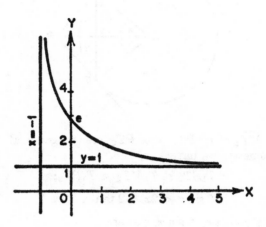

more advanced methods (Chapter Eleven), we shall learn how to compute this limit e to any number of decimal places. Note that y approaches e as a limit not only as x approaches 0 from the right, but also as x approaches 0 from the left as well.

1—19. The Limit of $\dfrac{\sin x}{x}$. Another useful limit is

$$\lim_{x \to 0} \frac{\sin x}{x} = 1. \tag{16}$$

Reference to a five-place (or more) table of values of trigonometric functions will reveal that for all angles less than 2°, the sine of the angle and the angle itself (expressed in radians) are very nearly equal. In fact, for all angles less than 10°, the angle in radians and its sine are equal to three decimal places. Thus for small angles, the sine and the angle are nearly equal; the smaller the angle, the less the difference between them becomes. Hence we see that as $x \to 0$, $\lim \dfrac{\sin x}{x}$ appears to be equal to 1.

That this actually is the limit can be proved from the following geometrical proof. Let O be the center of a circle whose radius is unity.

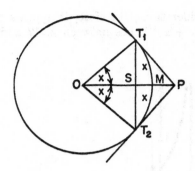

Let $\widehat{MT_1} = \widehat{MT_2} = x$; PT_1 and PT_2 are tangents at T_1 and T_2. From trigonometry, we have:

$$T_1ST_2 < T_1MT_2 < T_1PT_2;$$

hence $\qquad\qquad 2 \sin x < 2x < 2 \tan x.$ $\qquad\qquad$ (1)

Dividing (1) through by $2 \sin x$, we get:

$$1 < \frac{x}{\sin x} < \frac{1}{\cos x}. \qquad\qquad (2)$$

Now as $x \to 0$, the above inequality (2) holds true for all values of x, however small. At the same time, as $x \to 0$,

$$\lim_{x \to 0} \frac{1}{\cos x} = \frac{1}{\cos 0} = \frac{1}{1} = 1.$$

The $\lim\limits_{x \to 0} \dfrac{x}{\sin x}$ is therefore seen to lie between the two values 1, and the inequality (2) can hold only if $\lim\limits_{x \to 0} \dfrac{x}{\sin x} = 1$; so, $\lim\limits_{x \to 0} \dfrac{\sin x}{x} = 1$, by the properties of inequalities and reciprocals.

1—20. Continuity of a Function in an Interval. We are now in a position to formulate what is meant by the continuity of a function in an interval more precisely than was stated in §1—6.

A single-valued function $f(x)$ is said to be continuous at $x = a$ if:

(1) $f(a)$ is defined;
(2) $\lim\limits_{x \to a} f(x)$ exists;
(3) $\lim\limits_{x \to a} f(x) = f(a)$.

If any one of these three conditions is not satisfied, the function is said to be discontinuous at the point where $x = a$. If the reader should find this formal definition somewhat complicated at first, he will find that it will become more significant as he becomes more familiar with the subject.

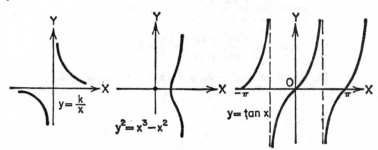

Examples of Discontinuity

In general, a function is said to be continuous throughout an interval if it is continuous at every point in the interval. More specifically, polynomial functions, such as

$$f(x) = a_0 x^n + a_1 x^{n-1} + a_2 x^{n-2} + \cdots + a_n,$$

are continuous for every value of x.

EXERCISE 1—3

Evaluate each of the following limits:

1. $\lim\limits_{x \to 2} (x^3 + 3x - 5).$

2. $\lim\limits_{x \to 0} \left(\dfrac{1}{x} + x \right).$

3. $\lim\limits_{x \to \infty} \left(\dfrac{3}{x^2 + 1} \right).$

4. $\lim\limits_{x \to 0} \left(\dfrac{x^2 + 3}{2x} \right).$

5. $\lim\limits_{x \to 3} \left(\dfrac{5}{2x - 6} \right).$

6. $\lim\limits_{x \to \infty} \left(\dfrac{2x + 3}{x} \right).$

7. $\lim\limits_{x \to 1} (x + 2)(x^2 - 3).$

8. $\lim\limits_{x \to \infty} \left(\dfrac{x^2 - 1}{x + 2} \right).$

9. $\lim\limits_{x \to 0} \left(\dfrac{4x^3 - 3x^2}{3x^4 - 2x^2} \right).$

10. $\lim\limits_{t \to \infty} \left(\dfrac{2t - 1}{4 - t} \right).$

11. $\lim\limits_{t \to -2} \left(\dfrac{t^2 + 3t + 2}{t^2 - t - 6} \right).$

12. $\lim\limits_{x \to \pi} \left(\dfrac{\sin 2x}{\tan x} \right).$

13. Prove $\lim\limits_{t \to -1} \left(\dfrac{t^3 + 1}{t^2 - 1} \right) = -\dfrac{3}{2}$.

14. Prove $\lim\limits_{x \to \infty} \left(\dfrac{ax^3 + bx + k}{mx^2 - nx + p} \right) = \infty$.

15. Prove $\lim\limits_{x \to \infty} \left(\dfrac{ax^2 + bx + k}{px^3 + qx + r} \right) = 0$.

16. Prove $\lim\limits_{y \to 0} (3x^3 - 5x^2y + 2xy^2 - 6) = 3x^3 - 6$.

The Derivative of a Function

CHAPTER TWO

INCREMENT NOTATION

2—1. A Convenient Symbolism. When discussing average and instantaneous rates in Chapter One, we frequently had occasion to refer to small "intervals." We shall now introduce a convenient notation for small intervals, or for small changes in the value of a variable or a function. When a variable changes from one numerical value to another, we refer to the difference of the values as the *increment* of the variable; for example, if x changes from x_1 to x_2, then the increment equals $x_2 - x_1$. The increment is represented by the symbol Δx, read "delta x"; thus $\Delta x = x_2 - x_1$. The reader should note carefully that the symbol Δx does *not* mean "delta times x"; the "Δ" is not a number or a quantity. But Δx, taken as a single symbol, is a quantity, with the meaning we have just given it.

An increment is always equal to the new value less the original value, never the other way about. If the variable is increasing ($x_2 > x_1$), then Δx is positive; if the variable is decreasing ($x_2 < x_1$), then Δx is negative.

If in the function $y = f(x)$ we let the independent variable x take on an increment Δx, then the corresponding increment in the function $f(x)$, that is, the corresponding increment in y, is called Δy. Thus $\Delta y = y_2 - y_1$, or $\Delta y = f(x_2) - f(x_1)$, where the initial values of x and y are x_1 and y_1, respectively, and the final values of x and y are x_2 and y_2, respectively. For example, if $y = x^3 + 5$, and we start with $x_1 = 2$, then $y_1 = 8 + 5 = 13$; now if x takes on an increment from $x_1 = 2$ to $x_2 = 3$, then $\Delta x = x_2 - x_1 = 3 - 2 = 1$; also, $y_2 = (3)^3 + 5 = 32$, and hence $\Delta y = y_2 - y_1 = 32 - 13 = 19$. In other words, an increment of 1 unit in the value of x from 2 to 3 gives a corresponding increment in y of 19 units, from 13 to 32.

Thus the quantity Δy is always computed from a specific initial value of y which corresponds to the particular fixed initial value taken for x when selecting the increment Δx.

2—2. Using the Increment Notation. We shall now show how the idea of increments may serve a useful purpose.

EXAMPLE 1. Consider the function $A = s^2$, where A is the area of a square of side s. Now let s take on a small positive increment, say Δs; the area will then take on a corresponding increment, say ΔA, and we have:

$$A = s^2. \tag{1}$$

$$A + \Delta A = (s + \Delta s)^2.$$

Expanding:

$$A + \Delta A = s^2 + 2s \cdot \Delta s + (\Delta s)^2. \tag{2}$$

Subtracting (1) from (2):

$$\Delta A = 2s \cdot \Delta s + (\Delta s)^2. \tag{3}$$

Dividing both sides of (3) by Δs:

$$\frac{\Delta A}{\Delta s} = 2s + \Delta s. \tag{4}$$

Suppose that the initial value of $s = 5$; then the initial area of the square is 25. By considering successively smaller values for Δs, beginning with $\Delta s = 1$, we obtain, from equation (4), the table shown

Δs	$\dfrac{\Delta A}{\Delta s}$ $(= 2s + \Delta s)$
1.0	11.00
.5	10.50
.2	10.20
.1	10.10
.01	10.01
.001	10.001

here. It will be seen from the table that as we assign smaller and smaller values to Δs, the nearer the quantity $\dfrac{\Delta A}{\Delta s}$ approaches the value 10. We thus see that the average rate of change in A is represented by the quantity $\dfrac{\Delta A}{\Delta s}$; and the smaller we take the interval, or increment, Δs, the closer the average rate of change $\dfrac{\Delta A}{\Delta s}$ approaches the limiting value 10.

Hence the limit 10 represents the instantaneous rate of change in A at the instant when $s = 5$; when the side of the square equals 5, the area of the square is changing 10 times as fast as a side. In other words, the instantaneous rate of change of the area of a square is $2s$ times as fast as the rate of change of a side s.

EXAMPLE 2. Let us consider another illustration. Suppose an object thrown vertically upward, and the height in feet which it attains after t seconds is given by the function

$$h = 150t - 16t^2.$$

We propose the question, how fast is it moving 4 seconds after its motion began, that is, when $t = 4$? Using the increment notation, and considering a small interval Δt beginning at $t_1 = 4$, we have:

At $t_1 = 4$, $\qquad h_1 = 150(4) - 16(4)^2 = 344.$

At $t_2 = 4 + \Delta t$, $\qquad h_2 = 150(4 + \Delta t) - 16(4 + \Delta t)^2$
$$= 344 + 22\Delta t - 16(\Delta t)^2.$$

Now the difference between the heights h_2 and h_1 is the distance that the object rose during the interval of Δt seconds; or

$$h_2 - h_1 = \Delta h = 22\Delta t - 16(\Delta t)^2. \tag{1}$$

Dividing equation (1) by Δt, we get:

$$\frac{\Delta h}{\Delta t} = 22 - 16(\Delta t).$$

In other words, the average rate of change of the height, h, during the interval of Δt seconds, equals $22 - 16(\Delta t)$ feet.

But the rate of change in height (distance) per unit of time is the speed; hence the average speed of the object during the interval Δt is

$22 - 16(\Delta t)$. For example, if

$$\Delta t = \frac{1}{10} \text{ sec.,}$$

then $\quad\quad \dfrac{\Delta h}{\Delta t} = 22 - 16\left(\dfrac{1}{10}\right) = 20.4 \text{ ft. per sec.;}$

if $\quad\quad\quad\quad\quad\quad \Delta t = \dfrac{1}{100} \text{ sec.,}$

then $\quad\quad \dfrac{\Delta h}{\Delta t} = 22 - 16(.01) = 21.84 \text{ ft. per sec.}$

Now as Δt approaches zero, the *limiting value* approached by the average speed $22 - 16(\Delta t)$ is exactly equal to 22, at the instant when $t = 4$. In short, the *instantaneous* speed at $t = 4$ is exactly equal to 22 ft. per sec.

THE MEANING OF THE DERIVATIVE

2—3. The Derivative as a Limiting Value. From what has been said above, the distinction between an average rate of change in a function, or $\dfrac{\Delta y}{\Delta x}$, and the instantaneous rate of change, or the limiting value of $\dfrac{\Delta y}{\Delta x}$, should begin to be clear. This limiting value of $\dfrac{\Delta y}{\Delta x}$, the value, that is, of $\dfrac{\Delta y}{\Delta x}$ *at a particular instant*, is called a *derivative*. We represent this limiting value by the symbol $\dfrac{dy}{dx}$. Thus,

$$\lim_{\Delta x \to 0} \frac{\Delta y}{\Delta x} = \frac{dy}{dx}.$$

The reader is urged, at this point, to note carefully that the symbols Δy and Δx each represent a definite quantity; hence the fraction $\dfrac{\Delta y}{\Delta x}$ represents the *ratio of two quantities*. But the symbol $\dfrac{dy}{dx}$ does *not* represent the ratio of two quantities; it should distinctly be regarded as a *single symbol*—as one quantity, *the limiting value* of an average rate. The symbols dy and dx have (as yet) no meaning as separate symbols.

2—4. Analytical Representation of the Derivative. Let us examine the derivative a little further. We take any point $P(x_1, y_1)$ on the curve $y = x^2$, and $Q(x_2, y_2)$, another point on the curve, nearby. By making

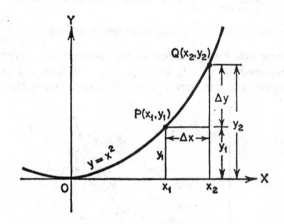

Δx, or $x_2 - x_1$, small enough, we can make the increment Δy, or $y_2 - y_1$, as small as we please. In short, as $\Delta x \rightarrow 0$, Δy also approaches zero.

Now, it is easy to show that the average rate of change of y during the interval from x_1 to x_2 is given by

$$\frac{\Delta y}{\Delta x} = 2x + \Delta x.$$

For, $$y = x^2. \tag{1}$$

$$y + \Delta y = (x + \Delta x)^2,$$

or $$y + \Delta y = x^2 + 2x \cdot \Delta x + (\Delta x)^2. \tag{2}$$

Subtracting (1) from (2):

$$\Delta y = 2x \cdot \Delta x + (\Delta x)^2. \tag{3}$$

Dividing (3) by Δx:

$$\frac{\Delta y}{\Delta x} = 2x + \Delta x. \tag{4}$$

We may now let Δx approach zero as a limit; the average rate $\frac{\Delta y}{\Delta x}$ for the interval Δx then becomes, in the limiting position, the instantaneous rate $\dfrac{dy}{dx}$ at the point x; or, in symbols,

$$\lim_{\Delta x \to 0} \frac{\Delta y}{\Delta x} = \frac{dy}{dx}, \quad \text{at some particular point.}$$

For the particular function under consideration, therefore, $\dfrac{dy}{dx} = 2x$.

2—5. Geometric Interpretation of the Derivative. Referring to the function discussed in §2—4, let $P(x_1, y_1)$ and $Q(x_2, y_2)$ be any two neighboring points on the curve. From the figure below, the slope of the

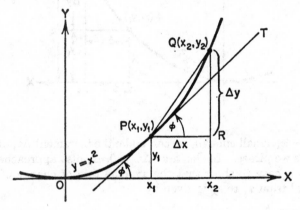

secant PQ is given by the ratio $\dfrac{\Delta y}{\Delta x}$. Now, as $\Delta x \to 0$, point Q will move along the curve toward P as a limit; the secant PQ will approach the position of the tangent PT as its limiting position; and the angle RPQ will approach the angle RPT, or ϕ, as its limiting position.

Inasmuch as $\dfrac{\Delta y}{\Delta x}$ gives the slope of the secant PQ, the limit of this ratio, as $\Delta x \to 0$, gives the slope of the tangent to the curve at the point P. In other words, the slope of the tangent to the curve at the point P is the value of

$$\lim_{\Delta x \to 0} \frac{\Delta y}{\Delta x} = \frac{dy}{dx}.$$

For this particular function, $y = x^2$, we see that the slope of the tangent at any point is given by

$$\lim_{\Delta x \to 0} \frac{\Delta y}{\Delta x} = \lim_{\Delta x \to 0} (2x + \Delta x),$$

or $$\frac{dy}{dx} = 2x.$$

For example, at the point where $x = 2$, the slope of the tangent equals $2(2)$, or 4; at the point where $x = 3$, the slope equals $2(3)$, or 6; at $x = 10$, the slope equals $2(10)$, or 20; etc.

2—6. The Derivative as a Slope. The conclusions of the preceding paragraph may now be generalized. Consider any function $y = f(x)$; $P(x_0,y_0)$ any particular point on the curve; and Δx an arbitrary incre-

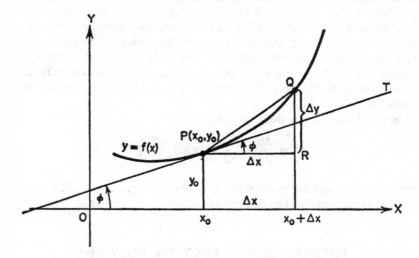

ment of the independent variable. Then the corresponding increment of the dependent variable, or of the function, is

$$\Delta y = f(x_0 + \Delta x) - f(x_0) = RQ.$$

The average rate of change within the interval Δx is given by

$$\frac{\Delta y}{\Delta x} = \frac{f(x_0 + \Delta x) - f(x_0)}{\Delta x} = \frac{RQ}{PR}.$$

But $\frac{RQ}{PR} = \frac{\Delta y}{\Delta x}$ is also the slope of the secant PQ.

Now as $\Delta x \to 0$, so that the interval Δx "closes down" about the point x, the limiting value of the ratio $\frac{\Delta y}{\Delta x}$ becomes the derivative $\frac{dy}{dx}$. The derivative thus represents not only the instantaneous rate of change

at the point where $x = x_0$, but also represents the slope of the tangent
to the curve at the point $x = x_0$. This, of course, means that the derivative represents the *slope of the curve* at that point, since we learned
that the slope of a curve at any point is given by the slope of the tangent
to the curve at that point.

The sense in which the tangent line is the "limit" of the secant lines
should be clearly understood. To begin with, the tangent is a *fixed* line;
secondly, the secant is a *variable* line, depending upon the value given
to Δx (that is, for every value of Δx, except 0, there is a corresponding
position of the secant); and in the third place, the angle between the
tangent and secant, angle *TPQ*, can be made to become and *remain* as
small as we please by taking Δx sufficiently small. The tangent line
itself does not, in general, belong to the series of secant lines; it is a
unique line, which bears a special relation to the series of secant lines.
The tangent is the only line through P that has the property here
described.

In conclusion, then, we may write:

$$\lim_{\Delta x \to 0} \frac{\Delta y}{\Delta x} = \frac{dy}{dx} = \tan \phi = f'(x),$$

where $f'(x)$ is merely an alternative symbolic form for $\frac{dy}{dx}$. Still a third
symbol often used for the derivative is $D_x y$.

DIFFERENTIATION: FINDING THE DERIVATIVE

2—7. The Process of Differentiation. The operation of finding the
derivative of a function is called *differentiation*. The derivative, as we
have already seen, is written $\frac{dy}{dx}$. This symbol is read: "the derivative
of y with respect to x," or, sometimes, the "x-derivative of y." It may
be written in a variety of ways, but they all denote the same thing: thus

$$\frac{dy}{dx} = \frac{d}{dx}(y); \quad \text{or} \quad \frac{dy}{dx} = \frac{d}{dx} f(x);$$

$$\text{or} \quad \frac{d}{dx} f(x) = f'(x).$$

We shall now illustrate the operation of differentiation, and then formulate a General Rule of procedure.

EXAMPLE 1. Differentiate: $y = 3x - 2$.

Solution.

$$y = 3x - 2. \tag{1}$$

Let x take on an increment Δx, and let the corresponding increment in y be Δy.

Then $\qquad\qquad y + \Delta y = 3(x + \Delta x) - 2,$

or $\qquad\qquad y + \Delta y = 3x + 3\Delta x - 2. \tag{2}$

Subtract equation (1) from equation (2):

$$\begin{aligned} y + \Delta y &= 3x + 3\Delta x - 2 \\ y &= 3x - 2 \\ \hline \Delta y &= 3\Delta x \end{aligned} \tag{3}$$

Now divide equation (3) by Δx:

$$\frac{\Delta y}{\Delta x} = \frac{3\Delta x}{\Delta x} = 3.$$

Finally, let $\Delta x \to 0$:

$$\lim_{\Delta x \to 0} \frac{\Delta y}{\Delta x} = 3,$$

or $\qquad\qquad \dfrac{dy}{dx} = 3.$

EXAMPLE 2. Differentiate: $y = x^2 + 5$.

Solution.

$$y = x^2 + 5. \tag{1}$$

$$y + \Delta y = (x + \Delta x)^2 + 5,$$

or $\qquad y + \Delta y = x^2 + 2x \cdot \Delta x + (\Delta x)^2 + 5. \tag{2}$

Subtracting (1) from (2):

$$\begin{aligned} y + \Delta y &= x^2 + 2x \cdot \Delta x + (\Delta x)^2 + 5 \\ y &= x^2 + 5 \\ \hline \Delta y &= 2x \cdot \Delta x + (\Delta x)^2 \end{aligned} \tag{3}$$

Dividing (3) by Δx:

$$\frac{\Delta y}{\Delta x} = 2x + \Delta x.$$

Let $\Delta x \to 0$:

$$\lim_{\Delta x \to 0} \frac{\Delta y}{\Delta x} = 2x,$$

or

$$\frac{dy}{dx} = 2x.$$

EXAMPLE 3. Differentiate: $y = x^2 + 3x + 8$.

Solution.

$$
\begin{aligned}
y + \Delta y &= (x + \Delta x)^2 + 3(x + \Delta x) + 8 \\
y + \Delta y &= x^2 + 2x \cdot \Delta x + (\Delta x)^2 + 3x + 3\Delta x + 8 \\
y &= x^2 \qquad\qquad\qquad\quad + 3x \qquad\quad + 8 \\
\hline
\Delta y &= 2x \cdot \Delta x + (\Delta x)^2 + 3\Delta x
\end{aligned}
$$

Dividing through by Δx:

$$\frac{\Delta y}{\Delta x} = 2x + \Delta x + 3.$$

Let $\Delta x \to 0$:

$$\lim_{\Delta x \to 0} \frac{\Delta y}{\Delta x} = 2x + 3,$$

or

$$\frac{dy}{dx} = 2x + 3.$$

EXAMPLE 4. Differentiate: $y = x^3$.

Solution. $y + \Delta y = (x + \Delta x)^3$.

$$y + \Delta y = x^3 + 3x^2 \cdot \Delta x + 3x \cdot (\Delta x)^2 + (\Delta x)^3.$$

Subtracting: $\quad \Delta y = 3x^2 \cdot \Delta x + 3x(\Delta x)^2 + (\Delta x)^3.$

Dividing: $\quad \dfrac{\Delta y}{\Delta x} = 3x^2 + 3x(\Delta x) + (\Delta x)^2.$

Passing to the limit:

$$\frac{dy}{dx} = 3x^2,$$

for, as $\Delta x \to 0$, the terms $3x(\Delta x)$ and $(\Delta x)^2$ vanish.

EXAMPLE 5. Differentiate: $y = \dfrac{x + 1}{x}$.

Solution. $y + \Delta y = \dfrac{x + \Delta x + 1}{x + \Delta x}$.

$$\Delta y = \frac{x + \Delta x + 1}{x + \Delta x} - \frac{x + 1}{x};$$

$$\Delta y = \frac{x(x + \Delta x + 1) - (x + \Delta x)(x + 1)}{x(x + \Delta x)}$$

$$= \frac{x^2 + x \cdot \Delta x + x - x^2 - x \cdot \Delta x - x - \Delta x}{x(x + \Delta x)}$$

$$= \frac{-\Delta x}{x(x + \Delta x)} = \frac{-\Delta x}{x^2 + x \cdot \Delta x}.$$

$$\frac{\Delta y}{\Delta x} = \frac{-1}{x^2 + x \cdot \Delta x},$$

or $\dfrac{dy}{dx} = \dfrac{-1}{x^2}$.

2—8. The General Rule for Differentiation.   We may now formulate the "four-step" rule for differentiating a function, as follows:

Step 1.   Substitute $(x + \Delta x)$ for x in the given equation, thus giving y a new corresponding value, $(y + \Delta y)$.

Step 2.   Subtract the given equation from the equation obtained in Step 1, thus obtaining an expression for Δy.

Step 3.   Divide the equation obtained in Step 2 by Δx, thus obtaining a value for $\dfrac{\Delta y}{\Delta x}$.

Step 4.   Find the limit of $\dfrac{\Delta y}{\Delta x}$ as Δx approaches zero as a limit.   This limit is the required derivative.

EXERCISE 2—1

Differentiate each of the following functions by the General Rule:

1. $y = 2x - 3$
2. $y = 4 - x^2$
3. $y = 2x^2 + 3x$
4. $y = 10x^2$
5. $y = x^2 - x + 6$

6. $y = 3x^2 - 2x + 1$
7. $y = 2x^3 + 3$
8. $y = x^3 - 2x$
9. $y = (x + 1)(x - 2)$
10. $y = (2x - 1)(3x + 2)$

11. $y = \dfrac{4x - 5}{2}$

14. $y = \dfrac{x}{x + 1}$

12. $y = \dfrac{1}{x}$

15. $y = \dfrac{x + 1}{x - 1}$

13. $y = \dfrac{x - 2}{x}$

16. $y = \dfrac{x + 1}{x^2}$

2—9. Finding the Tangent to a Curve. In the light of the discussion in §2—6, we are now able to find the slope of the tangent to a given curve at a given point, or the slope of a curve at any desired point, provided, of course, that the equation of the curve is given.

EXAMPLE 1. Find the slope of the curve $y = x^2 + 6$ at the point where $x = 5$.

Solution. $\qquad\qquad\qquad y = x^2 + 6.$

By the General Rule,

$$\frac{\Delta y}{\Delta x} = 2x + \Delta x,$$

and $\qquad\qquad \dfrac{dy}{dx} = 2x.$

Hence, when $x = 5$, $\dfrac{dy}{dx} = 2(5) = 10$. The slope of the tangent to $y = x^2 + 6$ at the point where $x = 5$ is 10, as is the slope of the curve at that point. The inclination of the tangent is $\phi =$ arc tan 10, or approximately $84°17'$.

EXAMPLE 2. Find the slope of the curve $y = 3x^2 - 4x + 8$ at the point where $x = 3$.

Solution. By the General Rule,

$$\frac{\Delta y}{\Delta x} = 6x - 4 + 3(\Delta x),$$

and $\qquad\qquad \dfrac{dy}{dx} = 6x - 4.$

Hence, when $x = 3$, $\dfrac{dy}{dx} = 6(3) - 4 = 14$, the required slope.

EXERCISE 2—2

By differentiating by the General Rule, find the slope of the tangent to each of the following curves at the point indicated:

1. $y = x^2 - 5$, where $x = 6$.
2. $y = 4x^2 + 3$, where $x = 1$.
3. $y = x^2 - 3x + 6$, where $x = 4$.
4. $y = 3x^2 + 4x$, where $x = -2$.
5. $y = (x + 3)^2$, where $x = 0$.
6. $y = x^3 - x^2$, where $x = -3$.
7. $y = x^3 + 8$, where $x = -2$.
8. $y = x^4 - 1$, where $x = \frac{1}{2}$.

Differentiation of Algebraic Functions

CHAPTER THREE

THE DERIVATIVE OF A CONSTANT, A VARIABLE, AND A SUM

3—1. The Derivative of a Constant. We shall now derive standard formulas for differentiating algebraic functions, using the General Rule developed in Chapter Two. With these formulas for finding the derivative of functions and various combinations of functions, it will then be possible to differentiate specific functions in later work when applying the calculus to practical problems.

The simplest algebraic function is that given by $y = c$, where c is any constant. Here, as x changes, y remains constant, being equal to c for all values of x. Hence, when x is given an increment Δx, the corresponding increment in y has the value zero; or, $\Delta y = 0$. Dividing by Δx:

$$\frac{\Delta y}{\Delta x} = 0;$$

and since the difference-quotient is always zero, its limit is likewise zero. Therefore

$$\frac{dy}{dx} = \lim_{\Delta x \to 0} \frac{\Delta y}{\Delta x} = 0.$$

In other words:

$$\frac{d}{dx}(c) = 0. \qquad [1]$$

RULE. *The derivative of a constant equals zero.*

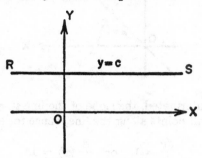

Geometrically, if the line RS represents the equation $y = c$, we get, by differentiating:

$$\frac{dy}{dx} = \frac{dc}{dx}.$$

But $\frac{dy}{dx}$ represents the slope of the line RS, or the function $y = c$; and since the line RS is parallel to OX, its slope equals zero. Therefore, as above,

$$\frac{d}{dx}(c) = 0.$$

3—2. The Derivative of a Variable with Respect to Itself. Consider the simple function, $y = x$. Applying the General Rule:

Step 1. $\qquad\qquad y + \Delta y = x + \Delta x.$

Step 2. $\qquad\qquad \Delta y = \Delta x.$

Step 3. $\qquad\qquad \frac{\Delta y}{\Delta x} = \frac{\Delta x}{\Delta x} = 1.$

Step 4. $\qquad\qquad \frac{dy}{dx} = \lim_{x \to 0} \frac{\Delta y}{\Delta x} = 1,$

or $\qquad\qquad \frac{d}{dx}(x) = 1. \qquad [2]$

RULE.   *The derivative of a variable with respect to itself equals unity.*

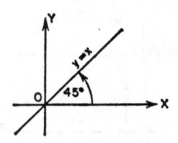

Geometrically interpreted, the slope of the line $y = x$ is constant and equal to unity at all points along the line; hence for all values of x,

$$\frac{dy}{dx} = 1, \quad \text{or} \quad \frac{d}{dx}(x) = 1.$$

3—3. The Derivative of the Algebraic Sum of Two or More Functions.
Consider the function

$$y = u + v - w,$$

where u, v, and w are functions of x. Applying the General Rule, we have:

Step 1. $y + \Delta y = u + \Delta u + v + \Delta v - w - \Delta w.$

Step 2. $\Delta y = \Delta u + \Delta v - \Delta w.$

Step 3. $\dfrac{\Delta y}{\Delta x} = \dfrac{\Delta u}{\Delta x} + \dfrac{\Delta v}{\Delta x} - \dfrac{\Delta w}{\Delta x}.$

Now, as $\Delta x \to 0$, $\Delta u \to 0$, $\Delta v \to 0$, and $\Delta w \to 0$; hence

Step 4. $\dfrac{dy}{dx} = \lim\limits_{\Delta x \to 0} \dfrac{\Delta y}{\Delta x} = \dfrac{du}{dx} + \dfrac{dv}{dx} - \dfrac{dw}{dx},$

or $\dfrac{d}{dx}(u + v - w) = \dfrac{du}{dx} + \dfrac{dv}{dx} - \dfrac{dw}{dx}.$ [3]

RULE.   *The derivative of the algebraic sum of any finite number of functions is equal to the algebraic sum of their individual derivatives.*

3—4. The Derivative of the Product of a Constant and a Variable.
Let the function be $y = cv$. By the General Rule:

Step 1. $\qquad y + \Delta y = c(v + \Delta v) = cv + c\Delta v.$

Step 2. $\qquad \Delta y = c\Delta v.$

Step 3. $\qquad \dfrac{\Delta y}{\Delta x} = c\,\dfrac{\Delta v}{\Delta x}.$

Now, as $\Delta x \to 0$, $\Delta v \to 0$; hence

Step 4. $\qquad \dfrac{dy}{dx} = \lim\limits_{\Delta x \to 0} \dfrac{\Delta y}{\Delta x} = c\,\dfrac{dv}{dx},$

or $\qquad \dfrac{d}{dx}\,(cv) = c\,\dfrac{dv}{dx}.$ $\hfill$ [4]

RULE. *The derivative of the product of a constant and a variable is equal to the product of the constant and the derivative of the variable.*

If the constant appears in the denominator, we have:

$$\frac{d}{dx}\left(\frac{v}{c}\right) = \frac{1}{c}\,\frac{dv}{dx}.$$ $\hfill$ [4a]

EXAMPLE 1. Find the derivative of $x + 5$.

Solution. $\dfrac{d}{dx}\,(x + 5) = \dfrac{d}{dx}\,(x) + \dfrac{d}{dx}\,(5) = 1 + 0 = 1.$

EXAMPLE 2. Differentiate $y = 3x + 2$.

Solution. $\dfrac{dy}{dx} = \dfrac{d}{dx}\,(3x) + \dfrac{d}{dx}\,(2) = 3 + 0 = 3.$

EXAMPLE 3. Differentiate $y = \dfrac{x}{4} + 5x - 1$.

Solution. $\dfrac{dy}{dx} = \dfrac{d}{dx}\left(\dfrac{x}{4}\right) + \dfrac{d}{dx}\,(5x) - \dfrac{d}{dx}\,(1)$

$$\frac{dy}{dx} = \frac{1}{4} + 5 - 0 = 5\tfrac{1}{4}.$$

EXAMPLE 4. Differentiate $y = 4(3x - 7)$.

Solution. $\dfrac{dy}{dx} = \dfrac{d}{dx}\,(cv) = c\,\dfrac{dv}{dx}$

Here $c = 4$, and $v = 3x - 7$

Therefore $\dfrac{dy}{dx} = (4) \dfrac{d}{dx} (3x - 7) = (4)(3) = 12.$

DERIVATIVE OF THE POWER FUNCTION

3—5. The Derivative of a Variable with a Constant Exponent.
Consider the power function $y = v^n$. Applying the General Rule, and
expanding by the Binomial Theorem, we obtain:

Step 1. $\qquad\qquad\qquad y + \Delta y = (v + \Delta v)^n.$

Step 2. $\qquad\qquad\qquad \Delta y = (v + \Delta v)^n - v^n.$

$$\Delta y = \left[v^n + n v^{n-1}(\Delta v) + \frac{n(n-1)}{2!} v^{n-2}(\Delta v)^2 + \cdots + (\Delta v)^n \right] - v^n,$$

or $\Delta y = n v^{n-1}(\Delta v) + \dfrac{n(n-1)}{2!} v^{n-2}(\Delta v)^2 + \cdots + (\Delta v)^n.$

Step 3.

$$\frac{\Delta y}{\Delta x} = n v^{n-1} \left(\frac{\Delta v}{\Delta x} \right) + \frac{n(n-1)}{2!} v^{n-2}(\Delta v) \left(\frac{\Delta v}{\Delta x} \right) + \cdots + (\Delta v)^{n-1} \left(\frac{\Delta v}{\Delta x} \right).$$

Now, as $\Delta x \to 0$, $\Delta v \to 0$; hence

Step 4.

$$\frac{dy}{dx} = \lim_{\Delta x \to 0} \frac{\Delta y}{\Delta x} = n v^{n-1} \frac{dv}{dx} + \frac{n(n-1)}{2!} v^{n-2}(0) \frac{dv}{dx} + \cdots ,$$

or $\qquad\qquad\qquad \dfrac{d}{dx} (v^n) = n v^{n-1} \dfrac{dv}{dx}.$ $\qquad\qquad$ [5]

RULE. *The derivative of a variable with a constant exponent is equal to
the product of that exponent, the variable raised to a power one less than the
original exponent, and the derivative of the variable.*

NOTE 1. It is assumed, in the above proof, that the exponent n is a
positive integer. If this were not so, we could not have applied the
binomial theorem, for when n is negative or fractional, the number of
terms in the binomial expansion is infinite; we could not pass to the limit,

therefore, since Principle (III) of §1—15 applies only to a finite number of terms. However, we shall find later that the relation

$$\frac{d}{dx}\,(v^n) = nv^{n-1}\frac{dv}{dx}$$

holds for any rational value of n, whether positive, negative, or fractional.

NOTE 2. As a special case of formula [5] above, we let $v = x$, and obtain:

$$\frac{d}{dx}\,(x) = nx^{n-1}\frac{dx}{dx},$$

or

$$\frac{d}{dx}\,(x) = nx^{n-1}. \tag{5a}$$

NOTE 3. Formulas [5] and [5a], in view of what was said in NOTE 1, may be applied to a power function when the exponent is negative or fractional, as shown in Examples 3 and 4 below; thus

$$\frac{d}{dx}\left(\frac{c}{v^n}\right) = \frac{d}{dx}\,(cv^{-n}) = c\frac{d}{dx}\,(v^{-n}) = -cnv^{-(n+1)}\frac{dv}{dx}.$$

EXAMPLE 1. Differentiate $y = x^3 + 5x^2 + 4$.

Solution. $\dfrac{dy}{dx} = 3x^2 + 10x$.

EXAMPLE 2. Differentiate $y = 5x^4 + 6x^3 - 3x^2 + 9x - 21$.

Solution. $\dfrac{dy}{dx} = 20x^3 + 18x^2 - 6x + 9$.

EXAMPLE 3. Differentiate $y = \dfrac{1}{x^3} + \dfrac{3}{x^2} - \dfrac{5}{x}$.

Solution. $y = x^{-3} + 3x^{-2} - 5x^{-1}$.

$$\frac{dy}{dx} = -3(x^{-4}) + (3)(-2)(x^{-3}) - 5(-1)(x^{-2}),$$

or

$$\frac{dy}{dx} = \frac{-3}{x^4} - \frac{6}{x^3} + \frac{5}{x^2}.$$

EXAMPLE 4. Differentiate $y = 2\sqrt{x} + \sqrt{3x} - \dfrac{4}{\sqrt[3]{x}}$.

Solution. $y = 2x^{\frac{1}{2}} + (3x)^{\frac{1}{2}} - 4(x^{-\frac{1}{3}})$.

$$\frac{dy}{dx} = 2(\tfrac{1}{2})(x^{-\frac{1}{2}}) + \tfrac{1}{2}(3x)^{-\frac{1}{2}}(3) - 4(-\tfrac{1}{3})(x^{-\frac{4}{3}}),$$

or $\qquad \dfrac{dy}{dx} = \dfrac{1}{\sqrt{x}} + \dfrac{3}{2\sqrt{3x}} + \dfrac{4}{3\sqrt[3]{x}}$.

EXAMPLE 5. Find $\dfrac{dy}{dx}$: $y = (5x + 3)^2$.

Solution. $v = 5x + 3$, and $y = v^2$.

Hence $\qquad \dfrac{dy}{dx} = nv^{n-1}\dfrac{dv}{dx} = 2v(5) = 10(5x + 3)$.

This result may be verified by expanding the given function before differentiating; thus

$$y = (5x + 3)^2 = 25x^2 + 30x + 9;$$

$$\frac{dy}{dx} = 50x + 30 = 10(5x + 3).$$

EXAMPLE 6. Differentiate with respect to x: $y = 4(x^2 + 6x)^3$.

Solution.
Here $v = x^2 + 6x$, and $y = 4v^3$.

$$\text{Hence} \quad \frac{dy}{dx} = nv^{n-1}\frac{dv}{dx} = (4)(3)v^2\frac{dv}{dx}$$

$$= (4)(3)(x^2 + 6x)^2(2x + 6)$$

$$= 12(x^2 + 6x)^2(2x + 6).$$

EXAMPLE 7. Differentiate $y = \dfrac{\sqrt{x^2 - 4}}{3}$.

Solution. $y = \dfrac{1}{3}(x^2 - 4)^{\frac{1}{2}}; \qquad v = x^2 - 4$

$$\frac{dy}{dx} = \frac{1}{3}\left(\frac{1}{2}\right)(x^2 - 4)^{-\frac{1}{2}}\frac{d}{dx}(x^2 - 4)$$

$$= \frac{1}{3}\left(\frac{1}{2}\right)(x^2 - 4)^{-\frac{1}{2}}(2x) = \frac{x}{3\sqrt{x^2 - 4}}.$$

EXERCISE 3—1

Differentiate:

1. $y = 2x^3 - 5x + 8$
2. $y = x^5 - 3x^4 + 2x^3 + x$
3. $y = \dfrac{1}{2}x + \dfrac{x^2}{3} - \dfrac{x^3}{6}$
4. $y = 5\sqrt{x} + 6$
5. $s = v_1 t + \frac{1}{2}gt^2$
6. $y = x^{-3} - 3x^{-2} + 4x^{-1}$
7. $y = \sqrt{6x + 3}$
8. $s = \dfrac{4t^3}{3} + 4t^2 - \dfrac{2}{t}$

9. $y = (x^2 - 5x)^2$
10. $s = k(a + 2t)^3$
11. $y = x^m - mx^3$
12. $y = \sqrt{x^2 - 3x}$
13. $y = (x^3 - 2x)^4$
14. $y = \sqrt{3x} + \dfrac{5}{x}$
15. $y = \sqrt{m^2 - x^2}$
16. $y = \dfrac{1}{\sqrt{m^2 + x^2}}$

DERIVATIVE OF PRODUCTS AND QUOTIENTS

3—6. The Derivative of a Product. Let the function to be differentiated be given by $y = uv$, where u and v are functions of x. If x is given an increment Δx, the functions u and v, and of course, y, also take on corresponding increments Δu, Δv, and Δy. We then have:

Step 1.
$$y + \Delta y = (u + \Delta u)(v + \Delta v)$$
$$= uv + u\Delta v + v\Delta u + \Delta u\Delta v.$$

Step 2.
$$\Delta y = u\Delta v + v\Delta u + \Delta u\Delta v.$$

Step 3.
$$\frac{\Delta y}{\Delta x} = u\frac{\Delta v}{\Delta x} + v\frac{\Delta u}{\Delta x} + \Delta u\frac{\Delta v}{\Delta x}.$$

Now as $\Delta x \to 0$, u and v remain unchanged, so that by Principle (II), §1—15, we have

$$\lim_{\Delta x \to 0}\left(u\frac{\Delta v}{\Delta x}\right) = u\left(\lim_{\Delta x \to 0}\frac{\Delta v}{\Delta x}\right) = u\frac{dv}{dx},$$

and
$$\lim_{\Delta x \to 0}\left(v\frac{\Delta u}{\Delta x}\right) = v\left(\lim_{\Delta x \to 0}\frac{\Delta u}{\Delta x}\right) = v\frac{du}{dx}.$$

Furthermore, as $\Delta x \to 0$, Δu and Δv both approach zero; hence

$$\lim_{\Delta x \to 0}\left(\Delta u\frac{\Delta v}{\Delta x}\right) = \left(\lim_{\Delta x \to 0}\Delta u\right)\left(\lim_{\Delta x \to 0}\frac{\Delta v}{\Delta x}\right) = (0)\left(\frac{dv}{dx}\right) = 0.$$

Step 4.
$$\frac{dy}{dx} = u\frac{dv}{dx} + v\frac{du}{dx},$$

or
$$\frac{d}{dx}(uv) = u\frac{dv}{dx} + v\frac{du}{dx}.$$ [6]

RULE. *The derivative of the product of two functions is equal to the product of the first function and the derivative of the second, plus the product of the second function and the derivative of the first.*

NOTE. In Step 3, when dividing the product $\Delta u \cdot \Delta v$ by Δx, we arbitrarily wrote $\Delta u\left(\dfrac{\Delta v}{\Delta x}\right)$; we could just as well have written $\Delta v\left(\dfrac{\Delta u}{\Delta x}\right)$ instead. Since Δv also approaches zero as $\Delta x \to 0$, the product $\Delta v\left(\dfrac{\Delta u}{\Delta x}\right)$ vanishes when we pass to the limit, and the final result is the same.

EXAMPLE 1. Find the derivative: $y = 5x^4(x^3 - 2x)$.

Solution. Let $u = 5x^4$, and $v = x^3 - 2x$.

Then
$$\frac{dy}{dx} = u\frac{dv}{dx} + v\frac{du}{dx}.$$

$$\frac{dy}{dx} = (5x^4)(3x^2 - 2) + (x^3 - 2x)(20x^3).$$

EXAMPLE 2. Differentiate $y = (3x^2 + 4)(3x^3 - 5x)$.

Solution. Here $u = 3x^2 + 4$, $v = 3x^3 - 5x$.

Hence
$$\frac{dy}{dx} = (3x^2 + 4)(9x^2 - 5) + (3x^3 - 5x)(6x).$$

EXAMPLE 3. Differentiate $y = (x^2 - 2)\sqrt{3x^2 + 5}$.

Solution. $u = x^2 - 2$, $v = (3x^2 + 5)^{\frac{1}{2}}$.

Hence
$$\frac{dy}{dx} = (x^2 - 2)(\tfrac{1}{2})(3x^2 + 5)^{-\frac{1}{2}} \cdot (6x) + (3x^2 + 5)^{\frac{1}{2}}(2x),$$

or
$$\frac{dy}{dx} = \frac{3x(x^2 - 2)}{\sqrt{3x^2 + 5}} + 2x\sqrt{3x^2 + 5}.$$

EXAMPLE 4. Prove that

$$\frac{d}{dx}(uvw) = vw\frac{du}{dx} + uw\frac{dv}{dx} + uv\frac{dw}{dx}.$$

Solution. Let $uv = z$.

Then

$$\frac{d}{dx}(zw) = z\frac{dw}{dx} + w\frac{dz}{dx}$$

$$= uv\frac{dw}{dx} + w\frac{d}{dx}(uv)$$

$$= uv\frac{dw}{dx} + w\left[u\frac{dv}{dx} + v\frac{du}{dx}\right]$$

$$= uv\frac{dw}{dx} + uw\frac{dv}{dx} + wv\frac{du}{dx}.$$

The method of solving Example 4 may be extended to find the derivative of the product of any finite number of functions.

EXERCISE 3—2

Differentiate:

1. $y = (x + 1)(x - 2)$
2. $y = (x^2 + 1)(3x + 4)$
3. $y = (x^2 - 1)(2x^2 - 3x + 1)$
4. $y = (x^3 - 1)^2$
5. $y = (x^2 + 1)^3$
6. $y = t(t - 1)(t^2 + 2)$
7. $y = (3x^2 + 2x)(4x^2 - x)$
8. $y = (x + 1)(x + 2)(x - 3)$
9. $y = (t^2 - 3t + 4)^3$
10. $y = (x + 2)^2(x + 3)^3$

3—7. The Derivative of a Quotient. We have already seen in §3—4, [4a] that

$$\frac{d}{dx}\left(\frac{v}{c}\right) = \frac{1}{c}\frac{dv}{dx}.$$

We have also seen (§3—5, NOTE 3) that

$$\frac{d}{dx}\left(\frac{c}{v}\right) = \frac{d}{dx}(cv^{-1}) = c\frac{d}{dx}(v^{-1}) = -cv^{-2}\frac{dv}{dx}.$$

We shall now derive the formula for the derivative of a quotient when both the numerator and the denominator contain variables.

Let $y = \dfrac{u}{v}$. Then, applying the General Rule once again:

Step 1. $y + \Delta y = \dfrac{u + \Delta u}{v + \Delta v}.$

Step 2. $\Delta y = \dfrac{u + \Delta u}{v + \Delta v} - \dfrac{u}{v} = \dfrac{v\Delta u - u\Delta v}{v(v + \Delta v)}.$

Step 3. $\dfrac{\Delta y}{\Delta x} = \dfrac{v\dfrac{\Delta u}{\Delta x} - u\dfrac{\Delta v}{\Delta x}}{v(v + \Delta v)} = \dfrac{v\dfrac{\Delta u}{\Delta x} - u\dfrac{\Delta v}{\Delta x}}{v^2 + v(\Delta v)}.$

Now as $\Delta x \to 0$, $\Delta u \to 0$, and $\Delta v \to 0$;

$$\lim_{\Delta x \to 0}\left(v\frac{\Delta u}{\Delta x}\right) = v\frac{du}{dx}; \quad \lim_{\Delta x \to 0}\left(u\frac{\Delta v}{\Delta x}\right) = u\frac{dv}{dx}; \quad \text{and} \quad v(\Delta v) \to 0.$$

Step 4. $\dfrac{dy}{dx} = \dfrac{v\dfrac{du}{dx} - u\dfrac{dv}{dx}}{v^2},$

or $\dfrac{d}{dx}\left(\dfrac{u}{v}\right) = \dfrac{v\dfrac{du}{dx} - u\dfrac{dv}{dx}}{v^2}.$ [7]

RULE. *The derivative of a fraction (i.e., a quotient) is equal to the denominator multiplied by the derivative of the numerator, minus the numerator multiplied by the derivative of the denominator, all divided by the square of the denominator.*

EXAMPLE 1.   Differentiate $y = \dfrac{2x}{x^2 + 1}.$

Solution.   Here $u = 2x$, and $v = x^2 + 1$.
From equation [7]:

$$\frac{dy}{dx} = \frac{(x^2 + 1)(2) - (2x)(2x)}{(x^2 + 1)^2},$$

$$\frac{dy}{dx} = \frac{2 - 2x^2}{(x^2 + 1)^2}.$$

EXAMPLE 2.   Differentiate $y = \dfrac{x + 3}{x - 2}.$

Solution. $u = x + 3$, and $v = x - 2$.
Therefore

$$\frac{dy}{dx} = \frac{(x-2)(1) - (x+3)(1)}{(x-2)^2} = \frac{-5}{(x-2)^2}.$$

EXAMPLE 3. Differentiate $y = \dfrac{x^2 - x + 2}{x^2 + 3}$.

Solution. $u = x^2 - x + 2$; $v = x^2 + 3$.

$$\frac{dy}{dx} = \frac{(x^2+3)(2x-1) - (x^2-x+2)(2x)}{(x^2+3)^2}$$

$$= \frac{x^2 + 2x - 3}{(x^2+3)^2}.$$

EXERCISE 3—3

Differentiate:

1. $y = \dfrac{x-1}{x}$

2. $y = \dfrac{x}{1+x}$

3. $y = \dfrac{x-k}{x+k}$

4. $y = \dfrac{a+x}{b+x}$

5. $y = \dfrac{1}{(x+k)^2}$

6. $y = \dfrac{x^3}{2-x}$

7. $y = \dfrac{x^2+9}{x^2-9}$

8. $y = \dfrac{3}{x^3+1}$

9. $y = \dfrac{1+x}{1+x^2}$

10. $y = \dfrac{2x^4}{m^2-x^2}$

3—8. Importance of Rapid, Accurate Differentiation. It need hardly be pointed out that to use the calculus skillfully requires considerable practice with standard formulas for differentiating various functions. Among the most commonly used formulas are those for the power function, for a product, and for a quotient. The exercise below affords further practice in the use of these formulas.

For the reader's convenience, we summarize below the standard formulas for differentiating algebraic functions.

[1]
$$\frac{d}{dx}(c) = 0$$

[2]
$$\frac{d}{dx}(x) = 1$$

[3]
$$\frac{d}{dx}(u + v - w) = \frac{du}{dx} + \frac{dv}{dx} - \frac{dw}{dx}$$

[4]
$$\frac{d}{dx}(cv) = c\frac{dv}{dx}$$

[4a]
$$\frac{d}{dx}\left(\frac{v}{c}\right) = \frac{1}{c}\frac{dv}{dx}$$

[5]
$$\frac{d}{dx}(v^n) = nv^{n-1}\frac{dv}{dx}$$

[5a]
$$\frac{d}{dx}(x^n) = nx^{n-1}$$

[6]
$$\frac{d}{dx}(uv) = u\frac{dv}{dx} + v\frac{du}{dx}$$

[7]
$$\frac{d}{dx}\left(\frac{u}{v}\right) = \frac{v\dfrac{du}{dx} - u\dfrac{dv}{dx}}{v^2}$$

EXERCISE 3—4

Differentiate:

1. $y = (x + 3)(x^2 - 2)$

2. $y = (a + x)(b - x)$

3. $y = x(k + x)^2$

4. $y = \dfrac{x^2 + 1}{x^3}$

5. $y = \dfrac{x - 5}{x + 2}$

6. $y = \dfrac{x - 1}{x}$

7. $y = \dfrac{x}{x - 1}$

8. $y = (3x + 2)\sqrt{3x - 2}$

9. $y = \sqrt{\dfrac{x - 1}{x + 1}}$

10. $y = \dfrac{ax^2 + bx + c}{x}$

11. $y = (1 - 2x^3)(1 + 5x^2)$

12. $y = (b - x)\sqrt{a + x}$

13. $y = (m^2 + x^2)\sqrt{m^2 - x^2}$

14. $y = \dfrac{x - a}{x + a}$

15. $y = \dfrac{2x^2 + 3}{x^3 - 2}$

16. $y = \dfrac{(x + 3)^2}{x + 1}$

17. $s = \dfrac{2t^3}{m^2 - t^2}$

18. $s = t(t + 1)(t - 2)$

19. $s = \dfrac{t^3}{(1 - t)^3}$

20. $y = \dfrac{\sqrt{x + 1}}{x^3}$

DIFFERENTIATION OF IMPLICIT FUNCTIONS

3—9. Explicit and Implicit Functions. Consider the function $y^2 = 2px$. When solved for y, we have $y = \pm\sqrt{2px}$, or $f(x) = \pm\sqrt{2px}$, where $y = f(x)$. When solved for x, we have $x = y^2/2p$, where $x = \phi(y)$. The functions $y = f(x)$ and $x = \phi(y)$ are *explicit* functions. In the first instance, y is expressed as a function of x; in the latter case, x is expressed as a function of y. Each function is the inverse of the other. But in the original form $y^2 = 2px$, each variable is defined *implicitly* as a function of the other. Additional examples of implicit functions are given herewith:

$$x^2 + y^2 + 2 = 0;$$

$$x^2 + xy + y^2 = 1;$$

$$x^3 + y^3 = kxy;$$

$$x + y = \sqrt{xy}.$$

Implicit functions, instead of being given as $y = f(x)$, or $x = \phi(y)$, or $s = f(t)$, are often written as $F(x,y) = 0$. It may be possible to solve an implicit function $F(x,y) = 0$ for one of the variables in terms of the other, yielding an explicit function such as $y = f(x)$; sometimes this cannot be done conveniently, and sometimes not at all.

3—10. Differentiation of Implicit Functions. Although it is not always simple or possible to obtain an explicit function from a given implicit function, nevertheless the derivative $\dfrac{dy}{dx}$ can be found, as shown by the following examples.

EXAMPLE 1. Find $\dfrac{dy}{dx}$ in the equation $x^2 + y^2 = 25$.

Solution. Differentiate each side of the equation with respect to x:

$$\frac{d}{dx}(x^2 + y^2) = \frac{d}{dx}(25),$$

or

$$\frac{d}{dx}(x^2) + \frac{d}{dx}(y^2) = \frac{d}{dx}(25),$$

$$2x + 2y\frac{dy}{dx} = 0.$$

Solve for $\frac{dy}{dx}$:

$$2y\frac{dy}{dx} = -2x,$$

$$\frac{dy}{dx} = -\frac{x}{y}.$$

EXAMPLE 2. Find $\frac{dy}{dx}$ in the equation $y^2 = x^3$.

Solution. Differentiating each side:

$$2y\frac{dy}{dx} = 3x^2.$$

Solving for $\frac{dy}{dx}$: $\qquad\qquad \dfrac{dy}{dx} = \dfrac{3x^2}{2y}.$

EXAMPLE 3. Find $\frac{dy}{dx}$ in the equation $xy = a^2$.

Solution.

$$\frac{d}{dx}(xy) = \frac{d}{dx}(a^2).$$

$$x\frac{dy}{dx} + y = 0,$$

$$\frac{dy}{dx} = -\frac{y}{x}.$$

EXAMPLE 4. Find $\frac{dy}{dx}$ in the equation $x^2 + y^2 - 2xy = 0$.

Solution. $2x + 2y\dfrac{dy}{dx} - 2x\dfrac{dy}{dx} - 2y = 0.$

$$\frac{dy}{dx}(2y - 2x) = 2y - 2x,$$

$$\frac{dy}{dx} = 1.$$

EXAMPLE 5. Find $\dfrac{dy}{dx}$ in the equation $x^3 - 2x^2y + 2y^3 = 0.$

Solution. $3x^2 - 2x^2\dfrac{dy}{dx} - 4xy + 6y^2\dfrac{dy}{dx} = 0.$

$$\frac{dy}{dx}(6y^2 - 2x^2) = 4xy - 3x^2,$$

$$\frac{dy}{dx} = \frac{4xy - 3x^2}{6y^2 - 2x^2}.$$

It should be observed that, in general, as in all the above examples except Example 4, the value obtained for $\dfrac{dy}{dx}$ contains both x and y. If we wish to obtain an expression for the derivative containing only x-terms, we may theoretically replace y by its value in terms of x as found from the original equation, $F(x,y) = 0$; however, this is sometimes very inconvenient, and is usually not necessary.

It is also worth noting that for most implicit functions, it is generally more convenient to find $\dfrac{dy}{dx}$ by the method shown above than it is first to express the given function explicitly and then to differentiate directly.

EXERCISE 3—5

Find $\dfrac{dy}{dx}$ for each of the following, leaving the result in terms of x and y:

1. $x^2 + y^2 = 36$
2. $x^2 = 4py$
3. $xy = -12$
4. $x^2 - y^2 + x - y = 0$
5. $x^2 + y^2 + xy = 0$

6. $2x + y = \sqrt{x + y}$
7. $x^3(x + a) = y^2$
8. $x + y = 2xy$
9. $b^2x^2 + a^2y^2 = a^2b^2$
10. $b^2x^2 - a^2y^2 = 1$

11. $xy^2 = 3(x + 2)$

12. $x^2 - y^2 + xy + 2y = 4$

13. $x - y + \sqrt{xy} = k$

14. $\sqrt{x} - \sqrt{y} = a + \dfrac{x}{2}$

15. $x^3 + y^3 = 3axy$

16. $x^3 + 2x^2y + 3y^3 = 0$

EXERCISE 3—6

Review

Differentiate:

1. $y = \sqrt{4px}$

2. $y = \sqrt{k^2 - 3x^2}$

3. $y = \dfrac{a}{b}\sqrt{x^2 + a^2}$

4. $y = \sqrt[3]{3x + 1}$

5. $y = \dfrac{k}{t} + at^3$

6. $y = (a + bt^2)^{3/2}$

7. $y = (x^3 + 3x^2 + 2)^3$

8. $y = (x^3 - 2)^5$

9. $y = \dfrac{1}{x^k}$

10. $y = \dfrac{x^2 + 3}{2x}$

11. $y = x(x - 2)(x + 1)$

12. $y = (2x^4 - 1)(2 + 3x^2)$

Find $\dfrac{dy}{dx}$ for each of the following:

13. $x^2 + y^2 + 2x - 4y = 0$

14. $x^3 - xy + y^2 = 1$

15. $x^2 + y - x^2y = 4$

16. $x^4 + x^2y^2 + y^4 = 16$

17. $a^2x^2 - b^2y^2 = a^2b^2$

18. $x^2y^2 = 2(x^2 - y^2)$

19. $Ax^2 + Cy^2 + Dx + Ey = 0$

20. $x^{2/3} + y^{2/3} = a^{2/3}$

Using the Derivative

CHAPTER FOUR

THE DERIVATIVE AS A TOOL

4—1. Interpretation of the Derivative. We have already seen that the derivative has several meanings:

(1) As a rate of change in one quantity which varies with another quantity.

If $\dfrac{\Delta y}{\Delta x}$ = the average rate of increase (grams per degree, square feet of area per foot of length, etc.),

then $\dfrac{dy}{dx}$ = the instantaneous rate of change.

(2) As a time rate of change.

If $\dfrac{\Delta y}{\Delta t}$ = the average speed during an arbitrary interval of time Δt,

then $\dfrac{dy}{dt}$ = the instantaneous speed.

(3) As a slope.

If $\dfrac{\Delta y}{\Delta x}$ = the average slope of a curve in an interval Δx,

then $\dfrac{dy}{dx}$ = the slope at a specified point on the curve.

Actually, these three meanings are simply various aspects of the same basic idea, namely, that of an *instantaneous rate of change as a limiting value*. Thus a *speed* is simply the *rate* at which the distance traveled is changing per unit of time; the slope of a curve is the *rate* at which a curve is rising per horizontal unit.

4—2. When Is a Function Increasing, When Decreasing? It is of great practical value to know whether, at some particular value of x, a given function $y = f(x)$ is increasing or decreasing. *If we adopt the convention that x is always increasing*, then we can see at once from the graph that the curve is rising at any point where its slope is positive, and falling where its slope is negative. In other words:

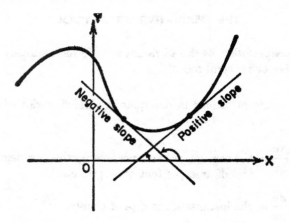

(1) y is increasing when $\dfrac{dy}{dx}$ is positive;

(2) y is decreasing when $\dfrac{dy}{dx}$ is negative.

The reader should note carefully that it is not a safe test to compare the *value of y* at the specified point with some near-by *value of y*; for y

might be decreasing at the point in question, but might have increased *before* reaching this point.

INSTANTANEOUS RATES OF CHANGE

4—3. Physical Changes. Many of the phenomena of physical science have to do with changing quantities. Thus *force, pressure, volume, temperature, work, power,* and the like are subject to change; indeed, it is not until we study such changes quantitatively that we really begin to understand the nature of our physical environment. The calculus, and particularly the derivative, furnishes the scientist, the technologist, and the engineer with a powerful tool to study phenomena involving change. The history of science affords eloquent testimony of the role played here by the calculus; for it was not until shortly after the invention of the calculus, roughly at the beginning of the eighteenth century, that modern science began to make tremendous advances in all fields, and notably in mechanics and astronomy.

We shall illustrate the application of the derivative to problems dealing with changing quantities.

EXAMPLE 1. A square metal plate, when heated, expands in area. Find the rate at which the area is increasing, per unit change in the length of the side s, at the instant when $s = 6$ cm.

Solution. $A = s^2$

$$\frac{dA}{ds} = 2s$$

When $s = 6$ cm., $\dfrac{dA}{ds} = 12$ sq. cm. per cm., or 12 sq. cm./cm.

EXAMPLE 2. The strength M of a certain beam varies with the thickness of the beam h according to the formula $M = 8.5h^2$. Find the rate at which the strength increases at the instant when $h = 5$ in.

Solution. $M = 8.5h^2$

$$\frac{dM}{dh} = 2(8.5)h$$

When $h = 5$, $\dfrac{dM}{dh} = 85$ units/inch.

EXAMPLE 3. The electromotive force developed by a thermoelectric couple is given by $E = 15t + 0.01t^2$, where t is the temperature in degrees, and E is in microvolts. How fast is E changing when $t = 1000$?

Solution. $\dfrac{dE}{dt} = 15 + .02t$

When $t = 1000$, $\dfrac{dE}{dt} = 15 + (.02)(1000) = 35$ microvolts per degree.

EXERCISE 4—1

1. The area of a circular metal plate increases as the temperature rises. Find the rate at which the area is increasing, per unit change in the radius, at the instant when $r = 10.3$ in.

2. The volume of a spherical rubber balloon increases as the gas pressure within increases. Find the rate at which the volume is increasing at the instant when $r = 12$ in.

3. The moment of inertia I of a square beam is given by

$$I = \frac{h^4}{12},$$

where h is the thickness of the beam. Find the rate of change in I when h equals 3.

4. The heat developed in an electric conductor in one second is equal to

$$H = 0.24I^2R,$$

where H is in calories and I is in amperes. For a fixed resistance of 20 ohms ($R = 20$), how fast is the heat increasing at the instant when $I = 5$?

5. The volume (V cu. in.) of a certain gas varies with the pressure (p lb. per sq. in.) as given by

$$V = \frac{600}{p}.$$

Find the rate at which V changes per unit change in p, at the instant when $p = 20$. Is the volume increasing or decreasing?

6. The theoretical discharge when water flows over a weir or a dam is

$$Q = 3.33lh^{\frac{3}{2}},$$

where Q is the amount of water in cu. ft. per sec., $l =$ the width of the weir in feet, and $h =$ the height of the water in feet (the head) above the weir. For a weir 200 feet wide, find the rate at which the amount of flow is increasing at the instant when $h = 9$ ft.

7. The heat of absorption of an ammonia solution in a refrigerating system is

$$Q = 887 - 350x - 400x^2,$$

where Q is the amount of heat in B.T.U., and x is the concentration of the ammonia solution. How fast is Q changing at the instant when $x = .18$?

8. The power required to propel a ship of a certain design is given by

$$H = \frac{d^{\frac{2}{3}}v^3}{200},$$

where H = horsepower, d = the ship's displacement in long tons, and v = the ship's speed in knots. For a ship with a displacement of 8000 long tons, how much power is required to propel the ship at 20 knots? How fast is the required power increasing at the instant when $v = 20$ knots?

9. The quantity of heat radiated by a surface varies directly as the fourth power of its absolute temperature:

$$Q = \frac{kAzT^4}{10^8},$$

where Q is the amount of heat in B.T.U.; A = the area of the surface in sq. ft.; z = the time in hours; T = the absolute temperature; and k is a constant depending upon the particular surface. Find how fast Q is changing when $T = 2000°$ for one square foot of surface for one hour; take $k = .15$.

10. The velocity with which a gas flows through a small hole varies inversely as the square root of the density d of the gas. (a) If $V = 50$ cu. in. per hr. for a gas whose density $d = .0225$, find the formula for the velocity of flow. (b) Find the rate at which V is changing at the instant when $d = .01$.

DISTANCE, VELOCITY, AND ACCELERATION

4—4. Distance and Speed. The distance traveled by a moving object is its *displacement*. If this displacement is achieved by motion in a straight line from its original position to a final position, it is referred to as *rectilinear motion*. The *speed* with which an object moves is the time rate of change—how far in a given unit of time. If in addition to the numerical value of the time rate of change the *direction* of motion is also specified, the rate is called the *velocity*. Thus, a plane flies with a speed of 150 miles per hour, but with a velocity of 150 miles per hour northeast.

If the distance is designated by s, the speed by v, and the time by t, then clearly, from our definitions,

$$v = \frac{ds}{dt},$$

or speed (velocity) equals the derivative of the distance with respect to the time. It is understood, of course, that the motion of a body may be *uniform* or *non-uniform*. In general, s is a function of t, or $s = f(t)$.

If the motion is uniform, $\frac{s}{t} = k$, or $s = kt$; and $\frac{ds}{dt} = k$. This is the familiar $R = \frac{D}{T}$ formula of elementary algebra, where R equals the rate, or $\frac{ds}{dt}$.

EXAMPLE 1. An object moves according to the formula $s = 30t + 5t^2$. Find (a) the distance traveled in 4 sec.; (b) the distance traveled in the fourth second; (c) its velocity at the end of the fourth second.

Solution.
  (a)   When $t = 4$, $s = (30)(4) + 5(4)^2 = 200$ ft.
  (b)   When $t = 3$, $s = (30)(3) + 5(3)^2 = 135$ ft.
        Hence, during fourth second, the distance covered equals $200 - 135 = 65$ ft.

  (c)   Velocity $= \frac{ds}{dt} = 30 + 10t$;

        when $t = 4$, $\frac{ds}{dt} = 30 + 10(4) = 70$ ft./sec.

EXAMPLE 2. A missile thrown vertically upward is moving in such a way that the height after t seconds is given by $h = 192t - 16t^2$. Find (a) the velocity at the end of 5 sec.; (b) the velocity at the end of 8 sec.; (c) the velocity at the end of 6 sec.; (d) what is the maximum height reached?

Solution.
$$h = 192t - 16t^2.$$

$$\frac{dh}{dt} = 192 - 32t.$$

  (a)   When $t = 5$, $\frac{dh}{dt} = 192 - (32)(5) = 32$ ft./sec.

  (b)   When $t = 8$, $\frac{dh}{dt} = 192 - (32)(8) = -64$ ft./sec.;

        the fact that the velocity is negative means that the missile is now moving *downward*, or falling.

(c)   When $t = 6$, $\dfrac{dh}{dt} = 192 - (32)(6) = 0$;

this means that the missile is moving neither upward nor downward.

(d)   The maximum height is attained when $\dfrac{dh}{dt} = 0$, or at $t = 6$; at $t = 6$, $h = (192)(6) - 16(6)^2 = 576$ ft.

4—5. Velocity and Acceleration. In motion that is not uniform, the velocity is either increasing or decreasing. The rate at which the velocity is changing is called *acceleration;* when the velocity is decreasing, that is, when the acceleration is negative, it is frequently called *deceleration.*

Since acceleration, usually denoted by a, is the rate of change of velocity, we may write

$$a = \frac{dv}{dt};$$

but since velocity is the rate of change of distance, or

$$v = \frac{ds}{dt},$$

we have

$$a = \frac{d}{dt}\left(\frac{ds}{dt}\right) = \frac{d^2s}{dt^2}.$$

The symbol $\dfrac{d^2s}{dt^2}$ is called the *second derivative* of the function $s = f(t)$. It should not be confused with $\left(\dfrac{ds}{dt}\right)^2$, which is the square of the derivative. A second derivative is thus *the derivative of a derivative.* This is to be expected, since it is the rate at which another rate is changing; or, the rate of change of a rate of change.

EXAMPLE.   A body is moving according to the formula $s = t^3 + 4t^2$. Find the distance traveled, the velocity, and the acceleration at the instant when $t = 3$.

Solution.

(a) $$s = t^3 + 4t^2,$$

$$s_3 = (3)^3 + 4(3)^2 = 63 \text{ ft.}$$

(b) $$v = \frac{ds}{dt} = 3t^2 + 8t,$$

$$v_3 = 3(3)^2 + 8(3) = 51 \text{ ft./sec.}$$

(c) $$a = \frac{d^2s}{dt^2} = 6t + 8,$$

$$a_3 = 6(3) + 8 = 26 \text{ ft./sec./sec.}$$

4—6. Laws of a Falling Body. It can be shown from physics that a body when falling freely from rest near the earth's surface, if air resistance is disregarded, follows the law

$$s = \tfrac{1}{2}gt^2, \tag{1}$$

where g is the constant of "gravitational acceleration," and equals about 32.2 feet per second per second when s is measured in feet and t is measured in seconds. From equation (1) we get:

$$v = \frac{ds}{dt} = 2\left(\frac{1}{2}g\right)t,$$

or $$v = gt; \tag{2}$$

and $$a = \frac{d}{dt}(v) = \frac{d}{dt}(gt) = g,$$

or $$a = g. \tag{3}$$

Also, from (1): $s = \tfrac{1}{2}gt^2 = \tfrac{1}{2}t(gt)$;

but from (2): $gt = v$;

hence $s = \tfrac{1}{2}tv$,   or $2s = vt$;

multiplying this last equation by the equation $gt = v$:

$$2sgt = v^2t,$$

or $$v^2 = 2gs. \tag{4}$$

Equations (1), (2), and (4) are often referred to as the *laws of a freely-falling body*.

EXAMPLE. If a body falls freely from rest, find (a) its velocity at the end of 5 sec.; (b) the distance fallen in 10 sec.; (c) its velocity after it has fallen 40 feet.

Solution.

(a) From (2), $v = gt = (32.2)(5) = 161$ ft./sec.

(b) From (1), $s = \frac{1}{2}gt^2 = \frac{1}{2}(32.2)(100) = 1610$ ft.

(c) From (4), $v^2 = 2gs$,

or $v = \sqrt{2gs} = \sqrt{2(32.2)(40)} = 50.8$ ft./sec.

4—7. Successive Derivatives. Differentiation may be repeated more than once. Thus, $\dfrac{d^2y}{dx^2}$, or $f''(x)$, is the derivative of the derivative of $y = f(x)$; that is,

$$f''(x) = \frac{d^2y}{dx^2} = \frac{d}{dx}\left(\frac{dy}{dx}\right).$$

Similarly,

$$f'''(x) = \frac{d^3y}{dx^3} = \frac{d}{dx}\left(\frac{d^2y}{dx^2}\right);$$

$$f^{\text{iv}}(x) = \frac{d^4y}{dx^4} = \frac{d}{dx}\left(\frac{d^3y}{dx^3}\right); \text{ etc.}$$

EXAMPLE 1. Find the fourth derivative of $y = 2x^5 + 3x^4 - 10x^2 + 5$.

Solution.
$$\frac{dy}{dx} = 10x^4 + 12x^3 - 20x;$$

$$\frac{d^2y}{dx^2} = 40x^3 + 36x^2 - 20;$$

$$\frac{d^3y}{dx^3} = 120x^2 + 72x;$$

$$\frac{d^4y}{dx^4} = 240x + 72.$$

EXAMPLE 2. Find the third derivative of $y = \dfrac{3}{x^5}$.

Solution.
$$y = 3x^{-5}.$$

$$\frac{dy}{dx} = -15x^{-6};$$

$$\frac{d^2y}{dx^2} = 90x^{-7};$$

$$\frac{d^3y}{dx^3} = -630x^{-8} = -\frac{630}{x^8}.$$

Example 3.   Find the second derivative of $y = \dfrac{x+1}{x}$.

Solution.

$$\frac{dy}{dx} = \frac{(x)(1) - (x+1)(1)}{x^2} = \frac{x - x - 1}{x^2} = -x^2;$$

$$\frac{d^2y}{dx^2} = \frac{d}{dx}(-x^{-2}) = \frac{2}{x^3}.$$

EXERCISE 4—2

1. Find the second derivative of each of the following:

   (a) $y = x^5 + 3x^3 - 4x$ (c) $y = \dfrac{2}{x^4}$

   (b) $s = 50t^2 - 2t^5$ (d) $V = \dfrac{4}{3}\pi r^3$

2. Find the third derivative of

   (a) $y = \dfrac{x - 1}{x}$ (b) $z = t(10 - t^2)$

3. Find the fourth derivative of

   (a) $y = x^6 + 20 + \dfrac{10}{x^2}$ (b) $s = t^5 - t^2$

4. Given the following equations of rectilinear motion, find the distance, velocity, and acceleration at the instant indicated in each case:

   (a) $s = 3t + t^3,\ t = 2$
   (b) $s = 10t - 5t^2,\ t = 1$
   (c) $s = 2t^3 - 6t^2,\ t = 1$

5. An object falls from rest; using the laws of a falling body (§4—6), find (a) its velocity at the end of 10 sec.; (b) its velocity after it has fallen 20 ft.; (c) its acceleration at the instant when $t = 6$; and (d) the distance it has fallen at the end of 4 seconds.   (Use $g = 32.2$.)

6. If an object starts with an initial velocity of v_0 and moves in a straight line with any given constant acceleration a (positive or negative), the distance traveled during a time t elapsed since the beginning of its motion is given by the formula

$$s = v_0 t + \tfrac{1}{2}at^2.$$

Prove: (a) $v = v_0 + at$, and (b) $\tfrac{1}{2}(v^2 - v_0^2) = as$.

7. Using the formulas from Problem 6, find the acceleration (assumed constant) with which a train, starting from rest, acquires a velocity of 60 miles per hour in 4 minutes. How far does it travel in that time?

8. An automobile moving at 40 miles an hour is brought to rest uniformly in 2 minutes. Find (a) its constant retardation (negative acceleration), and (b) how far it moves in that time.

9. Neglecting air resistance, the height (h ft.) reached in t seconds by an object projected vertically upward with an initial velocity of v_0 ft. per sec. is equal to $h = v_0 t - 16.1t^2$. Find (a) its velocity and acceleration at any instant; (b) its velocity and acceleration at the end of 3 seconds when the initial velocity equals 200 ft. per sec.; and (c) its velocity and acceleration at the end of 12 seconds.

10. A projectile is shot vertically upward with an initial velocity of 1200 ft./sec. Find (a) its velocity at the end of 15 sec.; (b) for how long after it is fired will it continue to rise? (Use formula from Problem 9.)

MAXIMA AND MINIMA

4—8. Changing Values of a Function. If a function increases as the independent variable increases, or decreases as the independent variable decreases, the function is said to be increasing; if the function decreases as the independent variable increases, or increases as the independent variable decreases, the function is decreasing. In other words, as we move left to right from a to b, $f(x)$ is increasing, or the

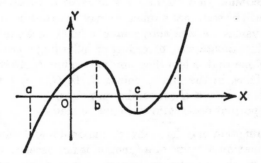

curve is rising; as we move from b to c, the function is decreasing, or the curve is falling; from c to d, the function is again increasing.

We have already noted that a function is

(a)   increasing when $\dfrac{dy}{dx} = f'(x)$ is positive (slope = $\tan \phi$ is $+$),

(b)   decreasing when $\dfrac{dy}{dx} = f'(x)$ is negative (slope = $\tan \phi$ is $-$).

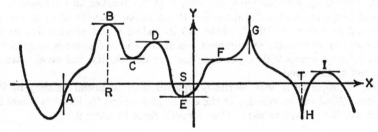

It will be seen, therefore, in order that a function shall change from an increasing to a decreasing function, or vice versa, the sign of its first derivative must change from $+$ to $-$, or from $-$ to $+$. In other words, at some point *the first derivative must have the value zero.* Such a point is called an *ordinary turning point,* and is illustrated by points B, C, D, E, and I. At ordinary turning points such as these,

$$\frac{dy}{dx} = f'(x) = 0,$$

the inclination of the tangent is zero, and the tangent at the turning point is parallel to the X-axis.

4—9. Maximum and Minimum Values of a Function. From the foregoing it will be seen that a function may have one or more maximum or minimum values. A *maximum value* of a function is one that is greater than any values *immediately* preceding or following; a *minimum value* of a function is one that is less than any values immediately preceding or following. Thus, in the figure, points B, D, G, and I are maximum points; points C, E, and H are minimum points.

Several important observations should be made.

(1)   A function may have several maximum and minimum values.

(2)   A maximum value of a function is not necessarily the greatest value the function may have, nor is a minimum value the least value. For example, there are values of y to the left of B which are greater than RB; there are values of y to the right of H which are less than SE and TH.

(3) At a maximum and minimum turning point (e.g., B, C, D, E, and I) the tangent is parallel to the X-axis, that is,

$$\text{slope} = \frac{dy}{dx} = f'(x) = 0.$$

(4) At a point of inflection (e.g., A and F), the curve changes from concave upward to concave downward arc, or vice versa; the slope, $f'(x)$, does not change in sign, but the tangent passes through the curve.

(5) At special points (e.g., G and H), although the derivative $\frac{dy}{dx}$ may not exist, the tangent (or the curve) may be perpendicular to the X-axis, although this need not necessarily be the case.

It is apparent, therefore, that for a function to have a maximum or a minimum value, *it is necessary* for the value of $f'(x)$ to be zero or infinite, but that this alone *is not a sufficient condition*. In addition to $f'(x)$ having a zero or infinite value, $f'(x)$ must change in sign as it passes through zero (or infinity).

We may then say that, in general, a function will have a maximum or minimum value under the following conditions:

(a) $f(x)$ is a maximum if $f'(x) = 0$, and $f'(x)$ changes from $+$ to $-$.

(b) $f(x)$ is a minimum if $f'(x) = 0$, and $f'(x)$ changes from $-$ to $+$.

Values of the independent variable at maximum and minimum points, or at exceptional turning points, are called *critical values*.

4—10. Rule for Finding Maxima and Minima. From the above discussion, we arrive at the following rule for determining the maximum and minimum values of a function:

Step 1. Find the first derivative of the function.

Step 2. Set the first derivative equal to zero and solve the resulting equation for real roots; these roots are critical values of the variable.

Step 3. Test the first derivative, using one critical value at a time, for a value first a little less, then a little greater, than the critical value. If the sign of the derivative changes from $+$ to $-$, the function has a maximum value for that particular critical value of the variable; if the sign of the derivative changes from $-$ to $+$, there is a minimum value; if the sign of the derivative does not change, then there is neither a maximum nor a minimum at that point.

EXAMPLE 1. Test for maximum and minimum values:

$$y = x^2 - 9x - 6.$$

Solution. $\dfrac{dy}{dx} = 2x - 9.$

Put $2x - 9 = 0$.

Solving, $x = \frac{9}{2} = 4\frac{1}{2} = $ *critical* value.

Testing $\dfrac{dy}{dx}$ near $x = 4\frac{1}{2}$:

when $x = 4$, $\qquad\qquad \dfrac{dy}{dx} = -,$

when $x = 5$, $\qquad\qquad \dfrac{dy}{dx} = +.$

Hence the value $x = 4\frac{1}{2}$ is a *minimum* value.

EXAMPLE 2. Test for maxima and minima:

$$y = x^3 - 27x + 2.$$

Solution. $\dfrac{dy}{dx} = 3x^2 - 27 = 0.$

$$3(x^2 - 9) = 0,$$
$$(x + 3)(x - 3) = 0,$$
$$x = -3, 3, \text{ critical values.}$$

Testing $x = -3$:

at $x = -4, f'(x) = +$
at $x = -2, f'(x) = -$ } hence $x = -3$ is a maximum point.

Testing $x = +3$:

at $x = +2, f'(x) = -$
at $x = +4, f'(x) = +$ } hence $x = +3$ is a minimum point.

EXAMPLE 3. Divide the number k into two parts such that their product is a maximum.

Solution. Let one part be x; then the other part is $(k - x)$.

Then
$$P = x(k - x),$$
$$P = kx - x^2.$$
$$\frac{dP}{dx} = k - 2x.$$

Let
$$k - 2x = 0,$$

then $x = \dfrac{k}{2}$, the value of x which makes the product of $x(k - x)$ a maximum.

EXAMPLE 4. Find the number which, when added to its square, yields a minimum sum.

Solution. Let x be the required number.

Then
$$S = x + x^2.$$
$$\frac{dS}{dx} = 1 + 2x = 0.$$
$$x = -\tfrac{1}{2}, \text{ the required number.}$$

4—11. An Alternative Method for Determining Maxima and Minima.
From the diagram it will be seen that in the vicinity of a maximum
value of $f(x)$, such as point R, as we pass along the graph from left to

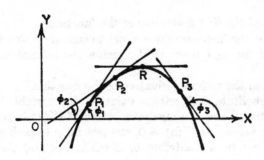

right, $f'(x)$, the slope of $f(x)$, changes from $+$ to 0 to $-$. Thus $f'(x)$
is a *decreasing* function; therefore, by §4—2 we know that *its* derivative
(that is, $f''(x)$, or the second derivative of the function itself) is negative
or zero.

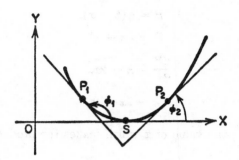

In the same way, in the vicinity of a minimum value of $f(x)$, the slope of $f(x)$, or $f'(x)$, changes from $-$ to 0 to $+$, and $f'(x)$ is an *increasing* function; hence by §4—2, $f''(x)$ is positive or zero.

At a maximum value, the curve is said to be *concave downwards;* at a minimum value, the curve is said to be *concave upwards*. These considerations lead to an alternative method of determining maxima and minima, based on the following principles:

(A) $f(x)$ is a maximum if $f'(x) = 0$ and $f''(x)$ is negative.
(B) $f(x)$ is a minimum if $f'(x) = 0$ and $f''(x)$ is positive.

Thus we see that if $\dfrac{dy}{dx} = 0$ and $\dfrac{d^2y}{dx^2}$ is $+$, the curve is concave upward; if $\dfrac{dy}{dx} = 0$ and $\dfrac{d^2y}{dx^2}$ is $-$, the curve is concave downward.

The rule of procedure is as follows:

Step 1. Find the first derivative of the function.

Step 2. Put the first derivative equal to zero and solve the equation thus obtained for real roots to determine the critical values of the variable.

Step 3. Find the second derivative of the function.

Step 4. Substitute each critical value for the variable in $f''(x)$. If $f''(x)$ is negative, there is a maximum value; if $f''(x)$ is positive, there is a minimum value. If $f''(x) = 0$, the test fails; in such a case, there may or may not be a maximum or a minimum, and the method of §15—10 must be used.

EXAMPLE. Test the function $y = x^3 - 48x + 8$ for maximum and minimum values.

Solution. $f'(x) = 3x^2 - 48$.

Put $$3x^2 - 48 = 0.$$

Hence $x = 4, -4$ are critical values.

$$f''(x) = 6x.$$

When $x = 4, f''(x) = +$; hence $x = 4$ is a minimum.
When $x = -4, f''(x) = -$; hence $x = -4$ is a maximum.

EXERCISE 4—3

Test each of the following for maximum and minimum values:

1. $y = x^2 - 4x$
2. $y = x^2 + 6x + 8$
3. $y = x^3 - 12x$
4. $y = x^4 - 18x^2 + 15$
5. $y = x^3 - 3x^2 + 6x - 2$

Find the following:

6. Two numbers whose sum is 40 and whose product will be as large as possible.

7. A number which, diminished by its square, will be a maximum.

8. A number which, when added to its reciprocal, will give the smallest possible sum.

4—12. Applications of Maxima and Minima.

Many practical situations arise in which the determination of maximum or minimum values is helpful.

EXAMPLE 1. A rectangle is to have a perimeter of 60 in.; find the dimensions which will give the maximum area.

Solution. $A = lw.$ \hfill (1)

$$2l + 2w = 60,$$

$$l + w = 30,$$

$$l = 30 - w. \hfill (2)$$

Substituting (2) in (1):

$$A = w(30 - w) = 30w - w^2,$$

$$\frac{dA}{dw} = 30 - 2w.$$

For A to be a maximum, $\dfrac{dA}{dw} = 0$.

Hence $30 - 2w = 0,$

$w = 15.$

From (2) $l = 15.$

Thus the rectangle, to have a maximum area, must be a square, 15×15.

NOTE. It is easy to prove that any rectangle of fixed perimeter, to have a maximum area, must be a square. This is left as an exercise for the reader.

EXAMPLE 2. A cylindrical tin can closed at both ends is to have a given fixed capacity, V. Prove that the amount of tin used (total surface of cylinder) will be a minimum when the height of the can equals the diameter of its base.

Solution.

$$V = \pi r^2 h; \qquad h = \frac{V}{\pi r^2}.$$

Total surface = $A = 2\pi r^2 + 2\pi r h$;

Substituting $\dfrac{V}{\pi r^2}$ for h:

$$A = 2\pi r^2 + \frac{2V}{r},$$

$$\frac{dA}{dr} = 4\pi r - \frac{2V}{r^2} = 0,$$

$$4\pi r^3 = 2V.$$

Substituting $\pi r^2 h$ for V:

$$4\pi r^3 = 2\pi r^2 h,$$

$$2r = h,$$

or $h = 2r = $ diameter.

EXAMPLE 3. Find the altitude of a cylinder of maximum volume inscribed in a sphere of radius k.

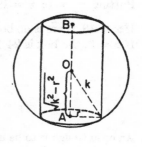

Solution. $OA = \sqrt{k^2 - r^2}$.

$h = AB = 2(OA) = 2\sqrt{k^2 - r^2}$.

$V = \pi r^2 h = 2\pi r^2 \sqrt{k^2 - r^2}$,

$\dfrac{dV}{dr} = 2\pi \left(\dfrac{r^2(\frac{1}{2})(-2r)}{\sqrt{k^2 - r^2}} + 2r\sqrt{k^2 - r^2} \right).$

Setting $\dfrac{dV}{dr} = 0$, and solving:

$$3r^2 = 2k^2,$$

$$r^2 = \frac{2k^2}{3} \text{, for max. value of } V.$$

Hence $\quad OA = \sqrt{k^2 - \dfrac{2k^2}{3}} = \sqrt{\dfrac{1}{3}k^2} = \dfrac{\sqrt{3}k}{3}$,

and $\quad h = 2(OA) = \dfrac{2\sqrt{3}}{3} k$, for max. value of V.

EXAMPLE 4. An open box of greatest possible capacity is to be made from a square piece of cardboard whose sides are each 36 in. long, by cutting equal small squares out of the corners and folding up the remaining piece as suggested by the diagram. What should be the length of each side of the small squares?

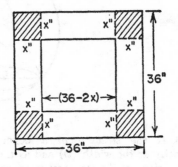

Solution. Let $x =$ side of small square; then $36 - 2x =$ side of square bottom of box.

$V = x(36 - 2x)^2$,

$\dfrac{dV}{dx} = -4x(36 - 2x) + (36 - 2x)^2$

$= 12(x - 18)(x - 6).$

Putting $\dfrac{dV}{dx} = 0$: $x = 18$, $x = 6$; critical values.

If $x = 18$, there is no bottom, and the box has a minimum capacity;
If $x = 6$, the box is $24 \times 24 \times 6$, and has a maximum capacity.

EXERCISE 4—4

1. An open trough is to be made from a long rectangular sheet of metal by bending up the long edges so as to give the trough a rectangular cross-section. If the width of the sheet is a inches, how deep should the trough be made so that it will have a maximum carrying capacity?

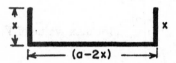

2. Solve illustrative Example 4 if the open box is made from a square a inches on a side.

3. Find the volume of the largest right circular cone that can be inscribed in a sphere of radius a inches.   *Hint:* $r^2 + (h - a)^2 = a^2$.

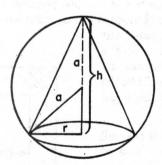

4. If the strength of a beam with rectangular cross-section varies directly as the width (w) and as the square of the height (h), what are the dimensions of the strongest beam that can be cut from a log of diameter d? *Hint:* Strength $= S = kwh^2$ is the function whose maximum we wish; $S = kw(d^2 - w^2)$.

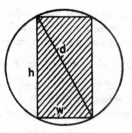

5. Find the altitude of the right circular cylinder having a maximum lateral area that can be inscribed in a sphere of radius a. *Hint:* $\dfrac{h}{2} = \sqrt{a^2 - r^2}$.

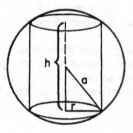

6. Find the volume of the largest right circular cone that can be generated by revolving a right triangle of hypotenuse a inches about one of its sides.

7. A Norman window consists of a rectangle surmounted by a semi-circle. For a given perimeter, find the height and width which will admit the maximum amount of light.

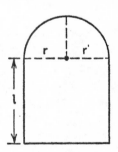

EXERCISE 4—5

Review

1. If a body moves in a fixed direction so that $s = \sqrt{t}$, prove that the acceleration is negative and proportional to the cube of the velocity.

2. At the end of t seconds, an object has a velocity of $(2t^2 + 8t)$ ft. per sec.; find its acceleration (a) in general, and (b) at the end of 5 seconds.

3. Determine the maximum and minimum values of

$$y = x^3 - 3x^2 - 24x + 10.$$

4. A circular filter paper of radius 9 inches is to be folded into a conical filter. Find the radius of the base of the filter if it has the maximum capacity.

5. The distance in feet covered in t seconds by a moving point is given by

$$s = 50t - 10t^2.$$

Find its velocity and acceleration at the end of $2\frac{1}{2}$ seconds.

6. Determine the maximum and minimum values of $y = x^3 - x^2 - 5x$.

7. Determine the maximum and minimum values of $y = x^4 - 2x^2 + 3$.

8. A freight train left the yards and in t hours was at a distance $s = t^3 - t^2 + 8t$ miles from the place where it started. Find its acceleration at the end of $1\frac{1}{2}$ hours.

9. If the sum of the length and girth of a cylindrical package must not exceed 90 inches, find the dimensions of the largest cylindrical package affording the largest volume and also meeting this requirement.

10. Find the area of the largest rectangle that can be inscribed in a circle of radius k.

Differentiation of Transcendental Functions

CHAPTER FIVE

DERIVATIVES OF LOGARITHMIC FUNCTIONS

5—1. Differentiating a Function of a Function. In this chapter we shall derive standard formulas for differentiating transcendental functions, that is, logarithmic, exponential, and trigonometric functions. Before we can do this, however, we must first discuss the problem of differentiating a function of a function, as well as that of differentiating an inverse function.

It frequently happens that the dependent variable y, instead of being defined directly as a function of x, is given as a function of another variable v, which, in turn, is defined as a function of x. In such a case, y is a function of x *through* v, and is called *a function of a function*. For example:

if
$$y = \frac{v^2}{1 + 3v},$$

and
$$v = 1 + 2x,$$

then y is a function of a function; that is, y is a function of v, or $y = f(v)$, and v is a function of x, or $v = \phi(x)$. By eliminating v, we may express

y directly as a function of x, or $y = F(x)$. In general, however, this is not convenient, nor is it necessary in order to find $\dfrac{dy}{dx}$.

Let $y = f(v)$, and $v = \phi(x)$; thus y is a function of x through the function v. If x takes on an increment Δx, then v will take on an increment Δv, and y will take on a corresponding increment Δy. Applying the General Rule to both functions together, we have:

$$y = f(v) \qquad\qquad\qquad v = \phi(x)$$

Step 1. $y + \Delta y = f(v + \Delta v)$ $\qquad v + \Delta v = \phi(x + \Delta x)$

Step 2. $\Delta y = f(v + \Delta v) - f(v)$ $\qquad \Delta v = \phi(x + \Delta x) - \phi(x)$

Step 3. $\dfrac{\Delta y}{\Delta v} = \dfrac{f(v + \Delta v) - f(v)}{\Delta v}$ $\qquad \dfrac{\Delta v}{\Delta x} = \dfrac{\phi(x + \Delta x) - \phi(x)}{\Delta x}$

In these last two equations in Step 3, the left-hand members express the ratio of the increment of each function to the increment of the corresponding variable; the right-hand members express the same ratios, respectively, in another form. Before passing to the limit, we may write the product

$$\frac{\Delta y}{\Delta v} \cdot \frac{\Delta v}{\Delta x} = \frac{\Delta y}{\Delta x},$$

or

$$\frac{\Delta y}{\Delta x} = \frac{\Delta y}{\Delta v} \cdot \frac{\Delta v}{\Delta x}.$$

Step 4. When $\Delta x \to 0$, both Δy and Δv also approach zero, and we have:

$$\frac{dy}{dx} = \frac{dy}{dv} \cdot \frac{dv}{dx}, \tag{1}$$

or

$$\frac{dy}{dx} = f'(v) \cdot \phi'(x). \tag{1a}$$

RULE. *If $y = f(v)$ and $v = \phi(x)$, the derivative of y with respect to x is equal to the product of the derivative of y with respect to v and the derivative of v with respect to x.*

5—2. Inverse Functions. Consider the pairs of functions given in the two columns below:

	(A)	(B)
(1)	$y = \dfrac{1}{x}$	$x = \dfrac{1}{y}$
(2)	$y = x^2 - 3$	$x = \pm\sqrt{y+3}$
(3)	$y = \pm\sqrt{ax}$	$x = \dfrac{y^2}{a}$
(4)	$y = x^n$	$x = \sqrt[n]{y}$
(5)	$y = k^x$	$x = \log_k y$
(6)	$y = \tan x$	$x = \text{arc tan } y^*$

It will be noted that each equation in column (A) has been solved for y, while those in column (B) have been solved for x. Thus if $y = \dfrac{1}{x}$ is called the *direct function*, as is customary, then $x = \dfrac{1}{y}$ is the *inverse function*. In other words, the functions in column (B) are the inverse functions, respectively, of those in column (A). In the direct functions, column (A), y is the dependent variable and x is the independent variable. In the inverse functions, however, y is the independent variable and x is the dependent variable.

5—3. Differentiating an Inverse Function. We shall now proceed to differentiate the inverse functions $y = f(x)$ and $x = \phi(y)$, together, by the General Rule.

Step 1. $\quad y + \Delta y = f(x + \Delta x)$ $\qquad\qquad x + \Delta x = \phi(y + \Delta y)$

Step 2. $\quad \Delta y = f(x + \Delta x) - f(x)$ $\qquad \Delta x = \phi(y + \Delta y) - \phi(y)$

Step 3. $\quad \dfrac{\Delta y}{\Delta x} = \dfrac{f(x + \Delta x) - f(x)}{\Delta x}$ $\qquad \dfrac{\Delta x}{\Delta y} = \dfrac{\phi(y + \Delta y) - \phi(y)}{\Delta y}$

Considering the product of the left-hand expressions, we have:

$$\frac{\Delta y}{\Delta x} \cdot \frac{\Delta x}{\Delta y} = 1, \quad \text{or} \quad \frac{\Delta y}{\Delta x} = \frac{1}{\dfrac{\Delta x}{\Delta y}}.$$

---

* The reader should know that the alternative forms arc tan y and $\tan^{-1} y$ are synonymous and interchangeable. In this book we shall consistently use the form arc sin θ for the inverse function of sin θ, arc cos θ for the inverse of cos θ, etc.

Step 4.
$$\frac{dy}{dx} = \frac{1}{\dfrac{dx}{dy}},$$ [2]

or
$$f'(x) = \frac{1}{\phi'(y)}.$$ [2a]

RULE. *The derivative of the inverse function is equal to the reciprocal of the derivative of the direct function.*

5—4. Logarithmic Functions. Before discussing the derivative of a logarithmic function, a word or two about logarithms may prove helpful. The reader will recall that, by definition,

if $N = a^x$,   then $x = \log_a N$.

He will also recall that

$$\log_{10} N = (\log_e N)(\log_{10} e),$$

where $e = 2.718\ldots$ is the base of the natural system of logarithms. It can be shown that $\log_{10} e = .434\ldots$; this is called the *modulus* of the system of *common logarithms*, where the base is 10. More generally, the logarithm of a number N to any base a may be found from the formula

$$\log_a N = \log_a e \cdot \log_e N = \frac{\log_e N}{\log_e a},$$

where $\log_a e$ is the modulus of the particular system whose base is a.

It should be emphasized that in higher mathematics, when we write $\log v$, without designating the base, the base is understood to be e. Hence, since $\log_e e = 1$, the modulus of the system of natural logarithms is unity.[*]

5—5. Differentiation of a Logarithm. We are now in a position to apply the General Rule in deriving a formula for differentiating a logarithm. Consider the direct function

$$y = \log_a v,$$

where a, the base, is any constant, and v is the independent variable.

Step 1. $y + \Delta y = \log_a (v + \Delta v).$

---

[*] In advanced mathematics, the notation $\ln x$ is used to designate $\log_e x$. Thus, $\ln e = 1$.

Step 2. $\Delta y = \log_a (v + \Delta v) - \log_a v.$

By algebra, we have:

$$\Delta y = \log_a \left(\frac{v + \Delta v}{v} \right) = \log_a \left(1 + \frac{\Delta v}{v} \right).$$

Step 3. $\dfrac{\Delta y}{\Delta v} = \dfrac{1}{\Delta v} \cdot \log_a \left(1 + \dfrac{\Delta v}{v} \right).$

By algebra, we may write this:

$$\frac{\Delta y}{\Delta v} = \log_a \left(1 + \frac{\Delta v}{v} \right)^{1/\Delta v},$$

or $\qquad\qquad \dfrac{\Delta y}{\Delta v} = \dfrac{1}{v} \log_a \left(1 + \dfrac{\Delta v}{v} \right)^{v/\Delta v}.$

It can be shown that $\lim\limits_{z \to 0} (1 + z)^{1/z} = e = 2.718 \ldots$ (See §1—17.)

Now, when $\Delta v \to 0, \dfrac{\Delta v}{v} \to 0.$ Hence,

$$\lim_{\Delta v \to 0} \left(1 + \frac{\Delta v}{v} \right)^{v/\Delta v} = e,$$

which is clear if we let $\dfrac{\Delta v}{v} = z$ in the expression $(1 + z)^{1/z}.$

Step 4. $\dfrac{dy}{dv} = \dfrac{1}{v} \log_a e,$ \hfill (1)

or $\qquad\qquad \dfrac{d}{dv} (\log_a v) = \dfrac{1}{v} \log_a e.$ \hfill (2)

NOTE 1. It should be remembered that the function $y = \log_a v$ is defined only for positive values of a and v.

NOTE 2. The algebraic transformation made after taking Step 2 and before taking Step 3 is necessary to avoid the occurrence of the indeterminate quotient $\dfrac{0}{0}$ later when passing to the limit in Step 4.

Since v is a function of x, and we wish to differentiate $\log v$ with respect to x, we now make use of §5—1, equation [1], to find the derivative of a function of a function. Since

$$\frac{dy}{dx} = \frac{dy}{dv} \cdot \frac{dv}{dx},\tag{3}$$

we may substitute the value of $\frac{dy}{dv}$ from equation (1) above in equation (3), giving:

$$\frac{dy}{dx} = \frac{1}{v} \log_a e \cdot \frac{dv}{dx},\tag{4}$$

or $$\frac{d}{dx} (\log_a v) = \log_a e \, \frac{\dfrac{dv}{dx}}{v}.\tag{[3]}$$

RULE. *The derivative of the logarithm of a function is equal to the product of the modulus of the system of logarithms and the derivative of the function, divided by the function.*

As a special case, which occurs very often in practice, we note that when $a = e$, $\log_a e = \log_e e = 1$, and formula [3] becomes

$$\frac{d}{dx} (\log v) = \frac{\dfrac{dv}{dx}}{v}.\tag{[3a]}$$

EXAMPLE 1. Differentiate $y = \log x$.

Solution. Here $v = x$.

$$\frac{d}{dx} (\log x) = \frac{\dfrac{dx}{dx}}{x} = \frac{1}{x}.$$

EXAMPLE 2. Differentiate $y = \log (x^3 + a)$.

Solution. Here $v = x^3 + a$.

$$\frac{dy}{dx} = \frac{\dfrac{d}{dx} (x^3 + a)}{x^3 + a} = \frac{3x^2}{x^3 + a}.$$

EXAMPLE 3. Differentiate $y = \log \sqrt{x^2 - 2}$.

Solution. Here $v = (x^2 - 2)^{\frac{1}{2}}$.

$$\frac{dy}{dx} = \frac{\dfrac{d}{dx}(x^2 - 2)^{\frac{1}{2}}}{(x^2 - 2)^{\frac{1}{2}}}$$

$$= \frac{\frac{1}{2}(x^2 - 2)^{-\frac{1}{2}}(2x)}{(x^2 - 2)^{\frac{1}{2}}} = \frac{x}{x^2 - 2}.$$

EXAMPLE 4. Differentiate $y = \log(ax^2 + bx + c)$.

Solution. Here $v = ax^2 + bx + c$.

$$\frac{dy}{dx} = \frac{2ax + b}{ax^2 + bx + c}.$$

EXAMPLE 5. Differentiate $y = \log \dfrac{x - 3}{x + 3}$.

Solution. Here $v = \dfrac{x - 3}{x + 3}$.

$$\frac{dy}{dx} = \frac{\dfrac{d}{dx}\left(\dfrac{x - 3}{x + 3}\right)}{\dfrac{x - 3}{x + 3}} = \frac{\dfrac{(x + 3) - (x - 3)}{(x + 3)^2}}{\dfrac{x - 3}{x + 3}} = \frac{6}{x^2 - 9}.$$

EXAMPLE 6. Differentiate $y = \log_a(x^3 - 4x)$.

Solution. Here $v = x^3 - 4x$, and the modulus is $\log_a e$; using formula [3] instead of [3a], we obtain:

$$\frac{dy}{dx} = \log_a e \frac{\dfrac{d}{dx}(x^3 - 4x)}{x^3 - 4x} = \log_a e \left(\frac{3x^2 - 4}{x^3 - 4x}\right).$$

EXERCISE 5—1

Differentiate:

1. $y = \log(k + x)$
2. $y = \log(x^2)$
3. $y = (\log x)^2$
4. $y = \log(ax^2)$
5. $y = \log(ax + b)$
6. $y = \log(x^3 + kx^2)$

7. $y = \log (x^2 + 3x - 2)$
8. $y = \log (ax) + \log (2x^3)$
9. $y = \log \sqrt[3]{x^3 + 1}$

10. $y = \log \dfrac{x + 1}{x - 1}$

11. $y = x \log x$
12. $y = \log (x^2 + ax)$
13. $y = \log x^3$

14. $y = \log^3 x$
15. $y = \log \sqrt{1 - x}$
16. $y = \log \sqrt{x^2 + 1}$
17. $y = \log_a (3x^2 - 2x)$
18. $y = \log_a (x^3 + k)$
19. $y = x^2 \log x$

20. $y = \log \dfrac{x}{x^2 - 1}$

DERIVATIVES OF EXPONENTIAL FUNCTIONS

5—6. Differentiating the Simple Exponential Function. Let us consider first the simple exponential function $y = a^v$, where a is a constant greater than zero (that is, positive), and v is a function of x. Taking the logarithm of both sides to the base e, we have:

$$\log y = v \log a,$$

or

$$v = \frac{\log y}{\log a} = \frac{1}{\log a} \cdot \log y.$$

Differentiating with respect to y:

$$\frac{dv}{dy} = \frac{1}{\log a} \cdot \frac{1}{y}.$$

Applying the formula for the derivative of an inverse function, §5—3, equation [2]:

$$\frac{dy}{dv} = \frac{1}{\dfrac{dv}{dy}},$$

we get

$$\frac{dy}{dv} = \log a \cdot y,$$

or

$$\frac{dy}{dv} = \log a \cdot a^v. \tag{1}$$

Now we apply the formula for the derivative of a function of a function, §5—1, equation [1], since v is a function of x; that is

$$\frac{dy}{dx} = \frac{dy}{dv} \cdot \frac{dv}{dx}.$$

Substituting the value of $\dfrac{dy}{dv}$ from (1) above, we get:

$$\frac{dy}{dx} = \log a \cdot a^v \cdot \frac{dv}{dx},$$

or $\qquad \dfrac{d}{dx}(a^v) = \log a \cdot a^v \cdot \dfrac{dv}{dx}.$ [4]

As a special case, when $a = e$, $\log a = \log e = 1$; hence

$$\frac{d}{dx}(e^v) = e^v \frac{dv}{dx}.$$ [4a]

RULE. *The derivative of a constant with a variable exponent is equal to the product of the natural logarithm of the constant, the constant with the variable exponent, and the derivative of the exponent.*

EXAMPLE 1. Find $\dfrac{d}{dx}(a^{4x+1})$.

Solution. $\dfrac{d}{dx}(a^{4x+1}) = \log a\,(a^{4x+1})(4)$, or $4 \log a\,(a^{4x+1})$.

EXAMPLE 2. Differentiate $y = e^{-2x}$.

Solution. $\dfrac{dy}{dx} = e^{-2x}(-2) = -2e^{-2x}$.

EXAMPLE 3. Differentiate $s = e^{t^3+1}$.

Solution. $\dfrac{ds}{dt} = e^{t^3+1}(3t^2)$.

EXAMPLE 4. Find $\dfrac{d\rho}{d\theta}$ in the equation $\rho = a^{\theta^2}$.

Solution. $\dfrac{d\rho}{d\theta} = \log a \cdot a^{\theta^2} \cdot 2\theta$.

EXAMPLE 5. Differentiate $y = e^{\log x}$.

Solution. $\dfrac{dy}{dx} = e^{\log x} \cdot \dfrac{1}{x}$.

But $e^{\log x} = x$, by the definition of a logarithm.

Therefore $$\frac{dy}{dx} = x\left(\frac{1}{x}\right) = 1.$$

EXAMPLE 6. Differentiate $y = xe^{2/x}$.

Solution.

$$\frac{dy}{dx} = x\frac{d}{dx}(e^{2/x}) + e^{2/x} \cdot \frac{d}{dx}(x)$$

$$= xe^{2/x}(-2x^{-2}) + e^{2/x}$$

$$= e^{2/x}\left(-\frac{2}{x} + 1\right).$$

EXERCISE 5—2

Differentiate:

1. $y = k^{4x}$

2. $y = a^{x^2-1}$

3. $y = e^{kx}$

4. $y = ke^x$

5. $s = e^{5-3t}$

6. $y = k^{m^2+x^2}$

7. $y = 5^{x^3+4x}$

8. $s = e^{a^2+t^2}$

9. $y = ke^{\sqrt{x}}$

10. $y = e^{-x^2}$

11. $\rho = a^\theta$

12. $\rho = a^{\log\theta}$

13. $\rho = e^{a\theta}$

14. $y = e^{1/x}$

15. $y = xe^x$

16. $y = x^2e^{2x}$

17. $y = x^n + n^x$

18. $y = a^x x^a$

19. $y = e^x(x^2 - 2x + 2)$

20. $y = e^x(x - 1)$

5—7. Differentiating the General Exponential Function. We now consider the more general exponential function $y = u^v$, that is, a function raised to a variable power instead of a constant raised to a variable power. The only restriction is that u shall assume only positive values.

Let $$y = u^v. \tag{1}$$

Take the logarithm of both sides to the base e:

$$\log_e y = v \log_e u,$$

or, by the definition of a logarithm,

$$y = e^{v \log u}. \tag{2}$$

Differentiating equation (2) by formula [4a]:

$$\frac{dy}{dx} = e^{v \log u} \cdot \frac{d}{dx}(v \log u). \tag{3}$$

But $e^{v \log u} = y$ from equation (2); and $y = u^v$ from equation (1); hence $e^{v \log u} = u^v$.

Also, by differentiating as a product,

$$\frac{d}{dx}(v \log u) = \frac{v}{u}\frac{du}{dx} + \log u \frac{dv}{dx}.$$

Making these substitutions in equation (3), we get:

$$\frac{dy}{dx} = u^v\left(\frac{v}{u}\frac{du}{dx} + \log u \frac{dv}{dx}\right),$$

or $$\frac{d}{dx}(u^v) = vu^{v-1}\frac{du}{dx} + \log u \cdot u^v \frac{dv}{dx}. \qquad [5]$$

RULE. *The derivative of a function with a variable exponent is equal to the sum of the two results obtained by first regarding the exponent as a constant and differentiating, and then by regarding the function as a constant and differentiating.*

EXAMPLE 1. Differentiate $y = x^x$.

Solution.

$$\frac{d}{dx}(u^v) = vu^{v-1}\frac{du}{dx} + \log u \cdot u^v \frac{dv}{dx}.$$

Hence $$\frac{d}{dx}(x^x) = x \cdot x^{x-1} + \log x \cdot x^x$$

$$= x^x(1 + \log x).$$

EXAMPLE 2. Differentiate $y = x^{e^x}$.

Solution. Here $u = x$, $v = e^x$.

Therefore $$\frac{d}{dx}(x^{e^x}) = e^x \cdot x^{e^x-1} + \log x \cdot x^{e^x} \cdot e^x$$

$$= e^x(x^{e^x-1} + \log x \cdot x^{e^x})$$

$$= e^x x^{e^x}\left(\frac{1}{x} + \log x\right).$$

EXAMPLE 3. Prove that the formula $\frac{d}{dx}(u^n) = nu^{n-1}\frac{du}{dx}$ holds for all values of the constant n by setting $v = n$ in formula [5].

Solution. If $v = n$, we have, from [5]:

$$\frac{d}{dx} (u^n) = nu^{n-1} \frac{du}{dx} + \log u \cdot u^n \cdot 0 = nu^{n-1} \frac{du}{dx}.$$

EXERCISE 5—3

Differentiate:

1. $y = x^{x+1}$ **4.** $y = x^{2x}$ **7.** $y = x^{1/x}$

2. $y = (x^2 + 1)^x$ **5.** $y = (2x)^{x^3}$ **8.** $y = x^{\log x}$

3. $y = (x^3)^{-x}$ **6.** $y = x^{x^2}$

5—8. Logarithmic Differentiation. When differentiating logarithmic and exponential functions, it is often more convenient to transform the given expression by making use of the properties of logarithms, namely:

$$\log AB = \log A + \log B; \qquad \log A^n = n \log A;$$

$$\log \left(\frac{A}{B}\right) = \log A - \log B; \qquad \log \sqrt[n]{A} = \frac{1}{n} \log A.$$

This is known as *logarithmic differentiation*, and is now illustrated.

Example 1. Differentiate $y = \log \sqrt{x^3 + 2}$.

Solution. First write

$$y = \tfrac{1}{2} \log (x^3 + 2),$$

thus eliminating the radical.

Then $$\frac{dy}{dx} = \frac{1}{2} \frac{\dfrac{d}{dx} (x^3 + 2)}{x^3 + 2} = \frac{3x^2}{2(x^3 + 2)}.$$

Example 2. Differentiate $y = \log \dfrac{x^2}{x + 1}$.

Solution. First write

$$y = \log x^2 - \log (x + 1).$$

Then
$$\frac{dy}{dx} = \frac{2x}{x^2} - \frac{1}{x+1}$$
$$= \frac{2}{x} - \frac{1}{x+1} = \frac{x+2}{x(x+1)}.$$

Example 3. Differentiate $y = \log\sqrt{\dfrac{x^2 - a}{x^2 + a}}$.

Solution. First write
$$y = \tfrac{1}{2}[\log (x^2 - a) - \log (x^2 + a)].$$

Then
$$\frac{dy}{dx} = \frac{1}{2}\left[\frac{2x}{x^2 - a} - \frac{2x}{x^2 + a}\right] = \frac{2ax}{x^4 - a^2}.$$

Example 4. Differentiate $y = \dfrac{(x+3)(x+2)}{x+1}$.

Solution. First take the logarithm of both sides:
$$\log y = \log (x+3) + \log (x+2) - \log (x+1).$$

Then, differentiating both sides with respect to x:
$$\frac{1}{y} \cdot \frac{dy}{dx} = \frac{1}{x+3} + \frac{1}{x+2} - \frac{1}{x+1};$$
$$\frac{dy}{dx} = \frac{x^2 + 2x - 1}{(x+1)^2}.$$

Example 5. Differentiate $y = x^{x+1}$.

Solution. First take the logarithm of both sides:
$$\log y = (x+1) \log x.$$

Differentiating both sides with respect to x:
$$\frac{1}{y} \cdot \frac{dy}{dx} = (x+1)\left(\frac{1}{x}\right) + \log x(1);$$
$$\frac{dy}{dx} = x^x(x+1) + \log x(x^{x+1}).$$

Example 6. Differentiate $y = x^{x^2}$.

Solution.   Taking the logarithm of both sides:

$$\log y = x^2 \log x.$$

Differentiating both sides with respect to x:

$$\frac{1}{y} \cdot \frac{dy}{dx} = x^2 \cdot \frac{1}{x} + \log x \cdot 2x$$

$$= x(1 + 2 \log x).$$

$$\frac{dy}{dx} = x^{x^2+1}(1 + 2 \log x).$$

EXERCISE 5—4

Differentiate, using the method of logarithmic differentiation:

1. $y = \left(\dfrac{k}{x}\right)^x$

2. $y = x^{x^n}$

3. $y = e^{x^x}$

4. $y = \dfrac{e^{3x}}{3x + 1}$

5. $y = \log \dfrac{x - 1}{x^3}$

6. $y = \log \sqrt{x^2 - 3x + 2}$

7. $y = \sqrt{\dfrac{x + 2}{x - 2}}$

8. $y = \dfrac{x + 3}{(x + 1)(x + 2)}$

DERIVATIVES OF TRIGONOMETRIC FUNCTIONS

5—9. The Derivative of the Sine.   To derive the formulas for the derivatives of the six trigonometric functions, we need only fall back upon the General Rule to find the derivative of the sine; the derivatives of the other five functions may be derived from this one by trigonometric transformations.   It should be recalled (§1—18) that

$$\lim_{x \to 0} \left(\frac{\sin x}{x}\right) = 1. \tag{1}$$

Now let

$$y = \sin v,$$

where v is any given function of the angle x expressed in radians.

Step 1. $y + \Delta y = \sin (v + \Delta v)$.

Step 2. $\Delta y = \sin (v + \Delta v) - \sin v$. $\qquad\qquad\qquad$ (2)

By a trigonometric transformation, the difference between two sines is given by

$$\sin A - \sin B = 2 \cos \tfrac{1}{2}(A + B) \sin \tfrac{1}{2}(A - B);$$

applying this to equation (2), we have:

$$\Delta y = 2 \cos \left(v + \frac{\Delta v}{2}\right) \sin \frac{\Delta v}{2}.$$

Step 3. $\dfrac{\Delta y}{\Delta v} = \cos \left(v + \dfrac{\Delta v}{2}\right) \cdot \left(\dfrac{\sin \dfrac{\Delta v}{2}}{\dfrac{\Delta v}{2}}\right).$

By equation (1) above,

$$\lim_{\Delta v \to 0} \left(\frac{\sin \dfrac{\Delta v}{2}}{\dfrac{\Delta v}{2}}\right) = 1.$$

Step 4. $\dfrac{dy}{dv} = \cos v.$

But v is a function of x; hence to find $\dfrac{dy}{dx}$, we use the relation

$$\frac{dy}{dx} = \frac{dy}{dv} \cdot \frac{dv}{dx}. \qquad\qquad\qquad (3)$$

Substituting the value of $\dfrac{dy}{dv}$ from Step 4 in equation (3), we obtain:

$$\frac{dy}{dx} = \cos v \, \frac{dv}{dx},$$

or $\qquad\qquad\qquad \dfrac{d}{dx} (\sin v) = \cos v \, \dfrac{dv}{dx}.$ $\qquad\qquad$ [6]

5—10. Derivative of the Cosine.

Let $\qquad\qquad\qquad\qquad y = \cos v.$

From trigonometry:

$$y = \sin\left(\frac{\pi}{2} - v\right). \tag{1}$$

Differentiating equation (1) by formula [6]:

$$\frac{dy}{dx} = \cos\left(\frac{\pi}{2} - v\right)\frac{d}{dx}\left(\frac{\pi}{2} - v\right)$$

$$= \cos\left(\frac{\pi}{2} - v\right)\left(-\frac{dv}{dx}\right). \tag{2}$$

But $\cos\left(\frac{\pi}{2} - v\right) = \sin v$, by trigonometry.

Hence equation (2) becomes:

$$\frac{dy}{dx} = -\sin v\,\frac{dv}{dx},$$

or

$$\frac{d}{dx}(\cos v) = -\sin v\,\frac{dv}{dx}. \tag{7}$$

5—11. Derivative of the Tangent and the Cotangent.

Let

$$y = \tan v.$$

Hence

$$y = \frac{\sin v}{\cos v}. \tag{1}$$

Differentiating equation (1) as a quotient:

$$\frac{dy}{dx} = \frac{\cos v\,\dfrac{d}{dx}(\sin v) - \sin v\,\dfrac{d}{dx}(\cos v)}{\cos^2 v}$$

$$= \frac{\cos^2 v\,\dfrac{dv}{dx} + \sin^2 v\,\dfrac{dv}{dx}}{\cos^2 v}$$

$$= \frac{\dfrac{dv}{dx}}{\cos^2 v} = \sec^2 v\,\frac{dv}{dx}.$$

Thus

$$\frac{d}{dx}(\tan v) = \sec^2 v\,\frac{dv}{dx}. \tag{8}$$

Again, let

$$y = \cot v.$$

Then

$$y = \frac{1}{\tan v}. \qquad (1)$$

Differentiating equation (1) as a quotient:

$$\frac{dy}{dx} = \frac{0 - \dfrac{d}{dx}(\tan v)}{\tan^2 v} = -\frac{\sec^2 v \dfrac{dv}{dx}}{\tan^2 v} = -\csc^2 v \frac{dv}{dx}.$$

Thus

$$\frac{d}{dx}(\cot v) = -\csc^2 v \frac{dv}{dx}. \qquad [9]$$

5—12. Derivative of the Secant and the Cosecant. These may be derived in a similar manner, and are left as an exercise for the reader. The formulas obtained are as follows:

$$\frac{d}{dx}(\sec v) = \sec v \tan v \frac{dv}{dx}, \qquad [10]$$

and

$$\frac{d}{dx}(\csc v) = -\csc v \cot v \frac{dv}{dx}. \qquad [11]$$

5—13. Differentiating Trigonometric Functions. The formulas for differentiating trigonometric functions may be used in conjunction with the formulas for differentiating algebraic exponential and logarithmic functions, as suggested by the following examples.

EXAMPLE 1. Differentiate $y = \sin^2 x$.

Solution. $y = \sin^2 x = (\sin x)^2$.
Here $v = \sin x$, and $n = 2$.

Then

$$\frac{dy}{dx} = 2(\sin x)(\cos x) = \sin 2x.$$

EXAMPLE 2. Differentiate $y = e^x \cdot \sin x$.

Solution.
Here $u = e^x$, and $v = \sin x$.

$$\frac{dy}{dx} = e^x \cdot \cos x + \sin x \cdot e^x$$

$$= e^x(\sin x + \cos x).$$

EXAMPLE 3.   Differentiate $y = (\sin x)^{\sin x}$.

Solution.
   Here $u = \sin x$, and $v = \sin x$.
   Using §5—7, formula [5]:

$$\frac{dy}{dx} = \sin x \cdot \sin x^{(\sin x - 1)} \cdot \cos x + \log \sin x \cdot (\sin x)^{\sin x} \cdot (\cos x),$$

or $$\frac{dy}{dx} = \cos x \, (\sin x)^{\sin x} \cdot (1 + \log \sin x).$$

EXAMPLE 4.   Differentiate $y = \cos x \sin^3 x$.

Solution.
   Here $u = \cos x$, and $v = \sin^3 x$.

$$\frac{dy}{dx} = \cos x (3 \sin^2 x \cos x) + \sin^3 x (-\sin x)$$

$$= \sin^2 x (3 \cos^2 x - \sin^2 x).$$

EXERCISE 5—5

Differentiate:

1. $y = \sin 3x$
2. $y = \cos 5ax$
3. $y = \sin^2 x \cos x$
4. $\rho = a \cos 2\theta$
5. $y = \sin (a + bx)$
6. $y = e^{\tan x}$
7. $\rho = \log \cos^2 \theta$
8. $y = (\sin x)^x$
9. $y = x^{\cos x}$
10. $y = e^x \cdot \log \sin x$
11. $\rho = \log \tan \theta$
12. $y = (\cos x)^x$

13. $y = \sin^2 \dfrac{x}{2}$
14. $y = \dfrac{1 + \sin x}{1 - \sin x}$
15. $y = x^m \log x$
16. $y = \dfrac{\sin x}{x}$
17. $\rho = \theta - \sin \theta \cos \theta$
18. $y = 2 \sin x \cos x$
19. $y = \log x + \tan x \cos x$
20. $\rho = \log \sin 2\theta$

DERIVATIVES OF THE INVERSE
TRIGONOMETRIC FUNCTIONS

5—14. The Derivative of Arc Sin v and Arc Cos v. It will be
recalled that the inverse trigonometric functions may be written as

follows:

if $\quad \alpha = \sin \beta, \quad$ then $\quad \beta = \text{arc sin } \alpha;$

if $\quad \alpha = \cos \beta, \quad$ then $\quad \beta = \text{arc cos } \alpha;$ etc.

Now let

$$y = \text{arc sin } v,$$

or $\qquad\qquad v = \sin y. \qquad\qquad (1)$

Differentiating equation (1) with respect to y by [6] (§5—9):

$$\frac{dv}{dy} = \cos y. \qquad\qquad (2)$$

From equation (2), by using the formula for the derivative of an inverse function (§5—2), we have:

$$\frac{dy}{dv} = \frac{1}{\cos y}. \qquad\qquad (3)$$

But v is a function of x; hence, using the formula for the derivative of a function of a function (§5—1):

$$\frac{dy}{dx} = \frac{dy}{dv} \cdot \frac{dv}{dx}. \qquad\qquad (4)$$

Substituting in (4) from (3):

$$\frac{dy}{dx} = \frac{1}{\cos y} \frac{dv}{dx}. \qquad\qquad (5)$$

From trigonometry

$$\cos y = \sqrt{1 - \sin^2 y} = \sqrt{1 - v^2}.$$

Substituting in (5):

$$\frac{dy}{dx} = \frac{1}{\sqrt{1 - v^2}} \frac{dv}{dx},$$

or $\qquad\qquad \dfrac{d}{dx}(\text{arc sin } v) = \dfrac{\dfrac{dv}{dx}}{\sqrt{1 - v^2}}. \qquad\qquad [12]$

By proceeding in the same manner, the reader may verify the formula for the derivative of arc cos v:

$$\frac{d}{dx}(\text{arc cos } v) = -\frac{\frac{dv}{dx}}{\sqrt{1-v^2}}. \tag{13}$$

5—15. Derivative of Arc Tan v and Arc Cot v. The general procedure is the same as for the derivation given in §5—14.

Let $\qquad\qquad\qquad y = \text{arc tan } v,$

or $\qquad\qquad\qquad v = \tan y.$

Differentiating with respect to y:

$$\frac{dv}{dy} = \sec^2 y.$$

Hence $\qquad\qquad\qquad \dfrac{dy}{dv} = \dfrac{1}{\sec^2 y}.$

Since $\qquad\qquad\qquad \dfrac{dy}{dx} = \dfrac{dy}{dv} \cdot \dfrac{dv}{dx},$

$$\frac{dy}{dx} = \frac{1}{\sec^2 y}\frac{dv}{dx}.$$

By trigonometry, $\sec^2 y = 1 + \tan^2 y = 1 + v^2$. Therefore

$$\frac{dy}{dx} = \frac{1}{1+v^2}\frac{dv}{dx},$$

or $\qquad\qquad \dfrac{d}{dx}(\text{arc tan } v) = \dfrac{\frac{dv}{dx}}{1+v^2}. \tag{14}$

In the same manner, the reader may verify

$$\frac{d}{dx}(\text{arc cot } v) = -\frac{\frac{dv}{dx}}{1+v^2}. \tag{15}$$

5—16. Derivative of Arc Sec v and Arc Csc v.

Let $\qquad\qquad\qquad y = \text{arc sec } v,$

or $\qquad\qquad\qquad v = \sec y.$

Differentiating with respect to y:

$$\frac{dv}{dy} = \sec y \tan y,$$

and

$$\frac{dy}{dv} = \frac{1}{\sec y \tan y}.$$

Since

$$\frac{dy}{dx} = \frac{dy}{dv} \frac{dv}{dx},$$

$$\frac{dy}{dx} = \frac{1}{\sec y \tan y} \cdot \frac{dv}{dx}.$$

By trigonometry,

$$\tan y = \sqrt{\sec^2 y - 1} = \sqrt{v^2 - 1};$$

and $\quad\sec y = v,\quad$ by hypothesis.

Therefore

$$\frac{dy}{dx} = \frac{1}{v\sqrt{v^2 - 1}} \frac{dv}{dx},$$

or

$$\frac{d}{dx} (\text{arc sec } v) = \frac{\dfrac{dv}{dx}}{v\sqrt{v^2 - 1}}. \qquad [16]$$

In the same manner, the reader may verify

$$\frac{d}{dx} (\text{arc csc } v) = -\frac{\dfrac{dv}{dx}}{v\sqrt{v^2 - 1}}. \qquad [17]$$

EXAMPLE 1. Differentiate $y = \text{arc sin } 2x$.

Solution. $\quad\dfrac{dy}{dx} = \dfrac{\dfrac{d}{dx}(2x)}{\sqrt{1 - 4x^2}} = \dfrac{2}{\sqrt{1 - 4x^2}}.$

EXAMPLE 2. Differentiate $y = \text{arc tan } \dfrac{x^2}{a^2}$.

Solution. $\quad\dfrac{dy}{dx} = \dfrac{\dfrac{d}{dx}\left(\dfrac{x^2}{a^2}\right)}{1 + \left(\dfrac{x^2}{a^2}\right)} = \dfrac{\dfrac{2x}{a^2}}{\dfrac{a^4 + x^4}{a^4}} = \dfrac{2a^2 x}{a^4 + x^4}.$

Differentiate:

1. $y = \text{arc sin } x^2$

5. $y = \text{arc cot } \dfrac{3x}{x-1}$

2. $y = \text{arc cos } \dfrac{x}{2}$

6. $y = \text{arc cos } \dfrac{2x+1}{2}$

3. $y = \text{arc tan } \dfrac{x-1}{x}$

7. $y = \text{arc cot } \dfrac{x^2}{a}$

4. $y = x \text{ arc sin } x$

8. $y = x^2 \text{ arc tan } \dfrac{a}{x}$

5—17. Summary of Formulas. For the reader's convenience, we summarize below the formulas for differentiating transcendental functions, as was done for algebraic functions in §3—8.

[1] $\quad \dfrac{dy}{dx} = \dfrac{dy}{dv} \cdot \dfrac{dv}{dx}$

[2] $\quad \dfrac{dy}{dx} = \dfrac{1}{\dfrac{dx}{dy}}$

[3] $\quad \dfrac{d}{dx}(\log_a v) = \log_a e \dfrac{\dfrac{dv}{dx}}{v}$

[3a] $\quad \dfrac{d}{dx}(\log v) = \dfrac{\dfrac{dv}{dx}}{v}$

[4] $\quad \dfrac{d}{dx}(a^v) = \log a \cdot a^v \cdot \dfrac{dv}{dx}$

[5] $\quad \dfrac{d}{dx}(u^v) = vu^{v-1}\dfrac{du}{dx} + \log u \cdot u^v \cdot \dfrac{dv}{dx}$

[6] $\quad \dfrac{d}{dx}(\sin v) = \cos v \dfrac{dv}{dx}$

[7] $\dfrac{d}{dx}(\cos v) = -\sin v\,\dfrac{dv}{dx}$

[8] $\dfrac{d}{dx}(\tan v) = \sec^2 v\,\dfrac{dv}{dx}$

[9] $\dfrac{d}{dx}(\cot v) = -\csc^2 v\,\dfrac{dv}{dx}$

[10] $\dfrac{d}{dx}(\sec v) = \sec v \tan v\,\dfrac{dv}{dx}$

[11] $\dfrac{d}{dx}(\csc v) = -\csc v \cot v\,\dfrac{dv}{dx}$

[12] $\dfrac{d}{dx}(\text{arc sin } v) = \dfrac{\dfrac{dv}{dx}}{\sqrt{1-v^2}}$

[13] $\dfrac{d}{dx}(\text{arc cos } v) = -\dfrac{\dfrac{dv}{dx}}{\sqrt{1-v^2}}$

[14] $\dfrac{d}{dx}(\text{arc tan } v) = \dfrac{\dfrac{dv}{dx}}{1+v^2}$

[15] $\dfrac{d}{dx}(\text{arc cot } v) = -\dfrac{\dfrac{dv}{dx}}{1+v^2}$

[16] $\dfrac{d}{dx}(\text{arc sec } v) = \dfrac{\dfrac{dv}{dx}}{v\sqrt{v^2-1}}$

[17] $\dfrac{d}{dx}(\text{arc csc } v) = -\dfrac{\dfrac{dv}{dx}}{v\sqrt{v^2-1}}$

5—18. Successive Differentiation. When discussing velocity and acceleration in §4—7, we learned how to find successive derivatives of algebraic functions. We shall now apply this idea to transcendental functions.

EXAMPLE 1. Find the fourth derivative of $y = e^x + 2x^3$.

Solution.

$$\frac{dy}{dx} = e^x + 6x^2,$$

$$\frac{d^2y}{dx^2} = e^x + 12x,$$

$$\frac{d^3y}{dx^3} = e^x + 12,$$

$$\frac{d^4y}{dx^4} = e^x.$$

EXAMPLE 2. Find the third derivative of $y = \sin 2x$.

Solution.

$$\frac{dy}{dx} = 2 \cos 2x,$$

$$\frac{d^2y}{dx^2} = -4 \sin 2x,$$

$$\frac{d^3y}{dx^3} = -8 \cos 2x.$$

5—19. Successive Differentiation of Implicit Functions. Suppose we wish to find the second derivative of y with respect to x in the equation of the ellipse $b^2x^2 + a^2y^2 = a^2b^2$. We proceed to differentiate with respect to x:

$$2b^2x + 2a^2y \frac{dy}{dx} = 0,$$

or

$$\frac{dy}{dx} = -\frac{b^2x}{a^2y}. \tag{1}$$

Now differentiate again, remembering that y is a function of x:

$$\frac{d^2y}{dx^2} = -\frac{a^2yb^2 - b^2xa^2 \dfrac{dy}{dx}}{a^4y^2}. \tag{2}$$

Substituting the value of $\frac{dy}{dx}$ from (1) in equation (2):

$$\frac{d^2y}{dr^2} = -\frac{a^2b^2y - a^2b^2x\left(-\dfrac{b^2x}{a^2y}\right)}{a^4y^2} = -\frac{b^2(a^2y^2 + b^2x^2)}{a^4y^3}.$$

But $a^2y^2 + b^2x^2 = a^2b^2$; therefore $\dfrac{d^2y}{dx^2} = -\dfrac{b^4}{a^2y^3}$.

EXAMPLE 1. Find the second derivative with respect to x of

$$x^2 + y = xy.$$

Solution. $\quad 2x + \dfrac{dy}{dx} = x\dfrac{dy}{dx} + y, \quad$ or $\quad \dfrac{dy}{dx} = \dfrac{y - 2x}{1 - x}.$ \hfill (1)

Differentiating again:

$$\frac{d^2y}{dx^2} = \frac{d}{dx}\left(\frac{y - 2x}{1 - x}\right) = \frac{(1 - x)\left(\dfrac{dy}{dx} - 2\right) + (y - 2x)}{(1 - x)^2}.$$

Substituting for $\dfrac{dy}{dx}$ from (1):

$$\frac{d^2y}{dx^2} = \frac{(1 - x)\left(\dfrac{y - 2x}{1 - x} - 2\right) + (y - 2x)}{(1 - x)^2} = \frac{2(y - x - 1)}{(1 - x)^2}.$$

EXAMPLE 2. Find the second derivative with respect to x of

$$e^x = \sin y.$$

Solution. $\quad \dfrac{d}{dx}(e^x) = \dfrac{d}{dx}(\sin y),$

or $\qquad\qquad\qquad\qquad e^x = \cos y\,\dfrac{dy}{dx}$

$$\frac{dy}{dx} = \frac{e^x}{\cos y}.$$ \hfill (1)

Differentiating again:

$$\frac{d^2y}{dx^2} = \frac{e^x\cos y + e^x\sin y\,\dfrac{dy}{dx}}{\cos^2 y}.$$

Substituting the value of $\dfrac{dy}{dx}$ from (1):

$$\frac{d^2y}{dx^2} = \frac{e^x\cos y + e^x\sin y\,\dfrac{e^x}{\cos y}}{\cos^2 y}$$

$$= \frac{e^x(\cos y + e^x\tan y)}{\cos^2 y}.$$

5—20. The nth Derivative of a Product. It is sometimes useful to express the nth derivative of the product of two variables in terms of the variables and their successive derivatives. For example, if u and v are functions of x, then

$$\frac{d}{dx}(uv) = u\frac{dv}{dx} + v\frac{du}{dx}.$$

By differentiating again, the reader can verify that

$$\frac{d^2}{dx^2}(uv) = v\frac{d^2u}{dx^2} + 2\frac{du}{dx}\cdot\frac{dv}{dx} + u\frac{d^2v}{dx^2},$$

and that

$$\frac{d^3}{dx^3}(uv) = v\frac{d^3u}{dx^3} + 3\frac{d^2u}{dx^2}\cdot\frac{dv}{dx} + 3\frac{du}{dx}\cdot\frac{d^2v}{dx^2} + u\frac{d^3v}{dx^3}.$$

By mathematical induction, it can be shown, in general, that

$$\frac{d^n}{dx^n}(uv) = v\frac{d^nu}{dx^n} + n\frac{d^{n-1}u}{dx^{n-1}}\cdot\frac{dv}{dx} + \frac{n(n-1)}{2!}\frac{d^{n-2}u}{dx^{n-2}}\cdot\frac{d^2v}{dx^2} +$$

$$+ \cdots + n\frac{du}{dx}\frac{d^{n-1}v}{dx^{n-1}} + u\frac{d^nv}{dx^n}.$$

This is known as *Leibniz's Formula* for the nth derivative of a product. It will be seen that the numerical coefficients follow the same law as those of the binomial theorem, and that the indices of the derivatives correspond to the exponents in the binomial expansion. The correspondence can be made complete as follows: just as the first derivatives $\frac{du}{dx}$ and $\frac{dv}{dx}$ can be considered as $\frac{d^1u}{dx^1}$ and $\frac{d^1v}{dx^1}$, so u and v (the variables themselves) may be considered as $\frac{d^0u}{dx^0}$ and $\frac{d^0v}{dx^0}$.

EXERCISE 5—7

1. Find the second derivative of

$$y = e^x \sin x.$$

2. Find the third derivative of

$$y = \log \sin x.$$

3. Find the fourth derivative of
  (a) $y = x^3 \log x$
  (b) $y = \sin ax$

4. Find $\dfrac{d^2y}{dx^2}$ for the equation $x^2 + y^2 = k^2$.

5. Find $\dfrac{d^2y}{dx^2}$ for $y^2 = 4px$.

6. Find $\dfrac{d^2y}{dx^2}$ for $y^2 + x = y$.

7. Find, by Leibniz's formula, the third derivative of $y = e^x x^2$. (Let $u = e^x$, $v = x^2$.)

EXERCISE 5—8

Review

Differentiate:

1. $y = \dfrac{\log x}{e^x}$

2. $y = e^t(1 + t^2)$

3. $y = \log(2 - 3x^2)$

4. $y = c^{e^x}$

5. $y = \log(\log x)$

6. $y = e^{\log x^2}$

7. $y = \log \dfrac{e^x}{2 + e^x}$

8. $y = e^{x \log x}$

9. $y = e^{\cos 2x}$

10. $y = \log \sin x$

11. $y = \tan x + \tan^2 x$

12. $y = \arcsin \dfrac{x - 3}{x + 3}$

13. $y = e^{x \sin x}$

14. $y = \text{arc cot} \dfrac{2x}{(x^2 - 1)}$

15. $y = \dfrac{1}{a}\left(\arctan \dfrac{x}{a}\right)$

16. $y = \log \cos^2 x$

17. $y = \log \dfrac{x + 1}{x - 1}$

18. $y = \dfrac{\log x}{e^x}$

19. $y = \dfrac{e^x - 1}{e^x + 1}$

20. $y = \dfrac{a^x}{x^x}$

Further Applications
of the Derivative

CHAPTER SIX

SLOPES, TANGENTS, AND NORMALS

6—1. Slope of a Curve. We have seen that for any curve, $y = f(x)$, the slope m of the tangent to the curve at any particular point (x_1, y_1)

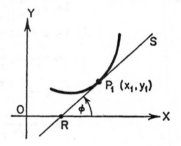

is given by $m = \tan \phi = \dfrac{dy_1}{dx_1}$, where the symbol $\dfrac{dy_1}{dx_1}$ indicates the particular value which the variable expression $\dfrac{dy}{dx}$ takes when $x = x_1$ and $y = y_1$. The reader is urgently cautioned not to think of $\dfrac{dy_1}{dx_1}$ as meaning the derivative of y_1 with respect to x_1; such an interpretation would be meaningless since both x_1 and y_1 are *constants*, not variables.

EXAMPLE 1. Find the slope of the tangent to $y = x^4 + 2x^3 - 5x + 3$ at the point where $x = 2$.

Solution.

$$\frac{dy}{dx} = 4x^3 + 6x^2 - 5,$$

and

$$\frac{dy_1}{dx_1} = 4(2)^3 + 6(2)^2 - 5 = 51.$$

EXAMPLE 2. Find the slope of the tangent to $y^2 = 12x + 13$ at the point $(3,7)$.

Solution. Differentiating:

$$2y\frac{dy}{dx} = 12;$$

$$\frac{dy}{dx} = \frac{6}{y}, \quad \text{and} \quad \frac{dy_1}{dx_1} = \frac{6}{7}.$$

EXAMPLE 3. Find the slope of the curve $2x^2 + y^2 = 4x + 42$ at the points where $x = 3$.

Solution. If $x_1 = 3$, then $y_1 = \pm 6$. Differentiating:

$$4x + 2y\frac{dy}{dx} = 4,$$

$$\frac{dy}{dx} = \frac{2 - 2x}{y}, \quad \text{and} \quad \frac{dy_1}{dx_1} = \frac{2 - 6}{\pm 6} = \mp\frac{2}{3}.$$

EXAMPLE 4. Find the slope of the curve

$$x^2y^2 + x^3 - 2x - y^4 - 6y = 0$$

at the point $(0,0)$.

Solution. Differentiating:

$$2xy^2 + 2x^2y\frac{dy}{dx} + 3x^2 - 2 - 4y^3\frac{dy}{dx} - 6\frac{dy}{dx} = 0.$$

$$\frac{dy}{dx} = \frac{2 - 2xy^2 - 3x^2}{2x^2y - 4y^3 - 6},$$

and

$$\frac{dy_1}{dx_1} = \frac{2 - 0 - 0}{0 - 0 - 6} = -\frac{1}{3}.$$

6—2. The Angle between Two Intersecting Curves. The reader
will recall that the angle between two intersecting curves is defined as
the angle between the tangents to the two curves, respectively, at their
point of intersection $P(x_1, y_1)$. Since

$$\tan \phi = \frac{m_2 - m_1}{1 + m_1 m_2},$$

it is merely necessary first to determine m_1 and m_2 by differentiation,

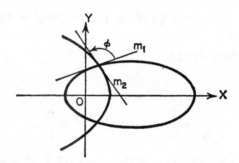

then to substitute the values of x_1 and y_1 in each of the two derivatives
so found, and then to substitute these values in the expression for
$\tan \phi$.

EXAMPLE. Find the angle of intersection between the circle

$$x^2 + y^2 = 25$$

and the parabola $4y^2 = 9x$.

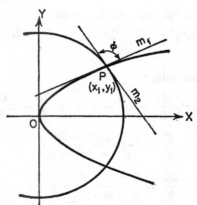

Solution. Solving these equations simultaneously we have:

$$x = 4, \quad y = \pm 3.$$

Let $$x_1 = 4, \quad y_1 = +3.$$

Differentiating:

(a) $2x + 2y \dfrac{dy}{dx} = 0;$

$$\frac{dy}{dx} = -\frac{x}{y}; \quad \frac{dy_1}{dx_1} = -\frac{4}{3} = m_2.$$

(b) $8y \dfrac{dy}{dx} = 9; \quad \dfrac{dy}{dx} = \dfrac{9}{8y};$

$$\frac{dy_1}{dx_1} = \frac{3}{8} = m_1.$$

$$\tan \phi = \frac{m_2 - m_1}{1 + m_1 m_2} = \frac{-\frac{4}{3} - \frac{3}{8}}{1 - \frac{1}{2}} = -\frac{41}{12};$$

hence, arc tan $\left(-\dfrac{41}{12}\right)$ = arc tan (-3.417) = $106°20'$ (approx.)

EXERCISE 6—1

1. Find the slope of the curve

$$\frac{x}{y} + \frac{y}{x} = 1$$

at any point.

2. Show that the tangent to the circle $x^2 + y^2 + 2x + 4y = 0$ at the origin is parallel to the line $x + 2y = 10$.

3. Find the slope of the curve

$$y = x(x^3 + 7)^{\frac{2}{3}}$$

at the point where $x = 1$.

4. At what point on the curve $y^2 = 3x^3$ is the slope equal to $2\frac{1}{4}$?

5. Find the angle of intersection between the curves

$$x^2 + y^2 + 2x - 3 = 0 \quad \text{and} \quad x^2 + y^2 = 7.$$

6. At what points on the circle $x^2 + y^2 = k^2$ is the slope of the tangent to the circle equal to $-\dfrac{5}{12}$?

7. Find the angle of intersection of the two parabolas $x^2 = 4py$ and $y^2 = 4px$.

8. Prove that for all values of k, the equation $kx = e^y$ has the same slope, that is, the slope is independent of the value of k.

6—3. Equations of Tangent and Normal. The equation of the tangent, ST, to a curve at a given point $P_1(x_1, y_1)$ is easily derived.

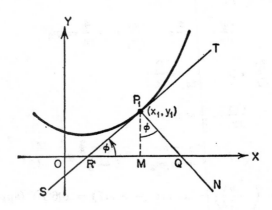

The slope of the tangent in question is given by $\dfrac{dy_1}{dx_1}$; since the point of tangency lies on the tangent as well as on the curve, we may use the point-slope formula for the equation of a line having a given slope and passing through a given point. Thus:

$$y - y_1 = \frac{dy_1}{dx_1}(x - x_1). \tag{1}$$

The *normal* is the line perpendicular to a tangent at the point of contact. Hence the equation of the normal, P_1N, given by

$$y - y_1 = -\frac{dx_1}{dy_1}(x - x_1). \tag{2}$$

EXAMPLE 1. Find the equations of the tangent and the normal to the curve $4x^2 + 9y^2 = 25$ at the point where $x = 2$ and y is positive.

Solution.

$$8x + 18y\frac{dy}{dx} = 0, \qquad \frac{dy}{dx} = -\frac{4x}{9y}.$$

At $x_1 = 2$, $y_1 = \pm 1$; taking y_1 positive:

$$\frac{dy_1}{dx_1} = -\frac{(4)(2)}{9(1)} = -\frac{8}{9}.$$

Hence, the equation of the tangent at $(2,1)$ is

$$y - 1 = -\frac{8}{9}(x - 2), \quad \text{or} \quad 8x + 9y - 25 = 0$$

and the equation of the normal at $(2,1)$ is

$$y - 1 = \frac{9}{8}(x - 2), \quad \text{or} \quad 9x - 8y - 10 = 0.$$

EXAMPLE 2. Find the equations of the tangent and the normal to the equation $2x^3 = y^2$ at the point where $x = 1$ and y is negative.

Solution.

$$6x^2 = 2y\frac{dy}{dx}; \quad \frac{dy}{dx} = \frac{3x^2}{y}.$$

At $x_1 = 1$, $y_1 = \pm\sqrt{2}$; taking y_1 negative:

$$\frac{dy_1}{dx_1} = -\frac{3}{\sqrt{2}} = -\frac{3\sqrt{2}}{2}.$$

Hence, the equation of the tangent is

$$y + \sqrt{2} = -\frac{3\sqrt{2}}{2}(x - 1), \quad \text{or} \quad 2y + 3\sqrt{2}\,x - \sqrt{2} = 0.$$

The equation of the normal:

$$y + \sqrt{2} = \frac{\sqrt{2}}{3}(x - 1), \quad \text{or} \quad 3y - \sqrt{2}\,x + 4\sqrt{2} = 0.$$

6—4. Length of Subtangent and Subnormal. Referring to the diagram in §6—3, the segment RM is called the *subtangent* of the point P_1; the length of the segment MQ is called the *subnormal* of P_1. The lengths of these segments, for any curve $f(x,y) = 0$, can be readily derived.

$$RM = \frac{MP_1}{\tan\phi} = \frac{y_1}{\dfrac{dy_1}{dx_1}} = \text{length of subtangent} \qquad [3]$$

$$MQ = MP_1 \cdot \tan \phi = y_1 \frac{dy_1}{dx_1} = \text{length of subnormal} \qquad [4]$$

EXAMPLE 1. Find the lengths of the subtangent and the subnormal to the curve $x^2 = 8y + 4$ when $x = 6$.

Solution. $2x = 8 \dfrac{dy}{dx}$,   or $\dfrac{dy}{dx} = \dfrac{x}{4}$.

At $x_1 = 6$, $y_1 = 4$;

hence, subtangent $= \dfrac{4}{\frac{6}{4}} = \dfrac{8}{3}$,

and subnormal $= 4 \left(\dfrac{6}{4} \right) = 6$.

EXAMPLE 2. Find the lengths of the subtangent and the subnormal to $x^2 = 2y^3$ at $x = 4$.

Solution. $2x = 6y^2 \dfrac{dy}{dx}$; $\dfrac{dy}{dx} = \dfrac{x}{3y^2}$.

At $x_1 = 4$, $y_1 = 2$; hence,

subtangent $= \dfrac{2}{\frac{4}{12}} = 6$,   subnormal $= 2 \left(\dfrac{4}{12} \right) = \dfrac{2}{3}$.

EXERCISE 6—2

1. Find the length of the subtangent to $3x^2 - y^2 = 12$ at $x = 4$.

2. Find the equation of the tangent and the normal to the curve $x^3 = y + xy$ at the point where $x = -2$.

3. Find the lengths of the subtangent and the subnormal to the curve $y^2 = 4px$ at any point (x,y).

4. Find the equation of the tangent, and the length of the subtangent, to the circle $x^2 + y^2 = 25$ at $x = 3$.

5. Find the equation of the tangent to the curve $a^2(x - y) = x^3 + x^2y$ at the origin.

6. Find the length of the subtangent to the curve $x = ky^n$ when $x = k$.

7. Find the length of the subtangent to the curve $y = k^x$.

8. Prove that the area of the triangle formed by the coordinate axes and any tangent drawn to the curve $2xy = a^2$ is a constant and equal to a^2.

POINTS OF INFLECTION AND CURVE TRACING

6—5. Points of Inflection. When a curve changes its nature in such a way that there is a point at which the curve changes from concave downwards to concave upwards, or vice versa, that point is called a *point of inflection*.

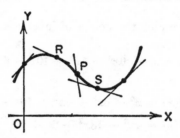

From the discussion of §4—11, it will be seen, from the figure above, that at R, $f''(x)$ is $-$, and at S, $f''(x)$ is $+$; for as we move from left to right, or from R to P, the slope, $f'(x)$, is decreasing, and when passing from P to S, the slope $f'(x)$ is increasing. Thus, as $f''(x)$ changes from $-$ to $+$ (or $+$ to $-$), it must pass through the value zero. Hence:

$$\text{at a point of inflection,} \quad f''(x) = 0. \tag{1}$$

Points of inflection may also be defined as points where

(1) $\dfrac{d^2y}{dx^2} = 0$ and $\dfrac{d^2y}{dx^2}$ changes sign,

(2) $\dfrac{d^2x}{dy^2} = 0$ and $\dfrac{d^2x}{dy^2}$ changes sign.

A general principle, given here without proof, is the following:

Given a function $f(x)$ and a specified point x_1 such that $f''(x_1) = 0$, but $f'''(x_1) \neq 0$, then x_1 is the abscissa of a point of inflection on the curve $y = f(x)$.

It should be noted that at a point of inflection, the tangent to the curve passes *through* the curve. Points of inflection are very common. For example, algebraic polynomials of the form

$$y = a_0x^n + a_1x^{n-1} + a_2x^{n-2} + \cdots + a_n$$

in general cannot have more than $(n - 2)$ points of inflection. A parabola cannot have a point of inflection; for if $f(x) = ax^2 + bx + c$, then $f'(x) = 2ax + b$, and $f''(x) = 2a$; thus there is no value of x which makes $f''(x) = 0$. Similarly, a straight line can have no point of inflection, since if $f(x) = ax + b$, then $f'(x) = a$, $f''(x) = 0$, and $f'''(x) = 0$. The trigonometric functions, such as $y = \sin x$, $y = \cos x$, $y = \tan x$, have an infinite number of points of inflection.

EXAMPLE 1. Determine the points of inflection of the curve whose equation is

$$y = x^4 - 6x^3 + 4x^2 + 10x - 5.$$

Solution.

$$f(x) = x^4 - 6x^3 + 4x^2 + 10x - 5,$$
$$f'(x) = 4x^3 - 18x^2 + 8x + 10,$$
$$f''(x) = 12x^2 - 36x + 8,$$
$$f'''(x) = 24x - 36.$$

The values of x for which $f''(x) = 0$ are therefore found from the equation

$$12x^2 - 36x + 8 = 0;$$

these values are

$$x = \frac{9 \pm \sqrt{57}}{6}.$$

EXAMPLE 2. Test for points of inflection the curve of

$$y = x^4 - 8x^3 + 24x^2 + 8x.$$

Solution.

$$f'(x) = 4x^3 - 24x^2 + 48x + 8,$$
$$f''(x) = 12x^2 - 48x + 48,$$
$$f'''(x) = 24x - 48.$$

Values of x for which $f''(x) = 0$ are found from the equation

$$12x^2 - 48x + 48 = 0,$$

which gives

$$(x - 2)^2 = 0, \quad \text{or} \quad x = +2, +2.$$

However, for the value $x = +2$,

$$f'''(x) = 24(2) - 48 = 0;$$

hence there are no points of inflection, since $f'''(x) = 0$.

EXAMPLE 3. Find the points of inflection in the curve of $y = \cos x$.

Solution.

$$f'(x) = -\sin x,$$
$$f''(x) = -\cos x,$$
$$f'''(x) = +\sin x.$$

Solving $f''(x) = 0$ for x:

$$-\cos x = 0, \quad x = \pm \frac{\pi}{2}, \pm \frac{3\pi}{2}, \pm \frac{5\pi}{2}, \text{ etc.}$$

For these values of x, $f'''(x) \neq 0$; hence points of inflection occur at

$$x = \pm \frac{2n+1}{2} \pi, \quad y = 0.$$

EXERCISE 6—3

Test the following functions for points of inflection:

1. $y = x^4 + 2x^3 - 36x^2 - x$
2. $y = x^4 + 4x^3 + 6x^2 + 60x$
3. $y = x^4 - 12x^2$
4. $y = 2x^3$
5. $y = x^4$
6. $y = x^5$
7. $y = e^x$
8. $y = xe^x$
9. $y = b + (x - a)^3$
10. $xy - x^3 = 1$

6—6. Curve Tracing. When discussing curves and equations, an examination of the equation for *intercepts, extent, symmetry,* and *asymptotes* constitutes a useful and economical approach to determining the general shape of a curve. For the reader who may have forgotten the tests for symmetry with respect to the origin and the axes, we remind him:

(1) If an equation is unchanged by the substitution of $-x$ for x, the curve is symmetrical with respect to the Y-axis.

(2) If an equation is unchanged by the substitution of $-y$ for y, the curve is symmetrical with respect to the X-axis.

(3) If an equation is unchanged by the substitution of $-x$ for x and $-y$ for y, the curve is symmetrical with respect to the origin.

We also remind the reader who may have forgotten how to determine horizontal and vertical asymptotes:

(1) To find a vertical asymptote, solve the equation for y; if the solution is a fraction, set the denominator equal to zero and solve for x.

(2) To find a horizontal asymptote, solve the equation for x; if the solution is a fraction, set the denominator equal to zero and solve for y.

Limiting our discussion to single-valued functions, we may now add the determination of *maximum and minimum values* and *points of inflection* as further tools for tracing a curve. When all these devices have been used, few additional points need to be calculated and plotted, and it will be possible to sketch the curve fairly accurately with a minimum of preliminary computation.

For convenience, the procedure for curve tracing is outlined below:

Step 1.   Find the intercepts on the coordinate axes, if any.

Step 2.   Test the curve for symmetry with respect to the axes and the origin.

Step 3.   Determine the extent of the curve; that is, what values, if any, of either variable must be excluded.

Step 4.   Find the asymptotes, if any, parallel to the coordinate axes; also, how y behaves for numerically $(+$ or $-)$ large values of x.

Step 5.   Locate the critical points, where $f'(x) = 0$, and determine all maxima and minima.

Step 6.   Locate the points at which $f''(x) = 0$, and determine all points of inflection.

EXAMPLE 1.   Trace the curve of $y = x^3 - 3x^2 + x + 1$.

Solution.

$$f(x) = x^3 - 3x^2 + x + 1,$$
$$f'(x) = 3x^2 - 6x + 1,$$
$$f''(x) = 6x - 6,$$
$$f'''(x) = 6.$$

(1) Intercept on Y-axis $= 1$. Intercepts on X-axis are $1, 1 + \sqrt{2}$, and $1 - \sqrt{2}$.

(2) Tests show that the curve is not symmetric to the origin or to either axis.

(3) Inspection shows that no values of x need be excluded.

(4) Inspection shows no asymptotes.

(5) Setting $f'(x)$ equal to zero and solving gives critical values at $x = \dfrac{3 \pm \sqrt{6}}{3}$. Testing these further, it is found that a minimum exists at $x = \dfrac{3 + \sqrt{6}}{3} = 1.82$; a maximum exists at $x = \dfrac{3 - \sqrt{6}}{3} = +.18$.

(6) Setting $f''(x) = 0$ and solving, $x = 1$; since $f'''(1) \neq 0$, there is a point of inflection at $x = 1$, i.e., the point (1,0). (That the abscissa of the point of inflection is also a root of $f(x) = 0$ is entirely accidental and of no significance.)

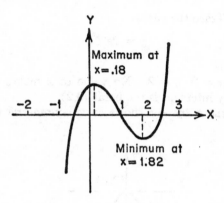

EXAMPLE 2. Trace the curve of

$$y = \frac{x}{x - 1}.$$

Solution.

(1) Intercept: $x = 0$, $y = 0$.

(2) No symmetry.

(3) Curve not defined when $x = 1$.

(4) Asymptotes: $x - 1 = 0$, and $y - 1 = 0$.

(5) $\dfrac{dy}{dx} = -\dfrac{1}{(x - 1)^2}$; no values of x for which $f'(x) = 0$; hence no maxima or minima.

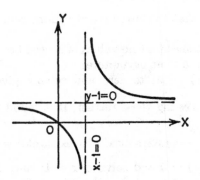

(6) $f''(x) = 2(x - 1)^{-3}$; no values of x for which $f''(x) = 0$; hence no points of inflection.

EXAMPLE 3.   Trace the curve:

$$y = 2e^{-x^2}.$$

Solution.

(1)   When $x = 0$, $y = 2$. No value of x makes $y = 0$. Hence the only intercept is the point (0,2).

(2)   Replacing x by $-x$ shows curve symmetric with respect to Y-axis.

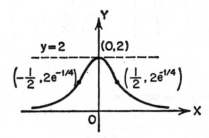

(3)   No values of x need be excluded; all values of x yield values of y greater than zero but not greater than 2. Thus the curve lies entirely between the line $y = 2$ and the X-axis.

(4)   As x becomes indefinitely large, positively or negatively, y approaches zero; thus the X-axis is an asymptote.

(5)   Differentiating, $f'(x) = -4xe^{-x^2}$; solving $f'(x) = 0$ for x, we get $x = 0$ as the only critical value. As x passes through zero, $f'(x)$ changes from $+$ to $-$, and so (0,2) is a maximum point.

(6) Setting $f''(x) = (16x^2 - 4)e^{-x_2} = 0$, we get $x = \pm\dfrac{1}{2}$. Since $f''(x)$ changes sign as x passes through either of these values, the points of inflection are $(\pm\dfrac{1}{2}, 2e^{-1/4})$.

EXERCISE 6—4

Trace the following curves and draw the sketch in each case:

1. $y = x^3 - 9x$

2. $y = x^3 - x^2 - 5x - 6$

3. $y = \dfrac{x}{x+1}$

4. $y = \dfrac{x}{x^2 - 1}$

5. $y = (x - 1)^3$

6. $y = e^{2x}$

7. $y = e^{-x^2}$

8. $y = x(4 - x^2)$

9. $y = x(\log x)$

10. $y = x + \sin x$

11. Prove that $(0,0)$ is not a point of inflection on the curve $y = x^6 - 2x^4$.

12. Prove that no conic can have a point of inflection.

PARAMETRIC EQUATIONS

6—7. Derivatives of Equations in Parametric Form. The reader will recall that the relation of two variables may be stated in terms of their relation to a third variable, called the *parameter*. Equations in parametric form may be differentiated as follows.

If $\qquad\qquad x = f(t), \qquad$ and $\qquad y = \phi(t),$

then, by §5—1,

$$\frac{dy}{dx} = \frac{\dfrac{dy}{dt}}{\dfrac{dx}{dt}}. \tag{1}$$

To find $\dfrac{d^2y}{dx^2}$, we proceed as shown:

$$\frac{d^2y}{dx^2} = \frac{d}{dx}\left(\frac{dy}{dx}\right) = \frac{\dfrac{d}{dt}\left(\dfrac{dy}{dx}\right)}{\dfrac{dx}{dt}}. \tag{2}$$

The numerator of the third member of equation [2] may be expanded; thus

$$\frac{d}{dt}\left(\frac{dy}{dx}\right) = \frac{d}{dt}\left(\frac{\dfrac{dy}{dt}}{\dfrac{dx}{dt}}\right) = \frac{\dfrac{d^2y}{dt^2}\cdot\dfrac{dx}{dt} - \dfrac{d^2x}{dt^2}\cdot\dfrac{dy}{dt}}{\left(\dfrac{dx}{dt}\right)^2}.$$

Hence:
$$\frac{d^2y}{dx^2} = \frac{\dfrac{d^2y}{dt^2}\dfrac{dx}{dt} - \dfrac{d^2x}{dt^2}\dfrac{dy}{dt}}{\left(\dfrac{dx}{dt}\right)^3}. \qquad [3]$$

Equation [3] may also be written in the form:

$$\frac{d^2y}{dx^2} = \frac{f'(t)\cdot\phi''(t) - \phi'(t)\cdot f''(t)}{[f'(t)]^3} \qquad [4]$$

With these expressions for $\dfrac{dy}{dx}$ and $\dfrac{d^2y}{dx^2}$, the slopes, maxima and minima, points of inflection, equations of tangents and normals, lengths of subtangents and subnormals, etc., for equations in parametric form are readily obtained.

EXAMPLE 1.   Find $\dfrac{dy}{dx}$ and $\dfrac{d^2y}{dx^2}$ for the curve whose parametric equations are $x = 2t^2$ and $y = 2t - t^3$; find the equation of the tangent to the curve at the point for which $t = 2$.

Solution.

$$\frac{dx}{dt} = 4t, \qquad \text{and} \qquad \frac{dy}{dt} = 2 - 3t^2;$$

hence
$$\frac{dy}{dx} = \frac{2 - 3t^2}{4t}.$$

Now
$$\frac{d^2x}{dt^2} = 4, \qquad \text{and} \qquad \frac{d^2y}{dt^2} = -6t;$$

hence
$$\frac{d^2y}{dx^2} = \frac{(-6t)(4t) - (4)(2 - 3t^2)}{(4t)^3} = \frac{-3t^2 - 2}{16t^3}.$$

When $t = 2$, $x = 8$, $y = -4$.

Equation of tangent at point (x_1, y_1) is

$$y - y_1 = \frac{dy_1}{dx_1}\,(x - x_1),$$

or, at (x_1, y_1), when $t = 2$, we have $\dfrac{dy_1}{dx_1} = \dfrac{2 - 3(2)^2}{4(2)} = -\dfrac{5}{4}\,;$

hence $\qquad\qquad y + 4 = -\dfrac{5}{4}\,(x - 8),$

or $\qquad\qquad 5x + 4y = 24.$

EXAMPLE 2.   Given the parametric equations $x = e^{2t}$ and $y = e^{t+1}$. Find (a) the values of $\dfrac{dy}{dx}$ and $\dfrac{d^2y}{dx^2}\,;$ (b) the length of the subtangent and the subnormal at the point for which $t = 0$.

Solution. $\quad \dfrac{dx}{dt} = 2e^{2t}, \quad$ and $\quad \dfrac{dy}{dt} = e^{t+1}\,;$

hence $\quad \dfrac{dy}{dx} = \dfrac{e^{t+1}}{2e^{2t}} = \dfrac{1}{2}\,e^{1-t}.$

Now, $\quad \dfrac{d^2x}{dt^2} = 4e^{2t}, \quad$ and $\quad \dfrac{d^2y}{dt^2} = e^{t+1}\,;$

hence $\quad \dfrac{d^2y}{dx^2} = \dfrac{(e^{t+1})(2e^{2t}) - (4e^{2t})(e^{t+1})}{8e^{6t}} = -\dfrac{1}{4}\,e^{1-3t}$

At $t = 0$, $\;\; x_1 = 1,\;$ and $\;\; y_1 = e;\;\; \dfrac{dy_1}{dx_1} = \dfrac{e}{2}\,.$

Hence, length of subtangent $= \dfrac{y_1}{\dfrac{dy_1}{dx_1}} = \dfrac{e}{\dfrac{e}{2}} = 2;$

and length of subnormal $= y_1 \cdot \dfrac{dy_1}{dx_1} = (e)\left(\dfrac{e}{2}\right) = \dfrac{1}{2}\,e^2.$

EXAMPLE 3.   Given the parametric equations of the cycloid

$$x = a(\theta - \sin \theta), \qquad y = a(1 - \cos \theta).$$

Find the lengths of the subtangent and the subnormal to the point on the curve for which $\theta = \dfrac{\pi}{2}\,;$ find also the value of $\dfrac{d^2y}{dx^2}$ at this point.

Solution.

$$\frac{dx}{d\theta} = a(1 - \cos \theta), \quad \text{and} \quad \frac{dy}{d\theta} = a \sin \theta;$$

hence

$$\frac{dy}{dx} = \frac{\sin \theta}{1 - \cos \theta}.$$

At the point for which $\theta = \frac{\pi}{2}$:

$$x_1 = a\left(\frac{\pi}{2} - \sin \frac{\pi}{2}\right) = \frac{\pi a}{2} - a,$$

$$y_1 = a\left(1 - \cos \frac{\pi}{2}\right) = a,$$

and thus

$$\frac{dy_1}{dx_1} = \frac{\sin \frac{\pi}{2}}{1 - \cos \frac{\pi}{2}} = 1.$$

Therefore

$$\text{subtangent} = \frac{y_1}{\dfrac{dy_1}{dx_1}} = \frac{a}{1} = a,$$

and

$$\text{subnormal} = y_1 \cdot \frac{dy_1}{dx_1} = a \cdot 1 = a.$$

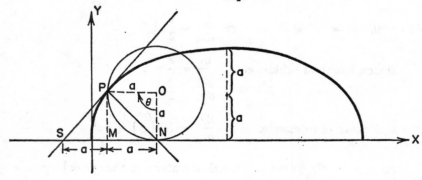

Further:

$$\frac{d^2y}{dx^2} = \frac{a(1 - \cos \theta) \cdot a \cos \theta - a \sin \theta \cdot a \sin \theta}{a^3(1 - \cos \theta)^3},$$

or $\quad \dfrac{d^2y_1}{dx_1{}^2} = \dfrac{a^2\left(1 - \cos\dfrac{\pi}{2}\right)\cos\dfrac{\pi}{2} - a^2\sin^2\dfrac{\pi}{2}}{a^3\left(1 - \cos\dfrac{\pi}{2}\right)^3} = -\dfrac{1}{a}.$

EXERCISE 6—5

1. Find (a) the equation of the tangent to the curve whose parametric equations are $x = 40t$ and $y = 40t - \frac{1}{2}gt^2$, at the point on the curve given by $t = 1$; (b) find the rate at which the slope is changing at any point on the curve.

2. Given the parametric equations $x = 4 + t$ and $y = \log 4t$. Find (a) the equation of the tangent at the point on the curve given by $t = \frac{1}{4}$; (b) find the value of $\dfrac{d^2y}{dx^2}$ when $t = 3$.

3. In the parametric equations of the circle $x = r \cos \phi$ and $y = r \sin \phi$, prove analytically (a) that the equation of the tangent at the point on the circle given by $\phi = \dfrac{\pi}{2}$ is $y = r$; (b) that the length of the subnormal to the point given by $\phi = \dfrac{\pi}{4}$ is equal to $- r \dfrac{\sqrt{2}}{2}$.

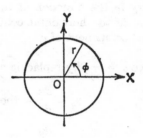

4. Find $\dfrac{d^2y}{dx^2}$ for the ellipse: $x = a \cos \theta$, $y = b \cos \theta$.

5. Find $\dfrac{d^2y}{dx^2}$ for the hypocycloid of four cusps: $x = a \cos^3 \theta$, $y = a \sin^3 \theta$.

6. Find the acceleration $\left(\dfrac{d^2y}{dx^2}\right)$ of the projectile whose path is given by:

$$x = (v_0 \cos \phi)t,$$
$$y = (v_0 \sin \phi)t - \tfrac{1}{2}gt^2.$$

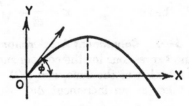

RECTILINEAR AND CIRCULAR MOTION

6—8. Component Velocities. Let us consider the path of a curve generated by a moving point following the law $y = f(x)$. The coordinates of x and y of P may be regarded as functions of the time; consequently the motion of $y = f(x)$ may also be defined by means

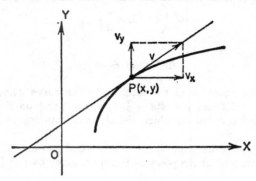

of the parametric equations $x = f(t)$, $y = \phi(t)$. If v represents the velocity in the direction of the tangent to the path of $y = f(x)$ at P, then v_x is the horizontal component of the velocity v, and v_y is the vertical component of v.

Since $v = \dfrac{ds}{dt}$, by replacing s by x and y, respectively, we have:

$$v_x = \frac{dx}{dt} = \text{the time rate of change of } x,$$

$$v_y = \frac{dy}{dt} = \text{the time rate of change of } y.$$

From the figure,

$$|v| = \sqrt{v_x^2 + v_y^2};$$

hence
$$v = \frac{ds}{dt} = \sqrt{\left(\frac{dx}{dt}\right)^2 + \left(\frac{dy}{dt}\right)^2}. \qquad [1]$$

6—9. Component Accelerations. In a similar manner, it is easy to find expressions for the component accelerations. Since v is, in general, a function of the time, $v = F(t)$. If t takes on an increment Δt, then v takes on an increment Δv. Hence the average acceleration during the interval $\Delta t = \dfrac{\Delta v}{\Delta t}$; and the acceleration at any instant is the limit of $\dfrac{\Delta v}{\Delta t}$ as $\Delta t \to 0$; hence,

$$\text{acceleration} = a = \frac{dv}{dt}.$$

By reasoning as in §6—8, the component accelerations are readily found to be

$$a_x = \frac{dv_x}{dt}, \qquad a_y = \frac{dv_y}{dt};$$

and

$$a = \frac{dv}{dt} = \sqrt{\left(\frac{dv_x}{dt}\right)^2 + \left(\frac{dv_y}{dt}\right)^2}.$$

6—10. Circular Motion. Another common type of motion, in addition to rectilinear motion already briefly described, is circular motion. A body moving in a circular pathway may move at a uniform or at a non-uniform rate. If the motion is uniform, the central angle θ changes at a constant rate, known as the *angular velocity* of the moving point.

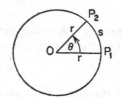

If, on the other hand, the motion is non-uniform, say $\theta = f(t)$, then the angular velocity will change for various positions of P; its value at any point P is given by

$$\omega = \frac{d\theta}{dt} = f'(t). \tag{1}$$

The rate of change of the angular velocity for non-uniform circular motion is known as the *angular acceleration*, and is designated by α. Thus

$$\alpha = \frac{d\omega}{dt} = \frac{d^2\theta}{dt^2} = f''(t). \tag{2}$$

If s (in feet, inches, etc.) represents the displacement, or distance traveled, as measured algebraically along the circular arc from P_1 to P_2, then

$$s = r\theta, \tag{3}$$

and the velocity (ft./sec., in./sec., etc.) *along the circle* is given by

$$v = \frac{ds}{dt} = r\frac{d\theta}{dt} = r\omega, \tag{4}$$

it being understood throughout that θ is measured in radians.

EXAMPLE 1. A motor 4 feet in diameter is revolving uniformly at 2400 revolutions per minute. Find (a) the angular velocity; (b) the velocity of a point on the rim; (c) the distance traveled by a point on the rim in 10 seconds.

Solution.

(a) Angular velocity $= \omega = 2400$ rev. per min.
$$= \frac{2\pi(2400)}{60} = 80\pi \text{ radians/sec.}$$

(b) Velocity of a point on rim $= r\omega = 2(80\pi) = 160\pi$ ft./sec.

(c) Distance traveled by a point on rim in 10 sec. is

$$(160\pi)(10) = 1600\pi \text{ ft.}$$

EXAMPLE 2. A point moves on a circle of diameter 24 inches in such a way that its motion follows the law

$$\theta = t^3 - 4t^2 + 6t.$$

Find (a) the angular velocity at $t = 6$ sec.; (b) the angular acceleration when $t = 2$ sec.; (c) the velocity of the point along the circle when $t = 4$ sec.; (d) the distance along the circle the point has moved when $t = 10$ sec.

Solution.

(a) $\omega = \dfrac{d\theta}{dt} = 3t^2 - 8t + 6;$

when $t = 6$, $\omega = 3(6)^2 - 8(6) + 6 = 66$ radians per sec.

(b) $\alpha = \dfrac{d^2\theta}{dt^2} = 6t - 8;$

when $t = 2$, $\alpha = 6(2) - 8 = 4$ radians/sec./sec.

(c) Velocity along the circle is given by

$v = r\dfrac{d\theta}{dt} = (12)(3t^2 - 8t + 6);$

when $t = 4$, $v = 12(48 - 32 + 6) = 22$ inches/sec.

(d) Distance traveled along the circle is given by

$s = r\theta$, where $\theta = t^3 - 4t^2 + 6t;$
when $t = 10$, $\theta = 1000 - 400 + 60 = 660$ radians;
hence $s = 12(660) = 7920$ inches.

6—11. Circular Motion with Constant Acceleration. If a body moving in a circular path starts from rest with an initial angular velocity

of ω_0, so that $\omega = \omega_0$ when $t = 0$ and $\theta = 0$, and if thereafter the motion is constantly accelerated by an amount α, positive or negative, it is shown in physics that throughout the motion the following relations hold:

$$(1) \quad \theta = \omega_0 t + \tfrac{1}{2}\alpha t^2;$$
$$(2) \quad \omega = \omega_0 + \alpha t;$$
$$(3) \quad \tfrac{1}{2}(\omega^2 - \omega_0^2) = \alpha\theta.$$

The reader should compare these three equations with the analogous relations for rectilinear motion, given in Exercise 4—2, problem 6.

EXAMPLE 1. A pulley on a shaft is revolving at 120 revolutions per minute. The diameter of the pulley is 3 ft. If the pulley is brought to rest at a constant (uniform) retardation in $\tfrac{3}{4}$ minute, find (a) the constant retardation, and (b) the number of revolutions made by the pulley before coming to rest.

Solution.
(a) $\omega_0 = 120$ r.p.m. $= 240\pi$ radians/min.;
$\omega = 0$, the final velocity.

$$\omega = \omega_0 + \alpha t,$$
$$0 = 240\pi + \alpha(\tfrac{3}{4}),$$
$$\alpha = -320\pi \text{ radians/min./min.}$$

(b) $\theta = \omega_0 t + \tfrac{1}{2}\alpha t^2,$

$$\theta = (240\pi)(\tfrac{3}{4}) + \tfrac{1}{2}(-320\pi)(\tfrac{3}{4})^2 = 180\pi - 90\pi = 90\pi \text{ radians.}$$

$$90\pi \div 2\pi = 45 \text{ revolutions.}$$

EXAMPLE 2. The armature of an electric motor is revolving at the rate of 1500 r.p.m. It is brought to rest at a uniform retardation of 5 radians per second. Find (a) the time required to bring it to rest; (b) how many revolutions it will make in coming to rest; and (c) how far a point on the rim will travel during this period, if the radius of the armature is 6 inches.

Solution.

(a) $\omega_0 = 1500$ r.p.m. $= 25$ rev./sec. $= 50\pi$ radians/sec.
$\alpha = -5$ radians/sec.

$$\omega = \omega_0 + \alpha t,$$

or $\qquad t = \dfrac{\omega - \omega_0}{\alpha} = \dfrac{0 - 50\pi}{-5} = 10\pi \text{ sec.}$

(b) $\qquad \alpha\theta = \dfrac{1}{2}(\omega^2 - \omega_0{}^2),$

$\qquad -5\theta = \dfrac{1}{2}(0 - 2500\pi),$

$\qquad \theta = 250\pi \text{ radians,} \qquad \text{or} \qquad \dfrac{250\pi}{2\pi} = 125 \text{ revolutions.}$

(c) $\quad s = r\theta = \dfrac{6}{12}(250\pi) = 125\pi \text{ feet.}$

RELATED TIME RATES

6—12. The Derivative as the Ratio of Two Rates. Consider the path generated by a moving point whose motion follows the law $y = f(x)$. The coordinates of point P in any position may be regarded

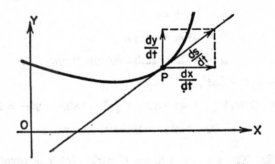

as functions of the time, as indicated in §6—8. Differentiating with respect to t:

$$\frac{dy}{dt} = f'(x)\,\frac{dx}{dt}\,. \qquad [1]$$

Equation [1] may be stated in words as follows:

At any instant, the time rate of change of a function (y) equals the derivative of the function multiplied by the time rate of change of the independent variable.

This equation [1] may also be written as:

$$\frac{\frac{dy}{dt}}{\frac{dx}{dt}} = f'(x) = \frac{dy}{dx}. \tag{2}$$

In words, equation [2] states:

The derivative is a measure of the time rate of change of y to the time rate of change of x.

The reader should note carefully that here we are regarding the derivative as *the ratio of two separate rates.*

Since $\frac{ds}{dt}$ represents the time rate of change of length of arc, we have, from §6—8:

$$\frac{ds}{dt} = \sqrt{\left(\frac{dx}{dt}\right)^2 + \left(\frac{dy}{dt}\right)^2}.$$

6—13. Related Time Rates. The principle represented by equations [1] and [2] finds ready application to many problems involving related time rates. Thus, suppose that two or more quantities, each of which varies with the time, are also functionally related to one another, i.e., connected by an equation; then the relation between their time rates of change may be determined by writing the functional relation between the variables, and differentiating with respect to the time. We illustrate the procedure to be followed.

EXAMPLE 1. A ladder 18 ft. long leans against the side of a building with one end resting on the horizontal ground. The foot of the ladder is drawn away from the building at the rate of 3 ft. per sec. (a) Find the rate at which the top of the ladder is descending when the foot is 4 ft. from the building. (b) Find how far the foot of the ladder will be from the building when the top is descending at the rate of 4 ft./sec.

Solution.

$$x^2 + y^2 = 18^2. \tag{1}$$

$$2x \frac{dx}{dt} + 2y \frac{dy}{dt} = 0. \tag{2}$$

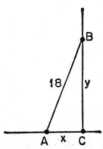

(a) To find the rate at which the top is descending $\left(\dfrac{dy}{dt}\right)$ when $x = 4$ ft.:

When $x = 4$,
$$y = \sqrt{324 - 16}$$
$$= \sqrt{308} = 2\sqrt{77};$$

substituting in (2), when $\dfrac{dx}{dt} = 3$ ft./sec. and $y = 2\sqrt{77}$:

$$(2)(4)(3) + 2(2\sqrt{77})\left(\frac{dy}{dt}\right) = 0,$$

$$\frac{dy}{dt} = -\frac{6}{\sqrt{77}} = .68 \text{ ft./sec.}$$

(b) To find the distance (x) of A from C when B is descending at the rate $\left(\dfrac{dy}{dx}\right)$ of 4 ft./sec., substitute in equation (2):

$$(2)(x)(3) + 2(y)(4) = 0,$$

or $3x + 4y = 0.$ (3)

Solving (1) and (3) simultaneously, $x = \dfrac{216}{15} = 14.4$ ft.

EXAMPLE 2. If a point moves on the parabola $8x = y^2$ so that when $y = 6$, the ordinate is increasing at the rate of $2\frac{1}{2}$ ft./sec.: (a) How fast is the abscissa moving at that instant? (b) At what point is the ordinate increasing 4 times as fast as the abscissa?

Solution.

(a) $8x = y^2.$ (1)

$$8\frac{dx}{dt} = 2y\frac{dy}{dt}.$$ (2)

Substituting in (2):

$$8\left(\frac{dx}{dt}\right) = 2(6)\left(\frac{5}{2}\right),$$

$$\frac{dx}{dt} = \frac{15}{4} = 3\frac{3}{4} \text{ ft./sec.}$$

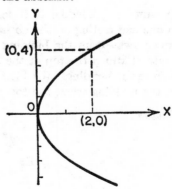

(b) $\dfrac{dy}{dt} = 4\left(\dfrac{dx}{dt}\right);$

hence, from (2): $\quad 8\left(\dfrac{dx}{dt}\right) = 2y(4)\left(\dfrac{dx}{dt}\right),$

$$y = 1,$$

and, from (1), if $\quad y = 1, \quad x = \dfrac{1}{8}.$

Therefore, at $(\frac{1}{8}, 1)$ the ordinate is increasing 4 times as fast as the abscissa.

EXAMPLE 3. A boy standing on a dock 8 ft. above the level of a lake is hauling in a canoe by means of a rope. If the rope is being pulled in at the rate of 2 ft./sec., how fast is the canoe moving when 17 ft. of tow rope are still paid out?

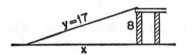

Solution.

$$x^2 + 8^2 = 17^2; \quad x = \sqrt{289 - 64} = 15 \text{ ft.}$$

$$y^2 - x^2 = 8^2; \quad 2y\frac{dy}{dt} - 2x\frac{dx}{dt} = 0.$$

Hence when $y = 17, \quad x = 15; \quad \dfrac{dy}{dt} = 2;$

therefore $\quad 2(17)(2) - 2(15)\dfrac{dx}{dt} = 0,$

or $\quad\quad\quad \dfrac{dx}{dt} = \dfrac{34}{15} = 2\dfrac{4}{15}\text{ ft.}$

EXAMPLE 4. A man 5 ft. tall walks towards a street lamp at the rate of 4 ft./sec. The lamp is 18 ft. high. (a) How fast is his shadow growing shorter? (b) How fast is the end of his shadow moving?

Solution.

Let BC represent the street lamp, DE the man, and let $x = AE$ be the length of his shadow.

Hence, by similar triangles:

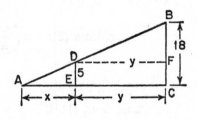

$$\frac{AE}{DE} = \frac{DF}{BF}, \text{ or } \frac{x}{5} = \frac{y}{18 - 5},$$

or $\qquad x = \dfrac{5y}{13}.$ $\qquad$ (1)

(a) By hypothesis, $\dfrac{dy}{dt} = -4$ ft./sec., since y decreases as t increases.

Differentiating (1):

$$\frac{dx}{dt} = \frac{5}{13}\left(\frac{dy}{dt}\right), \quad \text{or} \quad \frac{dx}{dt} = \frac{5}{13}(-4) = -\frac{20}{13} \text{ ft./sec.}$$

(b) To find how fast point A is moving, we must determine the rate of change of the distance $CA = y + x$, which is given by $\dfrac{dy}{dt} + \dfrac{dx}{dt}$. From (a):

$$-4 - \frac{20}{13} = -\frac{72}{13} = 5\frac{7}{13} \text{ ft./sec.,}$$

the rate at which the tip of the shadow (point A) is moving.

EXERCISE 6—6

1. A man is walking at the rate of 3 miles an hour toward the base of a tower 50 feet high. At what rate is he approaching the top of the tower when he is 120 ft. from the base of the tower?

2. The altitude of a cone and the diameter of its base are constantly equal. (a) If the volume of the cone is increasing at the rate of 8 cu. ft./min., how fast is the radius changing when the radius equals $\frac{1}{2}$ foot? (b) How fast is the lateral surface changing when the radius is increasing at $1\frac{1}{2}$ ft./min. and $h = 6$ ft.?

3. In the parabola $y^2 = 16x$, x increases uniformly at the rate of 4 in. per second. (a) At what rate is y increasing when $x = 9$ inches? (b) At what point on the parabola do x and y increase equally fast?

4. Two trains leave the same terminal at the same time, one traveling

60 mi./hr. southward, the other 80 mi./hr. westward. How fast are they separating after 3 hours?

5. In an expanding sphere, at any instant when the radius equals r, the rate of increase in volume is how many times as great as the rate of increase in area?

6. A kite is 90 ft. high, and 150 ft. of string are paid out. The kite drifts horizontally directly away from the boy flying the kite at the rate of 20 ft. per minute. How fast is the string being paid out at the instant the kite has drifted 60 ft. from K to D?

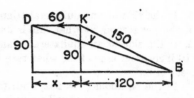

7. A plane flies horizontally at the rate of 240 mi. per hour and passes directly over a beacon light below at an elevation of 2 miles. How fast is its distance from the beacon increasing half a minute later?

8. A solution is being poured into a conical vessel at the rate of 40 cc. per minute. The vessel is 20 cm. across in diameter, and 30 cm. deep. Find the rate at which the surface of the liquid is rising at the instant the level is 6 cm. above the apex of the cone.

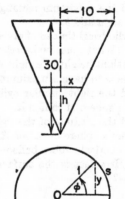

9. In the accompanying figure, where $y = \sin \phi$ and $\phi = s$, determine the value of ϕ, in the first quadrant, when the arc s is increasing twice as fast as the sin ϕ.

10. In each of the following, find the rate at which the length of arc is increasing, at the ordinate (or abscissa) given, when one of the variables is changing at the given rate:

(a)　$y^2 = 4x$;　$x = 4$;　$\dfrac{dx}{dt} = 8$.

(b)　$xy = 24$;　$y = 3$;　$\dfrac{dy}{dt} = 6$.

(c)　$y^2 = 2x^3$;　$x = 2$;　$\dfrac{dy}{dt} = 30$.

EXERCISE 6—7

Review

1. If the side of an equilateral triangle is increasing uniformly at the rate of 6 cm. per second, at what rate is its altitude increasing?

2. If the side of an equilateral triangle is increasing uniformly at the rate of 10 inches per second, at what rate is the area increasing when the side is 4 feet?

3. In the parabola $y^2 = 24x$, find: (a) the point at which the ordinate and abscissa are increasing equally rapidly; (b) the point at which the ordinate is increasing half as fast as the abscissa.

4. Determine the points of inflection of $y = \dfrac{x}{x^2 - 1}$.

5. Find the maximum rectangle that can be inscribed in the ellipse whose semiaxes are a and b.

6. A cylindrical tin can of given volume is to be made with the least amount of metal. What must be its relative dimensions?

7. The slant height of a right circular cone is S. What must be its altitude if the cone is to have a maximum volume?

8. Find the right circular cylinder of greatest convex surface that can be cut from a sphere of radius R.

9. Find the altitude of the right circular cone of greatest volume that can be cut from a sphere of radius R.

10. A spherical rubber balloon is being deflated at the rate of 20 cu. in. per second. How fast is the surface of the balloon decreasing when the radius is 4 inches?

Differentials

CHAPTER SEVEN

INCREMENTS AND INFINITESIMALS

7—1. The Meaning of a Differential. Let us consider again the function which was employed in §2—4 and §2—5 to develop the meaning of a derivative, namely, $y = x^2$. We learned that as x took on an increment Δx,

$$\Delta y = 2x \cdot \Delta x + (\Delta x)^2. \tag{1}$$

We know that, for this function, $\dfrac{dy}{dx} = 2x$; hence we may write equation (1) as follows, where $y' = \dfrac{dy}{dx}$:

$$\Delta y = y' \Delta x + (\Delta x)^2. \tag{2}$$

We wish to study the expression $y' \Delta x$, and thus introduce these new terms.

(1) The *differential of an independent variable* is the same as its increment. The symbol for the differential of x is dx; hence $dx = \Delta x$.

(2) The *differential of a function* (or dependent variable) is the product of its derivative and the differential of the independent variable. The symbol for the differential of a function is dy; hence $dy = y' dx$, or $y' = \dfrac{dy}{dx}$.

It will be seen that equation (2) can also be written as

$$\Delta y = y'dx + (dx)^2,$$

or
$$\Delta y = dy + (dx)^2. \tag{3}$$

Up to this point in our study of the calculus we have stressed the fact that $\frac{dy}{dx}$ was not to be regarded as the ratio of two quantities, but rather as a single symbol denoting the limiting value of $\frac{\Delta y}{\Delta x}$, where Δy and Δx represented two separately changing quantities. We now see that the meaning of the symbol $\frac{dy}{dx}$ may be extended. The derivative may now be thought of as a quotient of two differentials, or

$$\frac{dy}{dx} = dy \div dx.$$

7—2. Geometric Interpretation of *dx* and *dy*. This extension of meaning will become clearer from a study of the figure. For a given

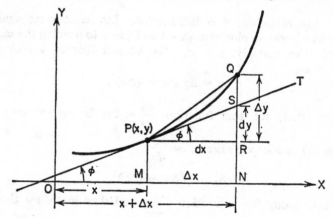

increment to x equal to Δx, the corresponding increment of y is $\Delta y = RQ$. The tangent PT, having an inclination ϕ, and a slope equal to $\frac{RS}{PR} = \frac{\Delta y}{\Delta x} = \tan \phi$, cuts off a part of Δy, namely, RS; this part of Δy equals dy. The segment $MN = \Delta x$, and $PR = dx$; thus $\Delta x = dx$.

We may regard dy as the value of Δy in the limit when the secant coincides with the tangent.

The relations between increments, differentials, and derivatives are extremely important and should be thoroughly understood. It should be noted that Δx and dx are quantities measured in the same unit as x, and Δy and dy are quantities measured in the same unit as y. The derivative, $\dfrac{dy}{dx}$, that is the slope, when considered as a single quantity, or rate (not a ratio of two quantities), is generally measured in a compound unit such as miles per hour or feet per second.

7—3. Relation between Δy and dy. The precise nature of the differential dy will be better understood from the following considerations. Originally, we defined the derivative with respect to x of a function $y = f(x)$ as the limit of the difference-quotient; that is

$$f'(x) = \lim_{x \to 0} \left(\frac{\Delta y}{\Delta x} \right).$$

Since the value of $\dfrac{\Delta y}{\Delta x}$ at each stage of the process of passing to the limit, when x is fixed, depends upon the corresponding value of Δx, the difference-quotient $\dfrac{\Delta y}{\Delta x}$ is a function of Δx.

By the definition of the limit of a function, the difference between $\dfrac{\Delta y}{\Delta x}$ and $f'(x)$ can be made as small as we please by taking Δx sufficiently small. This idea may be represented symbolically by

$$\frac{\Delta y}{\Delta x} - f'(x) = \epsilon, \tag{1}$$

where ϵ represents some variable quantity which is approaching zero in value. Equation (1) may be written as

$$\frac{\Delta y}{\Delta x} = f'(x) + \epsilon,$$

or $$\Delta y = f'(x)\Delta x + \epsilon \cdot \Delta x. \tag{2}$$

Now the product of a small quantity and another small quantity yields a much *smaller* quantity; for example, $\dfrac{1}{10} \times \dfrac{1}{11} = \dfrac{1}{110}$. To be sure, the terms "small" and "much smaller" are purely relative, and are admittedly used in a somewhat loose sense; but they may help us to understand the ideas involved. Thus, $\frac{1}{110}$ is, in a sense, of a different "degree of smallness" than $\frac{1}{10}$ or $\frac{1}{11}$.

Let us turn to equation (2) once more. As Δx becomes smaller, the quantity ϵ also becomes smaller; the product of ϵ and Δx is of a lesser degree of smallness than Δx. Hence, in the right member of the equation

$$\Delta y = f'(x)\Delta x + \epsilon \cdot \Delta x,$$

the sum usually consists of a relatively large part, $f'(x) \cdot \Delta x$, and a relatively small part, $\epsilon \cdot \Delta x$. The larger part, $f'(x) \cdot \Delta x$, may be called the *principal* part of Δy; and by disregarding the comparatively negligible part $\epsilon \cdot \Delta x$, we see that Δy is approximately equal to its principal part, $f'(x) \cdot \Delta x$. For example, if $y = f(x) = 10x^2$,

then $\Delta y = 10(x + \Delta x)^2 - 10x^2 = 20x \cdot \Delta x + 10(\Delta x)^2$;

also, $f'(x) = 20x.$

Hence, $\Delta y = f'(x)\Delta x + 10\Delta x \cdot \Delta x,$

which yields $\epsilon = 10\Delta x$. If we arbitrarily take $x = 1$ and $\Delta x = .01$, we have

$$\Delta y = (20)(.01) + (10)(.01)(.01) = .2 + .001 = .201.$$

Thus the value of Δy, .201, is approximately equal to its principal part, $f'(x)\Delta x = .200$.

The *differential* of a dependent variable may be defined as the principal part of Δy; or

$$dy = f'(x) \cdot \Delta x. \tag{3}$$

The value of dx is always taken as equal to Δx, for any given function, and is regarded as a constant. The value of dy, however, depends upon the values of both x and Δx; for a fixed value of x, the value of dy varies directly as the value of Δx.

We may therefore now finally write:

$$dy = f'(x)\, dx. \tag{4}$$

Equation (4) may be interpreted to mean that the derivative is considered as a ratio, or as the quotient of two differentials, for $\frac{dy}{dx}$ now becomes $dy \div dx$. In words, the differential of a function equals the product of its derivative and the differential of the independent variable.

From now on we may therefore regard the symbols dy and dx as separate quantities (differentials), and the symbol $\dfrac{dy}{dx}$ as a fraction.

7—4. Infinitesimals. We now see that an *infinitesimal* may be defined as a variable whose numerical value becomes and remains smaller than any preassigned value, however small. Or, an infinitesimal is a variable which approaches zero as a limit.

When comparing infinitesimals, we refer to their *order*. This is a relative term, suggesting comparative degree of smallness. If the limit of the quotient of two infinitesimals is a constant, not zero, they are said to be of the *same order*; if this limit is zero, the first differential (the numerator) is said to be of higher order than the second, and the second of *lower order* than the first. If the limit is infinite, the first differential (the numerator) is said to be of lower order than the second, and the second of higher order than the first.

EXAMPLE 1. In §7—3, when discussing the relation

$$\Delta y = 20x \cdot dx + 10(dx)^2,$$

we saw that, as $dx \to 0$, the quantity $10(dx)^2$ approached zero "faster" than did the quantity $20x \cdot dx$; in other words, $10(dx)^2$ is of a higher order than $20x \cdot dx$ because

$$\lim_{dx \to 0} \frac{10(dx)}{20x \cdot dx} = \lim_{dx \to 0} \left(\frac{dx}{2x}\right) = 0.$$

EXAMPLE 2. In $\lim\limits_{\theta \to 0} \left(\dfrac{\sin \theta}{\tan \theta}\right)$, the two infinitesimals $\sin \theta$ and $\tan \theta$ are of the same order, since

$$\lim_{\theta \to 0} \left(\frac{\sin \theta}{\tan \theta}\right) = \lim_{\theta \to 0} \left(\frac{\sin \theta}{\dfrac{\sin \theta}{\cos \theta}}\right) = \lim_{\theta \to 0} \cos \theta = 1.$$

Thus, in general,

$$\lim_{\Delta x \to 0} \frac{\Delta y}{\Delta x} = \lim_{\Delta x \to 0} \left(1 + \frac{\epsilon}{f'(x)}\right) = 1,$$

where $f'(x) \neq 0$; therefore Δy and dy are infinitesimals of the same order. On the other hand, from the relation

$$\Delta y - dy = \epsilon \cdot \Delta x,$$

it will be seen not only that the infinitesimal $\epsilon \cdot \Delta x$ is of a higher order than either Δy or dy, but also that dy is an approximation to Δy; the smaller the value of Δx, the more closely dy approximates Δy.

USING DIFFERENTIALS

7—5. Differential Notation. The standard formulas for differentiation can now be expressed in differential notation, as given below. For example, instead of writing

$$\frac{dc}{dx} = 0, \qquad \text{we write } dc = 0.$$

The relation, "the derivative with respect to x of a constant equals zero," may also be stated as "the differential of a constant equals zero."

Similarly,
$$\frac{d}{dx}(u + v) = \frac{du}{dx} + \frac{dv}{dx}$$

becomes
$$d(u + v) = du + dv.$$

SUMMARY OF DIFFERENTIAL NOTATION

$$dc = 0$$

$$d(cv) = c\, dv$$

$$d(u + v) = du + dv$$

$$d(uv) = u\, dv + v\, du$$

$$d\left(\frac{u}{v}\right) = \frac{v\, du - u\, dv}{v^2}$$

$$dy = \frac{dy}{du}\, du$$

$$d(v^n) = nv^{n-1}\, dv$$

$$d(x^n) = nx^{n-1}\, dx$$

$$d(\log_a v) = \frac{dv}{v} \log_a e$$

$$d(\log v) = \frac{dv}{v}$$

$$d(a^v) = a^v \log a\, dv$$

$$d(e^v) = e^v\, dv$$

$$d(u^v) = vu^{v-1}\, du + \log u \cdot u^v \cdot dv$$

$$d(\sin v) = \cos v \, dv$$

$$d(\cos v) = - \sin v \, dv$$

$$d(\tan v) = \sec^2 v \, dv$$

$$d(\cot v) = - \csc^2 v \, dv$$

$$d(\sec v) = \sec v \tan v \, dv$$

$$d(\csc v) = - \csc v \cot v \, dv$$

$$d(\arcsin v) = \frac{dv}{\sqrt{1 - v^2}}$$

$$d(\arccos v) = - \frac{dv}{\sqrt{1 - v^2}}$$

$$d(\arctan v) = \frac{dv}{1 + v^2}$$

7—6. Differentiation with Differentials. This may be illustrated by the following examples.

EXAMPLE 1. Find the differential of

$$y = x^5 + 3x^2.$$

Solution. $dy = 5x^4 \, dx + 6x \, dx,$

or $\qquad\qquad dy = (5x^4 + 6x) \, dx.$

EXAMPLE 2. Find the differential of

$$y = \frac{3x}{x^2 + 2}.$$

Solution. $dy = \dfrac{(x^2 + 2) \cdot 3dx - 3x \cdot 2x \, dx}{(x^2 + 2)^2},$

$$dy = \frac{3x^2 \, dx + 6dx - 6x^2 \, dx}{(x^2 + 2)^2} = \frac{-3x^2 \, dx + 6dx}{(x^2 + 2)^2},$$

or $\qquad dy = \dfrac{(-3x^2 + 6x) \, dx}{(x^2 + 2)^2}.$

EXAMPLE 3. Find dy for $y = \log \sin 2x.$

Solution. $d(\log v) = \dfrac{dv}{v},$

Hence $\quad dy = \dfrac{d(\sin 2x)}{\sin 2x} = \dfrac{2 \cos 2x \, dx}{\sin 2x} = (2 \cot 2x) \, dx.$

EXAMPLE 4. Find dy from

$$a^2x^2 + b^2y^2 = a^2b^2.$$

Solution. $2a^2x\,dx + 2b^2y\,dy = 0$,

$$dy = \left(-\frac{a^2x}{b^2y}\right)dx.$$

EXAMPLE 5.   Find $d\rho$ from

$$\rho^2 = a^3\sin 3\theta.$$

Solution. $2\rho\,d\rho = a^3\cos 3\theta\cdot 3d\theta$,

$$d\rho = \frac{3a^3\cos 3\theta}{2\rho}\,d\theta.$$

7—7. Successive Differentials. Consider the function $y = f(x)$. We may regard $d(dy)$ as the *second differential* of y, or the second differential of the function; it is represented by the symbol d^2y. In the same way,

$$d[d(dy)], \quad \text{or} \quad d^3y,$$

is the *third differential of y;* and d^ny is the *nth differential of y.*

The reader must be careful in interpreting the symbolism used here. Thus

d^2v is the *second differential* of v, and equals $d(dv)$.

dv^2 is the *square* of dv, and may also be written $(dv)^2$.

$d(v^2)$ is the *first differential* of v^2, and equals $2v\,dv$.

In general, in the function $y = f(x)$, the differential of the independent variable, namely, dx, is independent of x, and must be regarded as a constant when differentiating with respect to x. Consequently:

$$dy = f'(x)\,dx,$$

and, since dx is a constant,

$$d(dy) = d^2y = (dx)\cdot[f''(x)\,dx] = f''(x)(dx)^2.$$

Similarly, $d^3y = f'''(x)(dx)^3$,

and $d^ny = f^{(n)}(x)(dx)^n$.

The reader should note that, since $(dx)^2 = dx^2$, the differential notation

$$d^2y = f''(x)(dx)^2 \tag{1}$$

is equivalent to the familiar derivative notation

$$\frac{d^2y}{dx^2} = f''(x),$$

simply by dividing both sides of (1) by $(dx)^2$, or dx^2. Similarly,

$$\frac{d^ny}{dx^n} = f^{(n)}(x).$$

EXERCISE 7—1

Differentiate the following, using differentials:

1. $y = x^3 + 5x^2 - 3$
2. $y = (5x + 2)^3$
3. $y = \sqrt{x^2 - 1}$
4. $y = ke^{tx}$
5. $y = e^x \log x$
6. $\rho = \sin 2\theta + 2 \cos \theta$
7. $f(x) = (\log x)^4$
8. $\rho = a^2 \sec 3\theta$

Find dy for each of the following:

9. $x^2 + y^2 = r^2$
10. $b^2x^2 - a^2y^2 = a^2b^2$
11. $xy = k$

Find d^3y for each of the following:

12. $y = x^4 - 5x^3 + 4x^2 - 7$
13. $y = e^x \sin x$
14. $y = \sqrt{x^3}$
15. $y = \dfrac{x + 1}{x}$

APPROXIMATE CALCULATIONS

7—8. The Differential as an Approximation. In §7—4 we learned that for a very small change in the independent variable, the differential of the function is very nearly equal to the increment of the function; or, $dy = \Delta y$, approximately. This relationship is of practical value in determining the approximate change in a function corresponding to a particular value of the independent variable together with a small change in that variable.

EXAMPLE 1. What is the change in the area (y) of a square when the side of the square is increased from 8.0 to 8.2 inches?

Solution.

$$y = x^2,$$

$$\frac{dy}{dx} = 2x.$$

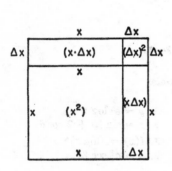

When $x = 8$, $\frac{dy}{dx} = 16$. Let the increase in x be Δx; thus $\Delta x = .2$. Let us assume that $\Delta y = dy$; then, since

$$\frac{\Delta y}{\Delta x} = \frac{dy}{dx}, \qquad \Delta y = \left(\frac{dy}{dx}\right)\Delta x;$$

hence $\Delta y = (16)(.2) = 3.2$.

The value of y when $x = 8.2$ is equal to 67.24; hence the *exact* value of the increment in y equals $67.24 - 64.00 = 3.24$, which is very close to the approximate value $\Delta y = 3.2$ found above.

NOTE 1. The reader should study the diagram and note the relation of the approximate formula $\Delta y = 2x \cdot \Delta x$ with the exact formula $\Delta y = 2x \cdot \Delta x + (\Delta x)^2$.

NOTE 2. The magnitude of the error introduced by using the value of the differential in lieu of the increment of the function will depend, in general, upon (1) the value taken for the independent variable, (2) the size of the increment given to the independent variable, and (3) the nature of the function.

EXAMPLE 2. In the function $y = x^8$, by how much will y increase when x changes from 3 to 3.02?

Solution. $\frac{dy}{dx} = 8x^7$; when $x = 3$, $\frac{dy}{dx} = 17{,}496$.

Hence $\Delta y = \left(\frac{dy}{dx}\right)\Delta x = (17{,}496)(.02) = 349.92 = 350$.

NOTE. To compute the exact value of Δy, or, for that matter, the exact value of $(3.02)^8$, involves considerable labor. By five-place logarithms, $(3.02)^8 = 6919.6$; hence the exact value of

$$\Delta y = 6919.6 - (3)^8 = 6919.6 - 6561 = 358.4,$$

as compared with the approximate value 350 found above.

EXAMPLE 3. The edge of a cubical block of metal is equal to 9.7 cm., but was measured as 10 cm. About how great is the error in the computed volume due to the error in measuring? What is the per cent of error in the volume?

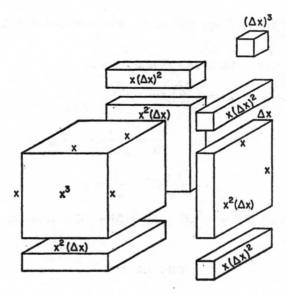

Solution.

$$V = x^3,$$

$$\frac{dV}{dx} = 3x^2;$$

when $x = 10$, $\dfrac{dV}{dx} = 300$.

Hence $\Delta V = \left(\dfrac{dV}{dx}\right)\Delta x = (300)(.3) = 90$ cu. cm., approx.

The per cent of error in the measurement of the edge is $\frac{.3}{10}$, or 3%; the per cent of error in the volume is about $\frac{90}{1000}$, or about 9%.

NOTE. The reader should note the interpretation, in the diagram, of the expression

$$V + \Delta V = x^3 + 3x^2 \cdot \Delta x + 3x(\Delta x)^2 + (\Delta x)^3;$$

in this way he will see the significance of the approximation formula $\Delta V = 3x^2(\Delta x)$.

EXAMPLE 4. Find, approximately, the square root of 402.

Solution. Let $y = \sqrt{x}$.
Here x may be taken as 400, and $\Delta x = 2$.

As before, $\quad \Delta y = \left(\dfrac{dy}{dx}\right)\Delta x; \quad \dfrac{dy}{dx} = \dfrac{1}{2\sqrt{x}}.$

When $\quad x = 400, \quad \dfrac{dy}{dx} = \dfrac{1}{2\sqrt{400}} = \dfrac{1}{40};$

hence $\quad y = \left(\dfrac{1}{40}\right)(2) = \dfrac{1}{20} = .05;$

therefore $\sqrt{402} = \sqrt{400} + \Delta y = 20 + .05 = 20.05$, approx.

EXERCISE 7—2

1. A metal cube when heated expands so that its edge increases from 20 cm. to 20.003 cm. Find (a) the approximate change in its area; (b) the approximate change in its volume.
2. Find the approximate change both in the circumference and in the area of a circle when the radius is decreased from 5 inches to 4.98 inches.
3. If the *diameter* of a sphere is actually 19.98 in., but is measured as 20.02 in., find the approximate error in computing both the surface and the volume of the sphere.
4. Find the change in the volume of a cylindrical metal bearing of length 10 cm. when it wears down from a radius of 2 cm. to a radius of 1.99 cm.
5. Find, by differentiation, the approximate square root of 291.
6. Find, by differentiation, the approximate value of $\sqrt[3]{60}$.

7. Approximately what error would be allowable in the side of a square, about 40 cm. on a side, if the error in the calculated area is not to exceed 2 sq. cm.?

8. The pressure p in lb. per sq. in. of a gas in a closed container at constant temperature varies according to the law $pv = k$. Derive (a) an expression for the approximate change in pressure if the volume is changed by a small amount; (b) an expression for the approximate change in volume if the pressure is changed by a small amount.

9. The inside diameter of a circular concrete conduit is 18 in.; the shell is $\frac{3}{4}$ in. thick. Find the approximate area of the cross-section of the concrete, i.e., the area of the "ring" between the inside and outside of the shell.

10. The heat generated by an electric current is given by the formula $H = 0.24I^2Rt$, where H is in calories, I = amperes, R is the resistance in ohms, and t = the time in seconds. Find the increase in the amount of heat generated per second by a constant resistance of 50 ohms when I changes from 20 to 22 amperes.

11. The horsepower required to propel a ship of a certain design is

$$H = \frac{D^{\frac{2}{3}}v^3}{200},$$

where D is the displacement in long tons and v the speed in knots. Find the additional power required to increase the speed of a 27,000 ton vessel from 25 to 27 knots.

12. Prove that the change in the circumference of a circle caused by a change in the radius is independent of the original circumference and radius. If a blacksmith inserted a piece of iron one inch long in the metal tire of a wagon wheel originally 42 in. in diameter, how much "space" would there be between the rim of the wheel and the tire? If a piece of metal 1 in. long were inserted in a metal bracelet which originally fitted the arm snugly, by how much would it stand away after lengthening?

EXERCISE 7—3

Review

1. Given the function $x^{\frac{1}{2}} + y^{\frac{1}{2}} = a^{\frac{1}{2}}$, find $\dfrac{dy}{dx}$; also, $\dfrac{dx}{dy}$.

2. Find $\dfrac{dy}{dx}$:

  (a) $x^2 + y^2 + 2x = 5y$
  (b) $(x^2 - y^2)^2 = x^2 + y^2$

3. (a) Find the slope of the curve $y = x(x^2 - 1)$ at the point of inflection.

  (b) Find the slope of the tangent to the curve $x^2 - xy + y^2 = 6$ at the point $(-2,1)$.

4. The distance S traveled by a body projected vertically upwards in time t is given by the equation $S = 192t - 16t^2$. Find the greatest height the body will reach, and how long it will take to reach it.

5. Differentiate:

(a) $\log \dfrac{e^x}{1 + e^x}$

(c) $\log (px^2 + qx + r)$

(b) $\log \dfrac{a + x}{a - x}$

(d) $\log \sqrt{x^2 - 1}$

6. Find $\dfrac{d^2y}{dx^2}$ for each of the following:

(a) $b^2x^2 + a^2y^2 = a^2b^2$
(b) $x^2 + y^2 = 2x$
(c) $x^2y^2 = x^2 + y^2$

Curvature

CHAPTER EIGHT

LENGTH OF ARC

8—1. Differential Arc Length. In order to understand the nature of curvature, it will be necessary to discuss what is meant by *differential arc length*. Let the curve in the figure represent the function $y = f(x)$,

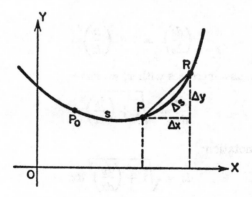

and let s represent the length of the arc measured from a definite, initial point P_0 to any point $P(x,y)$ of the curve. Since the length s will depend upon the position of P_0 on the curve, s is clearly a function of x. Thus suppose that as x takes on an increment Δx, the corresponding increment of s is Δs, and the corresponding point on the curve is $R(x + \Delta x, y + \Delta y)$. Intuitively, from the figure, we note that

$$\overline{PR}^2 = (\Delta x)^2 + (\Delta y)^2, \tag{1}$$

and hence

$$\left(\frac{PR}{\Delta x}\right)^2 = 1 + \left(\frac{\Delta y}{\Delta x}\right)^2. \tag{2}$$

Again intuitively, we see that, as Δx approaches zero, arc Δs and chord PR become more and more nearly equal, so that

$$\lim_{\Delta x \to 0} \frac{PR}{\Delta s} = 1. \tag{3}$$

Let us now transform the left-hand member of (2) by multiplying both numerator and denominator by $(\Delta s)^2$, and rewriting; thus

$$\left(\frac{PR}{\Delta x}\right)^2 = \left(\frac{PR}{\Delta s}\right)^2 \left(\frac{\Delta s}{\Delta x}\right)^2.$$

Hence,

$$\left(\frac{PR}{\Delta s}\right)^2 \left(\frac{\Delta s}{\Delta x}\right)^2 = 1 + \left(\frac{\Delta y}{\Delta x}\right)^2. \tag{4}$$

Now, as $\Delta x \to 0$, $\dfrac{\Delta y}{\Delta x} \to \dfrac{dy}{dx}$; $\dfrac{\Delta s}{\Delta x} \to \dfrac{ds}{dx}$; and, from (3), $\left(\dfrac{PR}{\Delta s}\right)^2 \to 1$.

Therefore

$$\left(\frac{ds}{dx}\right)^2 = 1 + \left(\frac{dy}{dx}\right)^2. \tag{5}$$

In other words, as s increases with x, we have

$$\frac{ds}{dx} = \sqrt{1 + \left(\frac{dy}{dx}\right)^2}. \tag{1}$$

In differential notation:

$$ds = \sqrt{1 + \left(\frac{dy}{dx}\right)^2}\, dx. \tag{1a}$$

The right-hand member of [1a] is an expression for the differential arc length ds.

It often happens that the given function is of the form $x = \phi(y)$, and s is an increasing function of y. In this case the corresponding formulas for differential arc length, by similar reasoning, are found to be

$$\frac{ds}{dy} = \sqrt{1 + \left(\frac{dx}{dy}\right)^2},$$ [2]

or
$$ds = \sqrt{1 + \left(\frac{dx}{dy}\right)^2}\, dy.$$ [2a]

8—2. Differential Arc Length for Equations in Parametric Form. If the function to be considered is given by a set of parametric equations, let us say

$$x = f(t), \qquad y = \phi(t),$$

where t is the parameter, we may proceed as follows.

From equation (5), §8—1,

$$\left(\frac{ds}{dx}\right)^2 = 1 + \left(\frac{dy}{dx}\right)^2;$$

therefore

$$\frac{(ds)^2}{(dx)^2} = 1 + \frac{(dy)^2}{(dx)^2},$$

or, clearing of fractions,

$$(ds)^2 = (dx)^2 + (dy)^2.$$ (1)

Since $x = f(t)$ and $y = \phi(t)$,

$$dx = f'(t)\, dt, \quad \text{and} \quad dy = \phi'(t)\, dt.$$ (2)

Substituting (2) in (1):

$$(ds)^2 = [f'(t)\, dt]^2 + [\phi'(t)\, dt]^2,$$

or
$$ds = (\sqrt{[f'(t)]^2 + [\phi'(t)]^2})\, dt.$$ [3]

8—3. Differential Arc Length in Polar Coordinates. If the function whose differential arc length is desired is expressed in polar coordinates, the corresponding formulas are easily derived by employing the usual transformation formulas from rectangular to polar coordinates, namely,

$$x = \rho \cos \theta, \qquad y = \rho \sin \theta.$$ (1)

Differentiating (1), using differential notation:

$$\left. \begin{array}{l} dx = -\rho \sin \theta\, d\theta + \cos \theta\, d\rho, \\ dy = \rho \cos \theta\, d\theta + \sin \theta\, d\rho. \end{array} \right\}$$ (2)

Substituting in [3], §19—2:

$$(ds)^2 = (-\rho \sin \theta \, d\theta + \cos \theta \, d\rho)^2 + (\rho \cos \theta \, d\theta + \sin \theta \, d\rho)^2.$$

Simplifying:

$$(ds)^2 = \rho^2 (\sin^2 \theta + \cos^2 \theta)(d\theta)^2 + (\sin^2 \theta + \cos^2 \theta)(d\rho)^2,$$

$$(ds)^2 = \rho^2 (d\theta)^2 + (d\rho)^2. \tag{3a}$$

Hence, $\qquad ds = \sqrt{(\rho \, d\theta)^2 + (d\rho)^2} = \sqrt{\rho^2 + \left(\dfrac{d\rho}{d\theta}\right)^2} \, d\theta,$

and $\qquad\qquad \dfrac{ds}{d\theta} = \sqrt{\rho^2 + \left(\dfrac{d\rho}{d\theta}\right)^2}. \tag{4}$

To sum up, the differential length of arc is given by the formulas:

[A]   for rectangular coordinates: $ds = \sqrt{1 + \left(\dfrac{dy}{dx}\right)^2} \, dx;$

[B]   for polar coordinates: $ds = \sqrt{\rho^2 + \left(\dfrac{d\rho}{d\theta}\right)^2} \, d\theta;$

[C]   for parametric equations: $ds = \sqrt{[f'(t)]^2 + [\phi'(t)]^2} \, dt.$

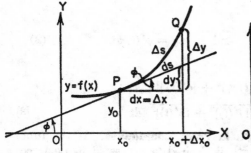 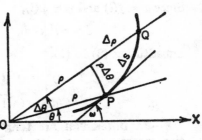

Thus the differentials dx, dy, and ds form the sides of a right triangle in which

$$(ds)^2 = (dx)^2 + (dy)^2. \tag{5}$$

Similarly, the differentials $d\rho$, $\rho \, d\theta$, and ds are related in such a way that they represent a right triangle whose hypotenuse is ds, and whose sides are $d\rho$ and $\rho \, d\theta$; or

$$(ds)^2 = (d\rho)^2 + (\rho \, d\theta)^2. \tag{6}$$

MEANING OF CURVATURE

8—4. Curvature. By observing their appearance, it is clear that various curves differ from one another in the degree of curvature, or their rate of turning; furthermore, on a given curve, the rate of turning varies at different points along the curve. For example, if we compare

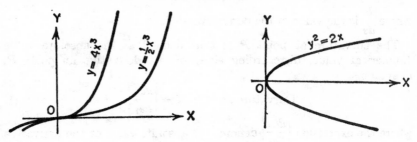

the curves, $y = \frac{1}{2}x^3$ and $y = 4x^3$, the latter curve is seen to "bend around" or curve more rapidly than the former as x increases from zero. Or again, the parabola $y^2 = 2x$ is more "sharply" curved in the neighborhood of the vertex than elsewhere, and flattens out, or has less curvature, as the curve recedes from the vertex.

8—5. Definition of Curvature. To define precisely the amount of curvature, we proceed as follows. Let P be a fixed point on the arc AB, and let Q be a point in the neighborhood of P, at a distance Δs from P, measured along AB. The tangents at P and Q make angles

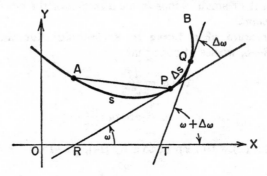

of ω and $\omega + \Delta\omega$, respectively, with the X-axis; thus the angle between the two tangents equals $\Delta\omega$. It will be appreciated that the entire change in direction of the curve in passing from P to Q is measured by

$\Delta\omega$; furthermore, the *average change of direction per unit arc length*, in passing from P to Q, is given by the ratio $\dfrac{\Delta\omega}{\Delta s}$.

Now let Q approach P; then $\Delta s \to 0$, and $\Delta\omega \to 0$. Therefore, in general,

$$\lim_{\Delta s \to 0} \left(\frac{\Delta\omega}{\Delta s}\right) = \frac{d\omega}{ds},$$

where $\dfrac{d\omega}{ds}$ is the value of the derivative at P.

The curvature at point P is thus defined as the absolute value (numerical value, disregarding sign) of the derivative at point P; in symbols,

$$\text{Curvature at } P = K = \left|\frac{d\omega_1}{ds_1}\right|, \tag{1}$$

where the expression $\left|\dfrac{d\omega_1}{ds_1}\right|$ denotes the absolute value of the derivative $\dfrac{d\omega}{ds}$ when evaluated for the particular values of ω and s that correspond to point P. Hence in order to find the value of $\dfrac{d\omega_1}{ds_1}$, the functional relation between ω and s must be known; but, for curves given in rectangular coordinates, such as $y = f(x)$, this is somewhat awkward. The procedure required will be explained shortly in §8—6.

The notion of curvature as just defined above will be seen to be consistent with our intuitive conception of curvature. In other words, at any point on a curve, the curvature is great or small according as there is a great or small *change in the direction* of the curve as we *pass through that point.*

8—6. Curvature of a Curve in Rectangular Coordinates. From the figure, §8—5, we see at once that

$$\tan \omega = \frac{dy}{dx},$$

or $\qquad\qquad\qquad \omega = \text{arc tan } \dfrac{dy}{dx}. \tag{1}$

Differentiating (1) with respect to x by [14], §5—15:

$$\frac{d\omega}{dx} = \frac{\dfrac{d^2y}{dx^2}}{1 + \left(\dfrac{dy}{dx}\right)^2}. \tag{2}$$

Also, from [1], §8—1, we recall:

$$\frac{ds}{dx} = \left[1 + \left(\frac{dy}{dx}\right)^2\right]^{\frac{1}{2}}. \tag{3}$$

Hence, dividing (1) by (2), we obtain:

$$\frac{\dfrac{d\omega}{dx}}{\dfrac{ds}{dx}} = \frac{\dfrac{d^2y}{dx^2}}{\left[1 + \left(\dfrac{dy}{dx}\right)^2\right]^{\frac{3}{2}}}. \tag{4}$$

But $\quad \dfrac{\dfrac{d\omega}{dx}}{\dfrac{ds}{dx}} = \dfrac{d\omega}{ds} = K.$

Therefore $\qquad K = \dfrac{\dfrac{d^2y}{dx^2}}{\left[1 + \left(\dfrac{dy}{dx}\right)^2\right]^{\frac{3}{2}}}. \qquad$ [1a]

8—7. Curvature of the Circle. Let us consider the circle so placed that its center is on the Y-axis, and at a distance from the origin equal to its radius r. Let the tangent at P make an angle ω with the X-axis, and let α be the central angle of the arc $OP = s$. Then, by trigonometry,

$$s = r\alpha.$$

But since the angles α and ω have their corresponding sides perpendicular, $\alpha = \omega$; hence

$$s = r\omega, \qquad \text{or} \qquad \omega = \frac{1}{r}s,$$

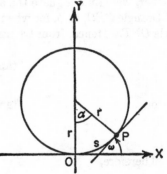

and, by differentiating,

$$d\omega = \frac{1}{r}ds, \qquad \text{or} \qquad \frac{d\omega}{ds} = \frac{1}{r}.$$

In other words, the curvature of a circle at any point P on the circle is constant, and equal to the reciprocal of the radius.

8—8. Curvature of a Curve in Polar Coordinates. Consider the polar curve, with the radius vector ρ to point P corresponding to θ, and the tangent at P, making an angle ω with the polar axis. From the figure,

$$\omega = \theta + \phi. \tag{1}$$

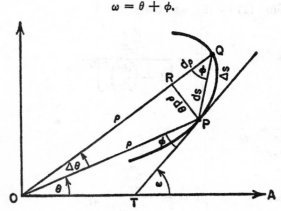

Differentiating (1) with respect to θ:

$$\frac{d\omega}{d\theta} = 1 + \frac{d\phi}{d\theta}. \tag{2}$$

Further, we may designate the angle between $d\rho$ and ds, in the differential triangle PRQ, as ϕ, for when passing to the limit, angle RQP equals angle OPT. Hence, from triangle PRQ, we have

$$\tan \phi = \frac{\rho\, d\theta}{d\rho},$$

or

$$\tan \phi = \frac{\rho}{\dfrac{d\rho}{d\theta}}. \tag{3}$$

Therefore,

$$\phi = \text{arc tan } \frac{\rho}{\dfrac{d\rho}{d\theta}}. \tag{4}$$

Differentiating (4) with respect to θ, by [14], §5—15, and simplifying:

$$\frac{d\phi}{d\theta} = \frac{\left(\dfrac{d\rho}{d\theta}\right)^2 - \rho\, \dfrac{d^2\rho}{d\theta^2}}{\rho^2 + \left(\dfrac{d\rho}{d\theta}\right)^2}. \tag{5}$$

Substituting (5) in (2):

$$\frac{d\omega}{d\theta} = \frac{\rho^2 - \rho\frac{d^2\rho}{d\theta^2} + 2\left(\frac{d\rho}{d\theta}\right)^2}{\rho^2 + \left(\frac{d\rho}{d\theta}\right)^2}. \tag{6}$$

But from §8—3, equation [4], we recall

$$\frac{ds}{d\theta} = \left[\rho^2 + \left(\frac{d\rho}{d\theta}\right)^2\right]^{\frac{1}{2}}. \tag{7}$$

Dividing (6) by (7), we obtain

$$\frac{\frac{d\omega}{d\theta}}{\frac{ds}{d\theta}} = \frac{\rho^2 - \rho\frac{d^2\rho}{d\theta^2} + 2\left(\frac{d\rho}{d\theta}\right)^2}{\left[\rho^2 + \left(\frac{d\rho}{d\theta}\right)^2\right]^{\frac{3}{2}}}.$$

And since $\dfrac{\frac{d\omega}{d\theta}}{\frac{ds}{d\theta}} = \dfrac{d\omega}{ds} = K$, therefore

$$K = \frac{\rho^2 - \rho\frac{d^2\rho}{d\theta^2} + 2\left(\frac{d\rho}{d\theta}\right)^2}{\left[\rho^2 + \left(\frac{d\rho}{d\theta}\right)^2\right]^{\frac{3}{2}}}. \tag{2}$$

8—9. Finding the Curvature at a Point on a Curve. This will now be illustrated by several examples.

EXAMPLE 1. Find the curvature of the parabola

$$y^2 = 12x \text{ at the point } (3,6).$$

Solution.

$$\frac{dy}{dx} = \frac{6}{y}; \qquad \frac{d^2y}{dx^2} = -\frac{6}{y^2}\frac{dy}{dx} = -\frac{36}{y^3}.$$

Substituting these values in [1a] of §8—6, for the point (3,6), we have

$$K = \frac{-\dfrac{36}{(6)^3}}{\left[1 + \left(\dfrac{6}{6}\right)^2\right]^{\frac{3}{2}}} = \frac{-\dfrac{1}{6}}{2^{\frac{3}{2}}} = -\frac{\sqrt{2}}{24};$$

or, considering the absolute value, $K = \dfrac{\sqrt{2}}{24}.$

EXAMPLE 2.   Find the curvature of the lemniscate

$$\rho^2 = \sin 2\theta \text{ at the point } \left(1, \frac{\pi}{4}\right).$$

Solution.

$$\frac{d\rho}{d\theta} = \frac{\cos 2\theta}{\rho};$$

$$\frac{d^2\rho}{d\theta^2} = \frac{-2\rho \sin 2\theta - \cos 2\theta \dfrac{d\rho}{d\theta}}{\rho^2}$$

$$= \frac{-2\rho^2 \sin 2\theta - \cos^2 (2\theta)}{\rho^3}.$$

At $\theta = \dfrac{\pi}{4}$, $\rho = 1$, $\sin 2\theta = 1$, and $\cos 2\theta = 0$.

Thus, at the point $\left(1, \dfrac{\pi}{4}\right)$, we have

$$\frac{d\rho}{d\theta} = 0, \qquad \frac{d^2\rho}{d\theta^2} = -2.$$

Substituting in [2], §8—8:

$$K = \frac{1 - (-2) + 2(0)}{[1 + 0]^{3⁄2}} = 3.$$

EXAMPLE 3.   Find the curvature of the hyperbola $xy = k$ at any point on the curve.   Where does the curve have the greatest curvature?

Solution. $\dfrac{dy}{dx} = -\dfrac{y}{x}$; $\dfrac{d^2y}{dx^2} = \dfrac{-x\dfrac{dy}{dx} + y}{x^2} = \dfrac{2y}{x^2}.$ (1)

Hence $K = \dfrac{\dfrac{2y}{x^2}}{\left[1 + \left(-\dfrac{y}{x}\right)^2\right]^{3⁄2}} = \dfrac{2xy}{(x^2 + y^2)^{3⁄2}}.$ (2)

Since $xy = k$, the numerator of this latter fraction in (2) is a constant; therefore K will have its greatest value when the denominator is least.   But the denominator will have the least value when $x = y$; hence the greatest curvature is at $x = y$, or where the 45°-axis cuts the hyperbola.

NOTE. To determine the minimum value of $x^2 + y^2$, let $u = x^2 + y^2$, or $u = x^2 + \dfrac{k^2}{x^2}$, since $xy = k$; find $\dfrac{du}{dx}$, set it equal to zero, and solve for critical values; they are $x = y = \pm\sqrt{k}$.

EXERCISE 8—1

Find the curvature of each of the following curves at the point indicated:

1. $y^2 = 2px$; $(2p, 2p)$
2. $a^2x^2 + b^2y^2 = a^2b^2$; $(0, a)$
3. $y = x^4 - 2x^3 + x^2$; $(0, 0)$
4. $y = \log x$; $(e, 1)$

Find the curvature of each of the following curves at the point where the slope of the curve is zero:

5. $y = 3x^2 - 6x + 5$
6. $y = x^3 + 9x^2 - 8$
7. $y = (x + 2)^2$
8. $y = \dfrac{x^2 + 1}{x}$
9. $y = \cos 2x$

Find the curvature at any point:

10. $x^2 - y^2 = a^2$
11. $\rho = e^{a\theta}$
12. $y = \frac{1}{4}x^4$
13. $\rho = a\theta$

Prove:

14. The curvature of a straight line, $y = mx + b$, is zero at every point on the line.
15. The curvature at a point of inflection on any curve equals zero.

CIRCLE OF CURVATURE

8—10. Radius of Curvature. The *radius of curvature* of a curve at a given point may be defined as the reciprocal of the curvature of the curve at that point. Thus, if the radius of curvature is represented by R, then

$$R = \frac{1}{K}.$$

Then we may at once write:

$$R = \frac{\left[1 + \left(\frac{dy}{dx}\right)^2\right]^{3/2}}{\frac{d^2y}{dx^2}}. \qquad [1]$$

$$R = \frac{\left[\rho^2 + \left(\frac{d\rho}{d\theta}\right)^2\right]^{3/2}}{\rho^2 - \rho\frac{d^2\rho}{d\theta^2} + 2\left(\frac{d\rho}{d\theta}\right)^2}. \qquad [2]$$

It should be noted that while the curvature K was defined as the absolute value of the respective fractions equivalent to $\frac{d\omega}{ds}$, K may be positive or negative. Hence R may also be positive or negative, and will have the same sign as K. If R is positive, the curve is concave upwards at the particular point in question; if R is negative, the curve is concave downwards at that point.

The radius of curvature may be thought of as the measure of the flatness or sharpness of a curve at a point; the smaller the radius of curvature, the sharper the curve.

The curvature of a curve at a point is the rate at which the inclination of the curve is changing with respect to the length of arc, that is, curvature $= \frac{d\omega}{ds}$.

The reader will see, from the following proof, why we take R equal to $\frac{1}{K}$. Let $R = CP$ be the required radius of curvature at point P on the curve AB; TU and MN are tangents to the curve at P and Q, respectively; $C'P$ and $C'Q$ are the respective normals; angle $\Delta\omega$ is the angle between the normals, and therefore equals $\Delta\omega$, the angle between the tangents (sides of the angles are respectively perpendicular); and arc $PQ = \Delta s$.

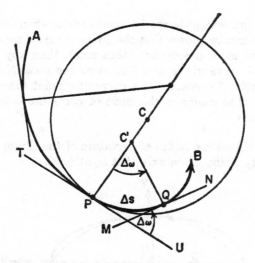

In triangle $PC'Q$, by the law of sines,

$$\frac{C'P}{PQ} = \frac{\sin Q}{\sin \Delta\omega}, \quad \text{or} \quad C'P = \frac{\overline{PQ}\sin Q}{\sin \Delta\omega}.$$

As Q approaches P, $\Delta s \rightarrow 0$; in passing to the limiting value of $C'P$ (that is CP, or R, the radius of curvature), we note that:

    (1)   chord PQ may be replaced by Δs,

    (2) $\sin \Delta\omega$ may be replaced by $\Delta\omega$, and

    (3)   limit of $\sin Q = 1$, since angle Q is approaching $90°$.

Therefore,

$$\lim_{\Delta s \rightarrow 0} C'P = \lim_{\Delta s \rightarrow 0} \left(\frac{PQ \cdot \sin Q}{\sin \Delta\omega}\right) = \lim_{\Delta s \rightarrow 0} \frac{\Delta s \cdot (1)}{\Delta\omega};$$

or

$$CP = R = \frac{ds}{d\omega}.$$

Since

$$K = \frac{d\omega}{ds}, \quad R = \frac{1}{K}.$$

8—11. Circle of Curvature. At a given point on a curve, there is, in general, but one tangent and one normal. An infinite number of circles, all having their centers lying on the normal, can be drawn through this point. Of these circles, the one whose radius equals the radius of curvature for the curve at that point is called the *circle of curvature for the point*. Obviously, each point on the curve has a different circle of curvature (except in special cases such as when the curve itself is a circle). It can be shown that the circle of curvature at any point "fits" the curve more closely, near that point, than any other circle.

The circle of curvature is also known as the *osculating circle*. In general, the circle of curvature of a curve at a point crosses the curve at that point. The center of the circle of curvature is known as the *center of curvature*.

EXAMPLE 1. Find the radius of curvature of the ellipse $\dfrac{x^2}{9} + \dfrac{y^2}{4} = 1$ at the extremity of the minor axis, that is, at (0,2).

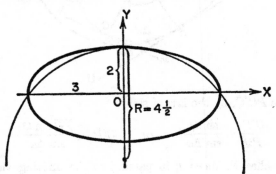

Solution. $\qquad\qquad 4x^2 + 9y^2 = 36.$

Differentiating:

$$8x + 18y\,\frac{dy}{dx} = 0; \qquad \frac{dy}{dx} = -\frac{4x}{9y};$$

hence, at point (0,2) the value of $\dfrac{dy}{dx} = 0$.

Differentiating again:

$$8 + 18y\,\frac{d^2y}{dx^2} + 18\left(\frac{dy}{dx}\right)^2 = 0;$$

since, at point (0,2) $y = 2$ and $\dfrac{dy}{dx} = 0$, the value of $\dfrac{d^2y}{dx^2}$ at this point is $-\dfrac{2}{9}$.

Therefore, $\quad R = \dfrac{[1 + 0]^{3/2}}{-\dfrac{2}{9}} = -\dfrac{9}{2} = -4\dfrac{1}{2}.$

NOTE 1. Since R is negative, the curve at (0,2) is concave downwards.

NOTE 2. It will be observed that the numerical value of R, or $4\frac{1}{2}$, is comparatively large, indicating that the curve at this point is fairly "flat."

NOTE 3. If we wish to determine the value of R at the extremity of the major axis, we should find that the value of $\dfrac{dy}{dx}$ at that point is infinite. In this case, we would *interchange the axes*, transforming the equation to

$$\frac{x^2}{4} + \frac{y^2}{9} = 1,$$

and the extremity in question becomes (0,3). See below, Exercise 8—2, Problem 11.

EXERCISE 8—2

Find the radius of curvature for each of the following at the point indicated; in each case sketch the circle of curvature:

1. $y = x^3 - 4x^2 + 3x$; (0,0)
2. $xy = 20$; (4,5)
3. $a^2x^2 + b^2y^2 = a^2b^2$; (0,a)
4. $x^2 = 2py$; (0,0)
5. $y = \sin x$; $\left(\dfrac{\pi}{2}, 1\right)$
6. $y = \tan x$; $\left(\dfrac{\pi}{4}, 1\right)$

Find the radius of curvature of the following curves at any point:

7. $y = x^3$
8. $\rho = a \cos \theta$
9. $y^2 = x^3$
10. $\rho = 1 - \cos \theta$

11. Find the radius of curvature of the ellipse $\dfrac{x^2}{9} + \dfrac{y^2}{4} = 1$ at the extremity of the major axis (see NOTE 3, Illustrative Example 1, §8—11). Compare the numerical values of the radii of curvature at the extremities of the two axes; what does this show about the comparative curvature at these two points?

12. Find the radius of curvature of the witch, $x^2y = 8 - 4y$, at the point $(0,2)$.

8—12. Radius of Curvature of Curves with Equations in Parametric Form.

We recall from §6—7 that when equations are given in parametric form, such as $x = f(t)$, $y = \phi(t)$, then

$$\frac{dy}{dx} = \frac{\dfrac{dy}{dt}}{\dfrac{dx}{dt}}, \tag{1}$$

and

$$\frac{d^2y}{dx^2} = \frac{\dfrac{d^2y}{dt^2}\dfrac{dx}{dt} - \dfrac{d^2x}{dt^2}\dfrac{dy}{dt}}{\left(\dfrac{dx}{dt}\right)^3}. \tag{2}$$

Substituting (1) and (2) in the formula for the radius of curvature, §8—10, equation [1], and simplifying, we obtain:

$$R = \frac{\left[\left(\dfrac{dx}{dt}\right)^2 + \left(\dfrac{dy}{dt}\right)^2\right]^{3/2}}{\dfrac{d^2y}{dt^2}\dfrac{dx}{dt} - \dfrac{d^2x}{dt^2}\dfrac{dy}{dt}}. \tag{1}$$

EXAMPLE 1. Find the value of the radius of curvature of the curve $x = t^2$, $y = 2t$, at the point where $t = 1$.

Solution.

$$\frac{dx}{dt} = 2t; \qquad \frac{dy}{dt} = 2.$$

$$\frac{d^2x}{dt^2} = 2; \qquad \frac{d^2y}{dt^2} = 0.$$

Therefore:

$$R = \frac{(4t^2 + 4)^{3/2}}{0 - (2)(2)} = -2(t^2 + 1)^{3/2}.$$

When $t = 1$, $\qquad\qquad R = -4\sqrt{2}.$

EXAMPLE 2. Find the radius of curvature of $x = 2 \sin t$, $y = \cos t$, at the point where $t = \dfrac{\pi}{4}$.

Solution.

$$\frac{dx}{dt} = 2 \cos t; \qquad \frac{dy}{dt} = -\sin t.$$

$$\frac{d^2x}{dt^2} = -2 \sin t; \qquad \frac{d^2y}{dt^2} = -\cos t.$$

At $t = \dfrac{\pi}{4} = 45°$:

$$\frac{dx}{dt} = \sqrt{2}; \qquad \frac{dy}{dt} = -\frac{\sqrt{2}}{2}.$$

$$\frac{d^2x}{dt^2} = -\sqrt{2}; \qquad \frac{d^2y}{dt^2} = -\frac{\sqrt{2}}{2}.$$

Therefore:

$$R = \frac{\left(2 + \dfrac{1}{2}\right)^{3/2}}{\left(-\dfrac{\sqrt{2}}{2}\right)\sqrt{2} - (-\sqrt{2})\left(-\dfrac{\sqrt{2}}{2}\right)} = \frac{\left(\dfrac{5}{2}\right)^{3/2}}{-1 - 1}$$

$$= -\frac{5\sqrt{5}}{4\sqrt{2}} = -\frac{5\sqrt{10}}{8}.$$

8—13. The Center of Curvature. Let $P(x,y)$ be any point on a given curve $y = f(x)$; let t be the tangent to the curve at P. Assume that the curve lies entirely on one side of the tangent. Along the

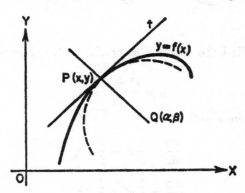

normal, toward the concave side of the curve, lay off the distance PQ, equal to the radius of curvature R at P. Thus Q, the center of the circle of curvature of the given curve for the point P, is called the *center of curvature* with respect to point P.

Without proof, we state the coordinates (α,β) of the center of curvature in terms of the coordinates (x,y) of P:

$$\alpha = x - \frac{\dfrac{dy}{dx}\left[1 + \left(\dfrac{dy}{dx}\right)^2\right]}{\dfrac{d^2y}{dx^2}}, \qquad [1]$$

$$\beta = y + \frac{1 + \left(\dfrac{dy}{dx}\right)^2}{\dfrac{d^2y}{dx^2}}. \qquad [2]$$

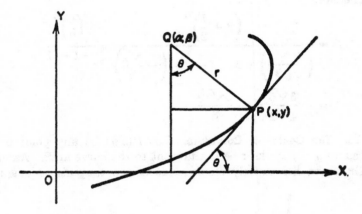

EXAMPLE. Find the coordinates of the center of curvature of the curve

$$y = \frac{e^x + e^{-x}}{2}$$

at the point where $x = 0$.

Solution.
$$\frac{dy}{dx} = \frac{1}{2}e^x - \frac{1}{2}e^{-x},$$

$$\frac{d^2y}{dx^2} = \frac{1}{2}e^x + \frac{1}{2}e^{-x}.$$

At the point where $x = 0$, $y = 1$.

Hence, substituting in equations [1] and [2],

$$\alpha = 0 - \frac{0}{1} = 0,$$

$$\beta = 1 + \frac{1}{1} = 2.$$

THE EVOLUTE

8—14. Definition and Properties of the Evolute. It is clear that every point of a given curve has its center of curvature. The locus of these centers of curvature is called the *evolute* of the curve. In other words, as a point P moves along a given curve, the center of curvature Q moves along another curve; this path is the evolute of the original curve. The original curve is the *involute* of its evolute.

If the given curve is given by $y = f(x)$, then the right-hand members of [1] and [2] of §8—13 may be stated in terms of x, whence they become

$$X = \phi(x), \qquad Y = \psi(x), \qquad [3]$$

where X and Y are the coordinates of any point on the curve $y = f(x)$, instead of the coordinates (α, β) of some particular point.

The evolute of a curve presents two interesting properties:

(1) The normals to a curve are tangent to the evolute, each at the center of curvature that lies upon it.

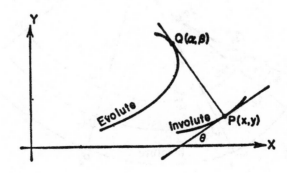

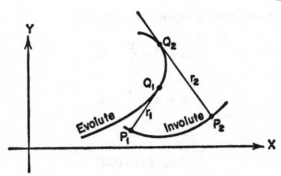

(2) The difference in length between any two radii of curvature is equal to the length of the arc of the evolute included by them; i.e., length of arc $Q_1Q_2 = r_2 - r_1$.

Property (2), sometimes referred to as the "string property" of the evolute, may be stated informally as follows. Let an inextensible but flexible string be attached to a fixed point on the evolute and wrapped tightly around a given arc AQ_1 of the evolute, and let the remaining portion of the string be held taut. Now as the free end (P) of the string is allowed to move so that the string remains tangent to the

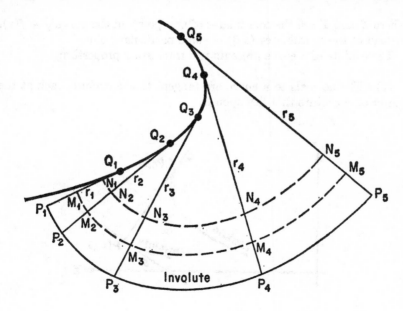

evolute, the free end P will trace the original curve or involute, and the point of tangency will be the corresponding center of curvature Q. From the diagram, it will be seen that the curve $P_1 \cdots P_5$ is not the only possible involute of the given evolute; for, as the string is unwrapped, similar curves $M_1 \cdots M_5$, $N_1 \cdots N_5$, etc. are traced. Thus, *while a curve has but one evolute, it has an infinite number of involutes.*

EXERCISE 8—3

Find the coordinates of the center of curvature of the following curves at the point indicated:

1. $y = e^{-x}$; point $(0,1)$.
2. $y = \log x$; at the point where $x = 2$.
3. $x^2 + y^2 = k^2$; point $(0,k)$.
4. $y^2 = 4x$; point $(1,2)$.
5. $y = \sin x$; point $\left(\dfrac{\pi}{2},1\right)$.

EXERCISE 8—4

Review

1. Find $\dfrac{dy}{dx}$:
   (a) $y^3 - 3y + x = 0$.
   (b) $x^3 + xy^2 + y^3 = 4$.

2. Find $\dfrac{d^2y}{dx^2}$:
   (a) $x^2 - y^2 = 1$.
   (b) $y^2 - 2xy + a^2 = 0$.

3. Find $\dfrac{dy}{dx}$:
   (a) $y = \arcsin \sqrt{x}$.
   (b) $y = \log \dfrac{e^x}{e^x - 1}$.

4. Find maximum or minimum values:
   (a) $y = x^{1/x}$.
   (b) $y = \sin x \, (1 + \cos x)$.
   (c) $y = x^3 - 3x^2 + 6x$.

5. Find the radius of curvature at the given point:

    (a) $y^2 = 4x$, at $(0,0)$.

    (b) $y = \dfrac{e^x + e^{-x}}{2}$, at $(0,1)$.

6. Find the equation of the tangent to the given curve at the point indicated:

    (a) $ay^2 = x^3$; (a,a).

    (b) $xy = 4$; $(4,1)$.

    (c) $y^2 = 4x$; $(1,2)$.

7. If the sides of a rectangle are a and b, find the rectangle of maximum area that can be drawn so as to have its sides pass through the vertices of the given rectangle.

8. Find the equation of the line tangent to the curve $x^2 + y^2 = x^3$ at the point $(2,2)$.

9. If $Z = \sqrt{x^2 - y^2}$, and $xy = a^2$, prove that $\dfrac{dZ}{dx} = \dfrac{x^2 + y^2}{xZ}$.

10. Assuming that a raindrop is a perfect sphere, and that it increases in volume at a rate proportional to its surface area, prove that the radius increases at a constant rate.

Indeterminate Forms

CHAPTER NINE

THEOREM OF MEAN VALUE

9—1. Rolle's Theorem. Let us consider a continuous single-valued function $y = f(x)$, with zero-values at $x = a$ and $x = b$. Let us assume that $f'(x)$ changes continuously as x varies from a to b. From the figure,

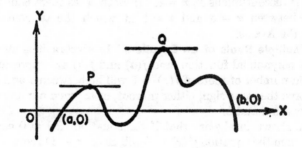

by intuition, it is clear that for at least one value of x between a and b the tangent to the curve must be parallel to the X-axis, as at P or Q; that is, the slope of the curve is zero for at least one value between $x = a$ and $x = b$.

This principle may be stated without formal proof as *Rolle's Theorem*, as follows:

If $f(x) = 0$ when $x = a$ and when $x = b$, and if both $f(x)$ and $f'(x)$

are continuous for all values of x from $x = a$ to $x = b$, then $f'(x)$ will equal zero for at least one value of x between a and b.

What this means is that, as x increases from a to b, $f(x)$ cannot *always* increase, since $f(a) = f(b) = 0$; similarly, $f(x)$ cannot *always* decrease as x increases from a to b. Therefore, for at least one value of x between a and b, $f(x)$ must stop increasing and begin to decrease, or stop decreasing and begin to increase. We have already learned that at a turning point in a curve, the value of $f'(x)$ is zero.

From a consideration of the following figures, it will be seen that Rolle's Theorem does not hold if either $f(x)$ or $f'(x)$ is discontinuous.

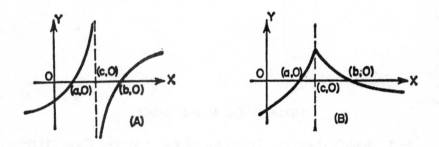

Thus in (A) the function is discontinuous at $x = c$; in (B), the first derivative is discontinuous at $x = c$. In both cases there is no point on the curve between $x = a$ and $x = b$ at which the tangent becomes parallel to the X-axis.

9—2. Multiple Roots of an Equation. In algebra it is shown that if $f(x)$ is a polynomial function, and $f(a)$ and $f(b)$ are opposite in sign, then an odd number of roots of $f(x) = 0$ will lie between a and b; if $f(a)$ and $f(b)$ have the same sign, either no root, or an even number of roots, will lie between a and b.

It is also shown in algebra that if the equation $f(x) = 0$ has r roots equal to a, then the equation $f'(x) = 0$ will have $(r - 1)$ roots equal to a. This is another way of saying that if $f(x)$ contains a factor $(x - a)^r$, then the equation $f'(x) = 0$ will have $(r - 1)$ roots equal to a; in short, $f(x)$ and $f'(x)$ have a common factor $(x - a)^{r-1}$. Thus an equation $f(x) = 0$ has or has not equal roots, according as $f(x)$ and $f'(x)$ have or do not have a common factor involving x. Such equal roots are also called *multiple roots* of an equation. To determine whether or not an equation has multiple roots, it is only necessary to determine whether $f(x)$ and $f'(x)$ have a common factor.

9—3. The Mean Value Theorem. The principle of Rolle's Theorem can be formulated analytically as follows. Let the abscissas of P_1 and

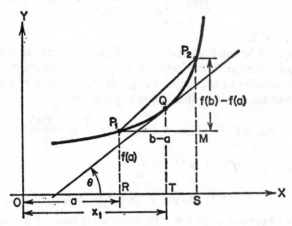

P_2 be a and b, respectively; then $RP_1 = f(a)$, $SP_2 = f(b)$, and $MP_2 = f(b) - f(a)$. The slope of the line P_1P_2 is seen to be

$$\frac{MP_2}{P_1M} = \frac{f(b) - f(a)}{b - a}.$$

Now let us denote this last fraction by the symbol Q, that is, let

$$Q = \frac{f(b) - f(a)}{b - a}. \tag{1}$$

Clearing of fractions:

$$f(b) - f(a) - (b - a)Q = 0. \tag{2}$$

Let $\phi(x)$ be a function formed by replacing b by x in equation (2):

$$\phi(x) = f(x) - f(a) - (x - a)Q. \tag{3}$$

From equation (2), we find $\phi(b) = 0$; and from equation (3), we find $\phi(a) = 0$. Therefore, from Rolle's Theorem, $\phi'(x)$ must vanish for at least one value of x between a and b; call this value x_1. By differentiating equation (3):

$$\phi'(x) = f'(x) - Q. \tag{4}$$

But $\phi'(x_1) = 0;$

hence $f'(x_1) - Q = 0,$

or $Q = f'(x_1). \tag{5}$

Substituting from equation (5) in equation (1):

$$\frac{f(b) - f(a)}{b - a} = f'(x_1),$$

where $a < x_1 < b$.

Referring now to the first paragraph of §9—3, we see that there is at least one point on the curve between P_1 and P_2 where the tangent to the curve is parallel to the chord P_1P_2. If the abscissa of this point is called x_1, then the slope at that point is given by

$$\tan MP_1P_2 = \tan \theta = f'(x_1) = \frac{f(b) - f(a)}{b - a}.$$

Hence: $\qquad\qquad f'(x_1) = \dfrac{f(b) - f(a)}{b - a},$ $\qquad\qquad$ [1]

or $\qquad\qquad f(b) = f(a) + (b - a)f'(x_1),$ $\qquad\qquad$ [2]

where x_1 lies between a and b. Equations [1] and [2] are alternative forms of stating the *law of the mean*, or the mean value theorem. It is a principle which has many uses in the calculus.

EVALUATION OF INDETERMINATE FORMS

9—4. Indeterminate Forms. It often happens that for a particular value of the independent variable, a function may take on a corresponding value which is indeterminate, such as $\dfrac{0}{0}$, $\dfrac{\infty}{\infty}$, $0 \cdot \infty$, $\infty - \infty$, 0^0, 1^∞, or ∞^0.

For example, what is the value of y in

$$y = \frac{x^2 - 9}{x + 3}$$

when $x = -3$? Direct substitution gives

$$\frac{9 - 9}{-3 - 3} = \frac{0}{0},$$

which is undefined. However, if we first write

$$y = \frac{x^2 - 9}{x + 3} = \frac{(x + 3)(x - 3)}{x + 3} = x - 3,$$

then $y = -3 - 3 = -6$ when $x = -3$.

Or, consider another example: find the value of

$$\lim_{x \to \infty} \left(\frac{x^3 + 3x^2 + 2}{4x - x^2 - 5x^3} \right).$$

Direct substitution gives the value $\frac{\infty}{\infty}$, which again is indeterminate, or undefined. By an algebraic transformation, however, we may write

$$\lim_{x \to \infty} \left(\frac{1 + \dfrac{3}{x} + \dfrac{2}{x^3}}{\dfrac{4}{x^2} - \dfrac{1}{x} - 5} \right) = \frac{1 + 0 + 0}{0 - 0 - 5} = -\frac{1}{5}.$$

Such algebraic transformations, however, do not constitute *general* methods for evaluating indeterminate forms. For this purpose we can use the process of differentiation, as shown in what follows.

9—5. The Indeterminate Form $\frac{0}{0}$. Let us consider a function of the form $\frac{f(x)}{F(x)}$ such that $f(a) = 0$ and $F(a) = 0$; in other words, suppose that the function takes on the indeterminate form $\frac{0}{0}$ when a is substituted for x. The problem, then, is to determine the value of

$$\lim_{x \to a} \frac{f(x)}{F(x)}.$$

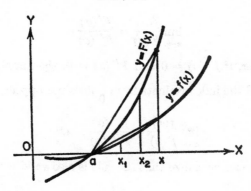

The curves $y = f(x)$ and $y = F(x)$ must intersect at $(a,0)$, since $f(a) = 0$ and $F(a) = 0$ by hypothesis. Let us now employ the law of the mean:

$$f(x) = f(a) + (x - a)f'(x_1), \qquad \text{where} \qquad a < x_1 < x;$$
$$\text{and} \quad F(x) = F(a) + (x - a)F'(x_2), \qquad \text{where} \qquad a < x_2 < x.$$

Dividing these two equations, we have:

$$\frac{f(x)}{F(x)} = \frac{f'(x_1)}{F'(x_2)},$$

the terms $f(a)$ and $F(a)$ vanishing, since each is equal to zero. Finally, let $x \to a$; then $x_1 \to a$ and $x_2 \to a$; hence

$$\lim_{x \to a} \frac{f(x)}{F(x)} = \frac{f'(a)}{F'(a)}, \qquad \text{where} \qquad F'(a) \neq 0.$$

This last relation, then, suggests a method of procedure to be used in evaluating a function which leads to the form $\frac{0}{0}$.

To evaluate the indeterminate form $\frac{0}{0}$, differentiate the numerator to obtain a new numerator, and the denominator to obtain a new denominator. The value of this new fraction for the assigned value of the variable will be the desired limiting value of the original fraction.

NOTE 1. If, after differentiation, it should turn out that both $f'(a)$ and $F'(a)$ are equal to zero, the rule may be applied again to *this* ratio; thus

$$\lim_{x \to a} \frac{f(x)}{F(x)} = \frac{f''(a)}{F''(a)}.$$

In like manner, if $f''(a) = 0$ and $F''(a) = 0$, then apply the rule repeatedly, until the indeterminate form $\frac{0}{0}$ does not appear; thus

$$\lim_{x \to a} \frac{f(x)}{F(x)} = \frac{f'''(a)}{F'''(a)}; \quad \text{etc.}$$

NOTE 2. It can be shown that the rule is also valid when $a = \infty$.

NOTE 3. When applying the rule, the expression $\frac{f(x)}{F(x)}$ should *not* be differentiated as a fraction; *each term* of the fraction should be differentiated *separately*.

EXAMPLE 1. Evaluate $\lim\limits_{x \to 3} \dfrac{x^2 - 9}{2x^2 + x - 21}$.

Solution. Let $f(x) = x^2 - 9$, and $F(x) = 2x^2 + x - 21$.

$$\frac{f(3)}{F(3)} = \frac{9 - 9}{18 + 3 - 21} = \frac{0}{0}, \quad \text{which is indeterminate.}$$

$$\frac{f'(x)}{F'(x)} = \frac{2x}{4x + 1}; \qquad \lim\limits_{x \to 3} \frac{2x}{4x + 1} = \frac{6}{13}.$$

EXAMPLE 2. Evaluate $\lim\limits_{x \to 0} \left(\dfrac{1 - \cos x}{\sin x} \right)$.

Solution. Let $f(x) = 1 - \cos x$, and $F(x) = \sin x$.

$$\frac{f(0)}{F(0)} = \frac{1 - 1}{0} = \frac{0}{0}, \quad \text{which is indeterminate.}$$

$$\frac{f'(x)}{F'(x)} = \frac{\sin x}{\cos x}; \qquad \lim\limits_{x \to 0} \frac{\sin x}{\cos x} = \frac{0}{1} = 0.$$

EXAMPLE 3. Evaluate $\lim\limits_{x \to 0} \left(\dfrac{e^x - 2 \sin x - e^{-x}}{x - \sin x} \right)$.

Solution. Let $f(x) = e^x - 2 \sin x - e^{-x}$, and $F(x) = x - \sin x$.
Substituting when $x = 0$:

$$\frac{f(0)}{F(0)} = \frac{1 - 0 - 1}{0 - 0} \bigg]_{x=0} = \frac{0}{0}; \quad \text{indeterminate.}$$

$$\frac{f'(0)}{F'(0)} = \frac{e^x - 2 \cos x + e^{-x}}{1 - \cos x} \bigg]_{x=0} = \frac{1 - 2 + 1}{1 - 1} = \frac{0}{0}.$$

$$\frac{f''(0)}{F''(0)} = \frac{e^x + 2 \sin x - e^{-x}}{\sin x} \bigg]_{x=0} = \frac{1 + 0 - 1}{0} = \frac{0}{0}.$$

$$\frac{f'''(0)}{F'''(0)} = \frac{e^x + 2 \cos x + e^{-x}}{\cos x} \bigg]_{x=0} = \frac{1 + 2 + 1}{1} = 4.$$

EXERCISE 9—1

Evaluate each of the following by differentiating:

1. $\lim\limits_{x \to 0} \dfrac{2x^3 - x^2}{x^3 + x^2}$

2. $\lim\limits_{x \to 1} \dfrac{x^n - 1}{x - 1}$

3. $\displaystyle\lim_{y \to -1} \frac{y^3 + 1}{y^2 - 1}$

7. $\displaystyle\lim_{x \to 0} \frac{1 - \cos x}{x^2}$

4. $\displaystyle\lim_{\theta \to 0} \frac{\cos \theta - 1}{\theta}$

8. $\displaystyle\lim_{x \to \pi} \frac{\cos x + 1}{(\pi - x)^2}$

5. $\displaystyle\lim_{x \to 0} \frac{x - \sin x}{x^3}$

9. $\displaystyle\lim_{\theta \to 0} \frac{\sin 2\theta}{\theta}$

6. $\displaystyle\lim_{y \to 1} \frac{\log y}{y - 1}$

10. $\displaystyle\lim_{x \to 0} \frac{e^x - e^{-x}}{\sin x}$

9—6. The Indeterminate Form $\dfrac{\infty}{\infty}$. To determine the value of an expression whose limit reduces to $\dfrac{\infty}{\infty}$, we proceed according to the same rule as used when evaluating the form $\dfrac{0}{0}$. A rigorous proof of the validity of the rule for the case $\dfrac{\infty}{\infty}$ must be left, however, for more advanced study.

EXAMPLE 1. Evaluate $\dfrac{\log x}{\dfrac{1}{x}}$ when $x = 0$.

Solution. Let $f(x) = \log x$, and $F(x) = \dfrac{1}{x}\cdot$

$$\frac{f(0)}{F(0)} = \frac{-\infty}{\infty}, \quad \text{which is indeterminate.}$$

$$\frac{f'(0)}{F'(0)} = \frac{\dfrac{1}{x}}{-\dfrac{1}{x^2}} = -x \Bigg]_{x=0} = 0.$$

EXAMPLE 2. Find the limit of $\dfrac{x^2}{e^x}$ when $x = \infty$.

Solution. Let $f(x) = x^2$, and $F(x) = e^x$.

$$\frac{f(0)}{F(0)} = \frac{\infty}{\infty}; \quad \text{indeterminate.}$$

$$\frac{f'(\infty)}{F'(\infty)} = \frac{2x}{e^x} \Bigg]_{x=\infty} = \frac{\infty}{\infty}; \quad \text{indeterminate.}$$

$$\frac{f''(\infty)}{F''(\infty)} = \frac{2}{e^x}\bigg]_{x=\infty} = \frac{2}{\infty} = 0.$$

9—7. Evaluation of the Form 0· ∞. When a function such as $f(x)\cdot\phi(x)$ takes the indeterminate form $0\cdot\infty$ when $x = a$, it may be evaluated by writing

$$f(x)\cdot\phi(x) = \frac{f(x)}{\dfrac{1}{\phi(x)}},$$

or

$$f(x)\cdot\phi(x) = \frac{\phi(x)}{\dfrac{1}{f(x)}}.$$

In this way it will take either the form $\dfrac{0}{0}$ or $\dfrac{\infty}{\infty}$, which may then be evaluated by the method of §9—5 or §9—6.

EXAMPLE 1. Evaluate $x^2 e^{-x^3}$ when $x = \infty$.

Solution. Let $f(x) = x^2$, and $\phi(x) = e^{-x^3}$.
When $x = \infty$, $f(x)\cdot\phi(x) = \infty\cdot 0$; indeterminate.

Writing $f(x)\cdot\phi(x)$ as $\dfrac{f(x)}{\dfrac{1}{\phi(x)}}$, the function becomes

$$\frac{f(x)}{F(x)} = \frac{x^2}{e^{x^3}}\bigg]_{x=\infty} = \frac{\infty}{\infty}; \text{ indeterminate.}$$

$$\frac{f'(\infty)}{F'(\infty)} = \frac{2x}{3x^2 e^{x^3}} = \frac{2}{3xe^{x^3}}\bigg]_{x=\infty} = \frac{2}{\infty} = 0.$$

EXAMPLE 2. Evaluate $\sec x \cos 3x$ when $x = \dfrac{\pi}{2}$.

Solution. $\lim\limits_{x\to\pi/2} (\sec x \cos 3x) = \infty\cdot 0$; indeterminate.

Put $\sec x = \dfrac{1}{\cos x}$; then $\dfrac{f(x)}{F(x)} = \dfrac{\cos 3x}{\cos x}$.

$$\frac{f\left(\dfrac{\pi}{2}\right)}{F\left(\dfrac{\pi}{2}\right)} = \frac{\cos 3x}{\cos x}\bigg]_{x=\pi/2} = \frac{0}{0}; \text{ indeterminate.}$$

$$\frac{f'\left(\dfrac{\pi}{2}\right)}{F'\left(\dfrac{\pi}{2}\right)} = \frac{-3\sin 3x}{-\sin x}\bigg]_{x=\pi/2} = \frac{-3(-1)}{-1} = -3.$$

9—8. Evaluation of the Form $\infty - \infty$. An expression which reduces to this form can usually be transformed into a fraction that will assume the form $\dfrac{0}{0}$ or $\dfrac{\infty}{\infty}$.

EXAMPLE 1. Find the value of

$$\lim_{x\to 0}\left(\csc x - \frac{1}{x}\right).$$

Solution. $\lim_{x\to 0}\left(\csc x - \dfrac{1}{x}\right) = \infty - \infty$;   indeterminate.

Write the given expression as

$$\frac{1}{\sin x} - \frac{1}{x}, \qquad \text{or} \qquad \frac{x - \sin x}{x \sin x}.$$

$$\frac{f(0)}{F(0)} = \frac{x - \sin x}{x \sin x}\bigg]_{x=0} = \frac{0}{0};\quad \text{indeterminate.}$$

$$\frac{f'(0)}{F'(0)} = \frac{1 - \cos x}{x \cos x + \sin x}\bigg]_{x=0} = \frac{0}{0};\quad \text{indeterminate.}$$

$$\frac{f''(0)}{F''(0)} = \frac{\sin x}{-x \sin x + 2 \cos x}\bigg]_{x=0} = \frac{0}{2} = 0.$$

EXAMPLE 2. Evaluate $\left(\dfrac{x}{x - 1} - \dfrac{1}{\log x}\right)$ when $x = 1$.

Solution. $\lim_{x\to 1}\left(\dfrac{x}{x - 1} - \dfrac{1}{\log x}\right) = \infty - \infty$;   indeterminate.

Write the given expression as

$$\frac{f(x)}{F(x)} = \frac{x \log x - x + 1}{(x - 1) \log x}\bigg]_{x=1} = \frac{0 - 1 + 1}{0} = \frac{0}{0}.$$

$$\frac{f'(1)}{F'(1)} = \frac{1 + \log x - 1}{\dfrac{x-1}{x} + \log x} = \frac{x \log x}{x - 1 + x \log x}\Big]_{x=1} = \frac{0}{0}.$$

$$\frac{f''(1)}{F''(1)} = \frac{1 + \log x}{2 + \log x}\Big]_{x=1} = \frac{1}{2}.$$

EXERCISE 9—2

Evaluate each of the following:

1. $\lim\limits_{x \to \infty} x e^{-x}$

2. $\lim\limits_{\theta \to 0} \dfrac{\log \theta}{\cot \theta}$

3. $\lim\limits_{x \to \infty} \dfrac{x^2}{e^x}$

4. $\lim\limits_{\theta \to 0} \theta \cot \theta$

5. $\lim\limits_{t \to \infty} \dfrac{t^2 - 1}{t^3 + 1}$

6. $\lim\limits_{x \to \infty} \dfrac{x^3}{e^{3x}}$

7. $\lim\limits_{x \to 0} x^2 \log x$

8. $\lim\limits_{x \to \infty} \left(\dfrac{e^{2x}}{x^2}\right)$

9. $\lim\limits_{x \to 0} (x \cot 2\pi x)$

10. $\lim\limits_{x \to \infty} \dfrac{x + x^2}{e^x + e^{-x}}$

11. $\lim\limits_{x \to \pi/2} (\sec x - \tan x)$

12. $\lim\limits_{x \to 0} (\sin x \log x)$

13. $\lim\limits_{x \to \pi/2} (\cos 3x \tan x)$

14. $\lim\limits_{x \to 0} \left(\csc x^2 - \dfrac{1}{x^2}\right)$

15. $\lim\limits_{x \to 1} \left(\dfrac{2}{x^2 - 1} - \dfrac{1}{x - 1}\right)$

9—9. Evaluation of the Forms 0^0, 1^∞, and ∞^0. A function of the form $f(x)^{\phi(x)}$ may yield indeterminate forms as follows:

(A)   if $f(x) = 0$, and $\phi(x) = 0$, we obtain 0^0;

(B)   if $f(x) = 1$, and $\phi(x) = \infty$, we obtain 1^∞;

(C)   if $f(x) = \infty$, and $\phi(x) = 0$, we obtain ∞^0.

In such cases, the indeterminate forms are evaluated by a logarithmic transformation. Thus, let

$$y = f(x)^{\phi(x)};$$

then $\qquad\qquad \log y = \phi(x) \log f(x).$

Now we have the indeterminate form $0 \cdot \infty$; for

in (A), $\log y = 0 \cdot (\log 0) = 0 \cdot (-\infty).$

in (B), $\log y = \infty \cdot (\log 1) = \infty \cdot 0$.

in (C), $\log y = 0 \cdot (\log \infty) = 0 \cdot \infty$.

The expression $\phi(x) \log f(x)$ may therefore be evaluated as in §9—7; the limit so found, however, is the *limit of the logarithm of the desired function*. But the limit of the logarithm of a function equals the logarithm of the limit of the function. Thus, if we know $\lim \log_e y = a$, then $y = e^a$.

EXAMPLE 1. Find the value of

$$\lim_{x \to 0} (1 + kx)^{1/x}.$$

Solution. $\lim_{x \to 0} (1 + kx)^{1/x} = 1^\infty$;   indeterminate.

Put $y = (1 + kx)^{1/x}$;

then $\log y = \frac{1}{x} \log (1 + kx)$,   and

$$\lim_{x \to 0} \frac{1}{x} \log (1 + kx) = \infty \cdot 0.$$

To find $\lim_{x \to 0} \frac{1}{x} \log (1 + kx)$, write the expression as

$$\frac{\log (1 + kx)}{x}.$$

$$\frac{f(x)}{F(x)} = \frac{\log (1 + kx)}{x} \Big]_{x=0} = \frac{0}{0}; \quad \text{indeterminate.}$$

$$\frac{f'(0)}{F'(0)} = \frac{\dfrac{k}{1 + kx}}{1} = \frac{k}{1 + kx} \Big]_{x=0} = k.$$

Since $\lim \log_e y = k$, then $y = e^k$; that is, $(1 + kx)^{1/x} = e^k$. In other words, since $y = (1 + kx)^{1/x}$, this gives $\log_e (1 + kx)^{1/x} = k$; that is, $(1 + kx)^{1/x} = e^k$.

EXAMPLE 2. Evaluate $\left(\dfrac{1}{x}\right)^{1/x}$ when $x = \infty$.

Solution. $\lim_{x \to \infty} \left(\dfrac{1}{x}\right)^{1/x} = 0^0$;   indeterminate.

Put $y = \left(\dfrac{1}{x}\right)^{1/x}$,   or $\log y = \dfrac{1}{x} \log \dfrac{1}{x}$.

Then $\lim\limits_{x \to \infty} \log y = \lim\limits_{x \to \infty} \left(\dfrac{1}{x} \log \dfrac{1}{x}\right) = 0 \cdot (-\infty);$ indeterminate.

Write $\dfrac{1}{x} \log \dfrac{1}{x}$ as $\dfrac{\log \dfrac{1}{x}}{x}$; let $f(x) = \log \dfrac{1}{x}$, and $F(x) = x$.

$$\frac{f(x)}{F(x)} = \frac{\log \dfrac{1}{x}}{x} \bigg]_{x=\infty} = \frac{-\infty}{\infty}; \quad \text{indeterminate.}$$

$$\frac{f'(x)}{F'(x)} = \frac{-\dfrac{1}{x}}{1} \bigg]_{x=\infty} = 0.$$

Since $\lim \log_e y = 0$, $y = e^0 = 1$; that is, $\left(\dfrac{1}{x}\right)^{1/x} = 1$. In other words,

since $y = \left(\dfrac{1}{x}\right)^{1/x}$, this gives $\log_e \left(\dfrac{1}{x}\right)^{1/x} = 0$; that is, $\left(\dfrac{1}{x}\right)^{1/x} = e^0 = 1$.

EXAMPLE 3. Evaluate $\lim\limits_{x \to 0} (\cot x)^x$.

Solution. $\lim\limits_{x \to 0} (\cot x)^x = \infty^0;$ indeterminate.

Put $\qquad\qquad\qquad\qquad y = (\cot x)^x;$

then $\qquad\qquad\qquad \log y = x \log \cot x.$

$$\lim\limits_{x \to 0} (x \log \cot x) = 0 \cdot \infty.$$

To find $\lim\limits_{x \to 0} (x \log \cot x)$, write the expression as $\dfrac{\log \cot x}{x^{-1}}$.

$$\frac{f(x)}{F(x)} = \frac{\log \cot x}{x^{-1}} \bigg]_{x=0} = \frac{\infty}{\infty}; \quad \text{indeterminate.}$$

$$\frac{f'(0)}{F'(0)} = \frac{\dfrac{-\csc^2 x}{\cot x}}{-\dfrac{1}{x^2}} = \frac{\dfrac{1}{\cos x}}{\dfrac{1}{x^2}} \bigg]_{x=0} = \frac{1}{\infty} = 0.$$

Hence, since $\lim \log_e y = 0$, then $y = e^0 = 1$; that is, $y = (\cot x)^x = 1$. In other words, since $y = (\cot x)^x$, this gives $\log_e (\cot x)^x = 1$; that is, $(\cot x)^x = 1$.

9—10. The Form 0^∞. It is interesting to note that the form 0^∞ is *not* indeterminate; it is always equal to zero. This can be explained as follows. In the function $y = f(x)^{F(x)}$, let $f(x) = 0$ and $F(x) = \infty$. Then,

$$\log y = F(x) \log f(x),$$

or $\qquad\qquad \log y = (\infty) \cdot (-\infty) = -\infty;$

since $\lim \log_e y = -\infty, y = e^{-\infty} = \dfrac{1}{e^\infty} = \dfrac{1}{\infty} = 0.$ Hence $0^\infty = 0.$

<hr>

EXERCISE 9—3

Evaluate each of the following:

1. $\lim\limits_{x \to 0} x^x$

2. $\lim\limits_{\theta \to \pi/2} (\sin \theta)^{\tan \theta}$

3. $\lim\limits_{y \to 0} \left(\dfrac{1}{y}\right)^{\tan y}$

4. $\lim\limits_{z \to 0} (1 + z)^{1/z}$

5. $\lim\limits_{y \to \infty} \left(1 + \dfrac{3}{y}\right)^y$

6. $\lim\limits_{x \to 0} (\log x)^x$

7. $\lim\limits_{z \to 0} (1 + mz)^{n/z}$

8. $\lim\limits_{y \to 1} y^{1/(1-y)}$

9. $\lim\limits_{y \to 0} (y + \cos y)^{1/y}$

10. $\lim\limits_{\theta \to 0} (\sin \theta + 1)^{\cot \theta}$

<hr>

EXERCISE 9—4

Review

1. Find the equation of the tangent to the curve $y^2 + 3x - 2y + 4 = 0$ at the point $(2,2)$.

2. Find the equation of the tangent line to the parabola $y^2 = 8x + 12$, parallel to the line $2x - 2y = 3$.

3. Find the value of:

    (a) $\lim\limits_{x \to 0} (\cot x)^{1/\log x}$.

    (b) $\lim\limits_{x \to 1} (x - 1)^{a/\log \sin \pi x}$.

    (c) $\lim\limits_{x \to 0} (1 - e^x)^x$.

4. Find $\dfrac{dy}{dx}$ and $\dfrac{d^2y}{dx^2}$ for the equation $x^2 + xy + y^2 = 0$.

5. A point moves so that its distance S at any instant t is given by $S = t^5 - 5t^4$. At what time t is the acceleration changing most rapidly?

6. Sand is falling onto a conical pile at the rate of 18 cu. ft. per minute. The radius of the pile is always one-third of its altitude. How fast is the altitude of the pile increasing when the pile is 6 feet high?

7. Find the point of inflection of the curve $y = x^3 + 3x^2 + 12$.

8. What is the curvature at any point of the curve $y = e^{2x}$?

9. Find the radius of curvature of:

(a) $y = \dfrac{e^x + e^{-x}}{2}$, at any point.

(b) $y = x^3$, at the point where $x = 1$.

10. Prove that the sum of any positive real number and its reciprocal is equal to or greater than 2.

Partial Differentiation

CHAPTER TEN

PARTIAL DERIVATIVES

10—1. Functions of Several Variables. Thus far we have been dealing with functions such as:

(1) $f(x) = y = x^3 + 2x^2 - 5x + 6$
(2) $y = x \log x$
(3) $y^2 = 2px$
(4) $x^2 + y^2 = r^2$

In these illustrations, (1) and (2) are *explicit* functions, while (3) and (4) are *implicit* functions. In all four instances, however, only two variables appear, either of which may be expressed as a function of the other; that is, each equation may be written either as $y = f(x)$, or as $x = F(y)$. These are functions involving only one independent variable.

Frequently, however, three or more variables may be related, as in the following examples:

(5) $z^2 = x^2 + y^2$
(6) $E = IR$
(7) $V = \frac{1}{3}\pi r^2 h$
(8) $PV = kT$, or $P = kT/V$
(9) $I = P(1 + i)^n$
(10) $H = .24I^2Rt$

Here (5), (6), (7), and (8) are each functions involving two independent variables; (9) and (10) are functions of three independent variables. When a quantity is a function of more than one variable, it is possible to "hold" one of the variables constant and investigate how the other variables are related to each other, and to the function. For example, in (7), if the altitude h of a cone remains constant, then doubling the value of r makes V four times as great; by holding r constant, and doubling h, the value of V becomes twice as great; doubling both r and h makes V eight times as great; etc. The various relations between the rates of change of variables with respect to one another, and of the function as a whole, are conveniently studied by means of *partial derivatives*.

A function whose value depends upon two variables is written as $F(x,y)$. For example:

(1)   if $z^2 = x^2 + y^2$, then $z = F(x,y) = \pm\sqrt{x^2 + y^2}$;

(2)   if $PV = kT$, then $V = F(P,T) = kTP^{-1}$.

Similarly, if a quantity depends upon three variables, the functional relation is expressed as

$$u = F(x,y,z).$$

10—2. Definition of a Partial Derivative.   Consider z as a function of two independent variables x and y, that is,

$$z = f(x,y).$$

Then the derivative of z with respect to x as x varies, but y *remains constant*, is called the *partial derivative of z with respect to x*; it is denoted by $\dfrac{\partial z}{\partial x}$. In like manner, if we hold x constant and allow y to vary, the partial derivative of z with respect to y is denoted by $\dfrac{\partial z}{\partial y}$. For example:

if $\qquad\qquad\qquad z = ax^2 + 3xy + y^2,$

$\dfrac{\partial z}{\partial x} = 2ax + 3y,$ since y is here regarded as a constant,

$\dfrac{\partial z}{\partial y} = 3x + 2y,$ since x is here regarded as a constant.

This notion of a partial derivative can be applied to *implicit* functions as well as to an explicit function as in the above illustration.

Thus, if $$2z^2 + x^2 - y^3 = 8,$$

then $\quad 4z \dfrac{\partial z}{\partial x} + 2x - 0 = 0,\quad$ where y is constant, $\hfill (1)$

and $\quad 4z \dfrac{\partial z}{\partial y} + 0 - 3y^2 = 0,\quad$ where x is constant. $\hfill (2)$

From (1), $\qquad\qquad \dfrac{\partial z}{\partial x} = -\dfrac{x}{2z};$

from (2), $\qquad\qquad \dfrac{\partial z}{\partial y} = \dfrac{3y^2}{4z}.$

The concept and notation of partial derivatives may also be extended to functions of more than two independent variables.

EXAMPLE 1. Find the partial derivatives of u, given

$$u = xy + xz - yz.$$

Solution.

$$\frac{\partial u}{\partial x} = y + z, \quad \text{where } y \text{ and } z \text{ are constants};$$

$$\frac{\partial u}{\partial y} = x - z, \quad \text{where } x \text{ and } z \text{ are constants};$$

$$\frac{\partial u}{\partial z} = x - y, \quad \text{where } x \text{ and } y \text{ are constants}.$$

EXAMPLE 2. If $x^3 - y^3 + z^3 = 3xyz$, find $\dfrac{\partial z}{\partial x}$ and $\dfrac{\partial z}{\partial y}$.

Solution.

Differentiating "implicitly" with respect to x:

$$3x^2 - 0 + 3z^2 \frac{\partial z}{\partial x} = 3yx \frac{\partial z}{\partial x} + 3yz,$$

or $\qquad\qquad\qquad \dfrac{\partial z}{\partial x} = \dfrac{yz - x^2}{z^2 - xy}.$

Differentiating with respect to y:

$$0 - 3y^2 + 3z^2 \frac{\partial z}{\partial y} = 3xy \frac{\partial z}{\partial y} + 3xz,$$

or
$$\frac{\partial z}{\partial y} = \frac{xz + y^2}{z^2 - xy}.$$

EXERCISE 10—1

In each of the following, find all the partial derivatives of z or u:

1. $z = x^3 + 5x^2y + y^3$

2. $z = e^x \log y$

3. $u = Ax^2 + Bxy + Cy^2 + F$

4. $z = e^{ax+y}$

5. $u = \sin \dfrac{x}{y}$

6. $u = x^y$

7. $\dfrac{x^2}{b^2} + \dfrac{y^2}{a^2} = 2cz$

8. $z = y \log x$

9. $x^2 + y^2 - z^2 = 36$

10. $x^2y^2 + y^2z^2 + x^2z^2 = 2a$

11. If $z = \dfrac{xy}{x + y}$, prove that $x \dfrac{\partial z}{\partial x} + y \dfrac{\partial z}{\partial y} = z$.

12. If $u = \sin (x + y + z)$, prove that $\dfrac{\partial u}{\partial x} = \dfrac{\partial u}{\partial y} = \dfrac{\partial u}{\partial z}$.

10—3. Higher Partial Derivatives. Inasmuch as the first partial derivatives of a function of several independent variables are also functions of those variables, the partial derivatives of these first partial derivatives may also be found; the results are known as *second-order partial derivatives.*

Thus, if $u = f(x,y)$, then, by partial differentiation, we obtain *two* first-order partial derivatives, $\dfrac{\partial u}{\partial x}$ and $\dfrac{\partial u}{\partial y}$, and *four* second-order partial derivatives, as follows:

$$\frac{\partial}{\partial x}\left(\frac{\partial u}{\partial x}\right) = \frac{\partial^2 u}{\partial x^2}; \qquad \frac{\partial}{\partial y}\left(\frac{\partial u}{\partial x}\right) = \frac{\partial^2 u}{\partial y\,\partial x};$$

$$\frac{\partial}{\partial x}\left(\frac{\partial u}{\partial y}\right) = \frac{\partial^2 u}{\partial x\,\partial y}; \qquad \frac{\partial}{\partial y}\left(\frac{\partial u}{\partial y}\right) = \frac{\partial^2 u}{\partial y^2}.$$

It can be shown that if the "cross-derivatives" $\dfrac{\partial^2 u}{\partial x\,\partial y}$ and $\dfrac{\partial^2 u}{\partial y\,\partial x}$ are continuous, they are identical; that is, *the order of differentiation does not affect the result.*

In a similar manner, partial differentiation of the four second-order derivatives yields *eight* third-order partial derivatives.

EXAMPLE. Find the second-order partial derivatives of

$$u = 2x^3y + x^2y^3.$$

Solution.

$$\frac{\partial u}{\partial x} = 6x^2y + 2xy^3; \qquad\qquad \frac{\partial u}{\partial y} = 2x^3 + 3x^2y^2;$$

$$\frac{\partial^2 u}{\partial x^2} = 12xy + 2y^3; \qquad\qquad \frac{\partial^2 u}{\partial y^2} = 6x^2y;$$

$$\frac{\partial^2 u}{\partial y\,\partial x} = 6x^2 + 6xy^2; \qquad\qquad \frac{\partial^2 u}{\partial x\,\partial y} = 6x^2 + 6xy^2.$$

EXERCISE 10—2

Find all the second-order partial derivatives:

1. $u = \frac{1}{3}(x^3 + y^3 + 3xy)$
2. $z = \sin xy$
3. $u = e^x + e^y$
4. $z = e^{xy}$
5. $x^3y^2 + 3xy^3 = z$
6. $z = x^3 + 4x^2y - y^3$

In each of the following, verify that $\dfrac{\partial^2 u}{\partial x\,\partial y} \equiv \dfrac{\partial^2 u}{\partial y\,\partial x}$:

7. $u = \sin (x + y)$
8. $u = y \log xy$
9. $u = \dfrac{x}{y} + \dfrac{y}{x}$
10. $u = x \cos y$

In each of the following, verify the relation given:

11. If $z = \log (x^2 + y^2)$, prove that

$$\frac{\partial^2 z}{\partial x^2} + \frac{\partial^2 z}{\partial y^2} = 0.$$

12. If $u = x^3 + y^3$, prove that

$$\frac{\partial^3 u}{\partial x^3} = \frac{\partial^3 u}{\partial y^3}.$$

THE TOTAL DERIVATIVE

10—4. Total Differential. We have already learned how to find the derivative of a function of a function; thus, if $y = f(u)$, and $u = F(x)$, then

$$\frac{du}{dx} = \frac{du}{dy} \cdot \frac{dy}{dx}.$$

In a similar way, it may be shown that if $z = f(x,y)$, and $x = F_1(t)$ and $y = F_2(t)$, then

$$\frac{dz}{dt} = \frac{\partial z}{\partial x}\frac{dx}{dt} + \frac{\partial z}{\partial y}\frac{dy}{dt}. \qquad [1]$$

Equation [1] gives the *total derivative* of z with respect to t. In the same way, if $u = f(x,y,z)$, and x, y, and z are functions of t, then

$$\frac{du}{dt} = \frac{\partial u}{\partial x}\frac{dx}{dt} + \frac{\partial u}{\partial y}\frac{dy}{dt} + \frac{\partial u}{\partial z}\frac{dz}{dt}. \qquad [2]$$

It should be understood that the meanings of $\dfrac{\partial u}{\partial x}$ and $\dfrac{du}{dx}$ are definitely different. Thus the partial derivative $\dfrac{\partial u}{\partial x}$ supposes that *only the particular variable x varies;* on the other hand, $\dfrac{du}{dx}$ is the limit of $\dfrac{\Delta u}{\Delta x}$, where Δu is the *total increment* in u brought about by *changes in all the variables due to an increment in the independent variable* x.

If we multiply equations [1] and [2], respectively, by dt, we obtain

$$du = \frac{\partial u}{\partial x}dx + \frac{\partial u}{\partial y}dy, \qquad [3]$$

$$du = \frac{\partial u}{\partial x}dx + \frac{\partial u}{\partial y}dy + \frac{\partial u}{\partial z}dz. \qquad [4]$$

The expression du is called the *total differential of u,* or the "complete differential."

The expressions $\dfrac{\partial u}{\partial x}dx$, $\dfrac{\partial u}{\partial y}dy$, etc. are called *partial differentials.*

10—5. Implicit Functions. The expression $f(x,y) = 0$ represents an equation in x and y where all the terms have been transposed to one side of the equation. In other words, y is an implicit function of x (or

x is an implicit function of y). Now, let us set $z = f(x,y)$, and find the total differential of z; from [3], §10—4, we have:

$$dz = \frac{\partial z}{\partial x}\,dx + \frac{\partial z}{\partial y}\,dy = \frac{\partial f}{\partial x}\,dx + \frac{\partial f}{\partial y}\,dy. \tag{1}$$

But by hypothesis, since $z = f(x,y) = 0$, then for all values of x, $z = 0$; hence $dz = 0$. Therefore, if $\frac{\partial f}{\partial y} \neq 0$:

$$\frac{\partial f}{\partial x}\,dx + \frac{\partial f}{\partial y}\,dy = 0; \tag{2}$$

or, dividing (2) through by dx and by $\frac{\partial f}{\partial y}$, we get:

$$\frac{dy}{dx} = -\frac{\dfrac{\partial f}{\partial x}}{\dfrac{\partial f}{\partial y}}, \quad \text{where } \frac{\partial f}{\partial y} \neq 0. \tag{1}$$

EXAMPLE. If $f(x,y) = x^2 + y^3 + xy = 0$, find $\frac{dy}{dx}$ by using partial derivatives.

Solution.

$$\frac{dy}{dx} = -\frac{\dfrac{\partial f}{\partial x}}{\dfrac{\partial f}{\partial y}} = -\frac{2x + y}{3y^2 + x}.$$

This result agrees, of course, with the value of $\frac{dy}{dx}$ found by the previously learned method; thus, differentiating "directly":

$$2x + 3y^2\frac{dy}{dx} + x\frac{dy}{dx} + y = 0,$$

$$\frac{dy}{dx} = \frac{-2x - y}{3y^2 + x}.$$

The relation above is quite general. Thus if

$$u = F(x,y,z) = 0,$$

then, by §10—4, equation [4]:

$$du = \frac{\partial F}{\partial x}\,dx + \frac{\partial F}{\partial y}\,dy + \frac{\partial F}{\partial z}\,dz. \tag{1}$$

Analogous to the derivation in the earlier part of the present section, we remember that z is a function of x and y, so that

$$dz = \frac{\partial z}{\partial x}\,dx + \frac{\partial z}{\partial y}\,dy. \tag{2}$$

Substituting the value of dz from (2) in equation (1), and remembering that $du = 0$, we obtain:

$$\frac{\partial F}{\partial x}\,dx + \frac{\partial F}{\partial y}\,dy + \frac{\partial F}{\partial z}\left(\frac{\partial z}{\partial x}\,dx + \frac{\partial z}{\partial y}\,dy\right) = 0,$$

or, by factoring:

$$\left(\frac{\partial F}{\partial x} + \frac{\partial F}{\partial z}\frac{\partial z}{\partial x}\right)dx + \left(\frac{\partial F}{\partial y} + \frac{\partial F}{\partial z}\frac{\partial z}{\partial y}\right)dy = 0. \tag{3}$$

Since z is a function of x and y, the variables x and y are independent variables, and so we may assign to the increments dx and dy such values as we please.

Let us first set $dy = 0$, $dx \neq 0$; then, from (3), we get:

$$\frac{\partial F}{\partial x} + \frac{\partial F}{\partial z}\frac{\partial z}{\partial x} = 0,$$

or
$$\frac{\partial z}{\partial x} = -\frac{\dfrac{\partial F}{\partial x}}{\dfrac{\partial F}{\partial z}}, \quad \text{where } \frac{\partial F}{\partial z} \neq 0. \tag{2}$$

Next, we set $dx = 0$, $dy \neq 0$; then, from (3) we get:

$$\frac{\partial F}{\partial y} + \frac{\partial F}{\partial z}\frac{\partial z}{\partial y} = 0,$$

or
$$\frac{\partial z}{\partial y} = -\frac{\dfrac{\partial F}{\partial y}}{\dfrac{\partial F}{\partial z}}, \quad \text{where } \frac{\partial F}{\partial z} \neq 0. \tag{3}$$

SIGNIFICANCE OF PARTIAL
AND TOTAL DERIVATIVES

10—6. Partial Differentials and Partial Derivatives. Some notion of the significance of partial differentials will be obtained from the following examples. Consider the changes in the area of a rectangle caused by variation in the lengths of the sides *separately*, and when varying *simultaneously*. Let the rectangle $ABCD$ have a variable base x and a variable altitude y; let its area equal $z = xy$. Now if we consider x constant

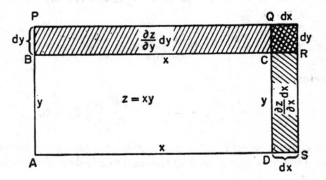

while y increases by an increment $BP = dy$, the corresponding change in z is $PQCB = \dfrac{\partial z}{\partial y} dy$; if we consider y constant while x increases by $DS = dx$, the corresponding change in z is $CRSD = \dfrac{\partial z}{\partial x} dx$; and the total differential of z is given by

$$PQCB + CRSD = \frac{\partial z}{\partial y} dy + \frac{\partial z}{\partial x} dx = dz.$$

Moreover, from the equation $z = xy$, we see that

$$\frac{\partial z}{\partial y} = x, \quad \text{and} \quad \frac{\partial z}{\partial x} = y;$$

hence,
$$\frac{\partial z}{\partial y} dy = x\, dy = \text{area } PQCB,$$

and
$$\frac{\partial z}{\partial x} dx = y\, dx = \text{area } CRSD.$$

It should also be carefully noted that the total increment Δz due to increments dx and dy is given by

$$\Delta z = x\, dy + y\, dx + dx\, dy;$$

the small rectangle RQ, whose area equals $dx\, dy$, represents the difference between Δz and dz. Thus the total increment and the total differential of a function of two or more variables are not, in general, equal.

EXAMPLE. The total area A of a right circular cone is given by $A = \pi rs + \pi r^2$, where s = slant height and r = radius of base. Find the rate of change of A with respect to r when s remains constant; the rate of change of A with respect to s when r is constant.

Solution.

$$A = \pi rs + \pi r^2;$$

$$\frac{\partial A}{\partial r} = \pi s + 2\pi r; \qquad \frac{\partial A}{\partial s} = \pi r.$$

10—7. The Total Derivative. As we have already learned, the total differential of $u = f(x,y)$ is

$$du = \frac{\partial u}{\partial x}\, dx + \frac{\partial u}{\partial y}\, dy; \tag{1}$$

if now x and y, and therefore u, are functions of another variable t, we may divide (1) by dt:

$$\frac{du}{dt} = \frac{\partial u}{\partial x}\frac{dx}{dt} + \frac{\partial u}{\partial y}\frac{dy}{dt}. \tag{2}$$

This, as we have seen, is the total derivative of u with respect to t; here $\dfrac{dx}{dt}$ and $\dfrac{dy}{dt}$ may be regarded as derivatives which represent the rates of change of x and y, respectively, with respect to t. In practical applications, the variable t often represents time.

EXAMPLE 1. The base of a rectangle equals 8 inches, and its altitude is 4 inches. At a certain instant the base is increasing at the rate of 3 in./sec., and the altitude at the rate of 2 in./sec. At what rate is the area changing at the same instant?

Solution. Let x = base, y = altitude, u = area.
Then $u = xy$; $\dfrac{\partial u}{\partial x} = y$; $\dfrac{\partial u}{\partial y} = x$.

By equation (2) above:

$$\frac{du}{dt} = y\frac{dx}{dt} + x\frac{dy}{dt}.$$

But $x = 8$ in., $y = 4$ in., $\dfrac{dx}{dt} = 3$ in./sec., $\dfrac{dy}{dt} = 2$ in./sec.;

hence $\dfrac{du}{dt} = (4)(3) + (8)(2) = 28$ sq. in./sec.

EXAMPLE 2. Assume that the formula for a gas is $pv = kT$, where p = pressure in lb./sq. unit, v = the volume in corresponding cubic units, T = absolute temperature, and k = a constant. If $k = 60$, and at a certain instant $v = 12$ cu. ft. and $T = 200°$, find p at this instant. If at the same instant v is increasing at the rate of 0.5 cu. ft./min., and T is increasing at 0.6 degree/min., at what rate is p changing at that instant?

Solution. Substituting $k = 60$, $v = 12$, $T = 200$ in the formula $pv = kT$, we find $p = 1000$. At the instant in question,

$$\frac{dv}{dt} = 0.5, \quad \text{and} \quad \frac{dT}{dt} = 0.6.$$

Now, since $p = \dfrac{60T}{v}$, we note that

$$\frac{\partial p}{\partial T} = \frac{60}{v}, \quad \frac{\partial p}{\partial v} = -\frac{60T}{v^2};$$

hence $\dfrac{dp}{dt} = \dfrac{\partial p}{\partial T}\dfrac{dT}{dt} + \dfrac{\partial p}{\partial v}\dfrac{dv}{dt}$

$$= \left(\frac{60}{12}\right)(0.6) + \left(-\frac{(60)(200)}{12^2}\right)(0.5) = -38.7.$$

In other words, the pressure is *decreasing* at the rate of 38.7 lb./sq. ft./minute.

SINGULAR POINTS OF A CURVE

10—8. Basic Ideas. In the light of what has been said thus far about the meaning of a continuous curve, we may say, without proof, that if a curve is continuous throughout an interval, then, in general, there will exist a tangent to the curve at every point of the curve in that interval. This is true for nearly all curves. There are continuous curves, however, for which no tangent exists, but such curves are beyond the scope of this book. Thus, if RS is an arc of a curve, and if at the point P on RS there exists one and only one tangent, AB, to the curve, then point P is known as an *ordinary point* of the curve.

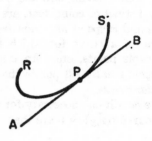

10—9. Singular Points of a Curve. If at a point P on a curve there exist two and only two distinct tangents, then that point is called a *node*. If the two tangents at a given point are not distinct, but coincide, we

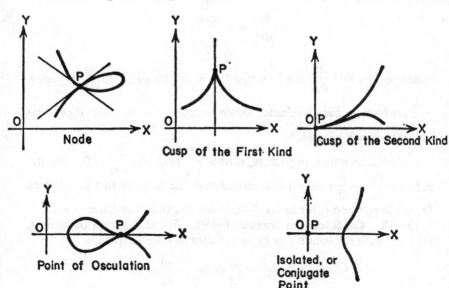

Node

Cusp of the First Kind

Cusp of the Second Kind

Point of Osculation

Isolated, or Conjugate Point

have what is called a *cusp*. There are several kinds of cusps: (1) if the curve in the neighborhood of a cusp lies partly on one side of the tangent and partly on the other side, the point is known as a *cusp of the first kind;* (2) if the curve lies entirely on one side of the common tangent (in the region of tangency), the point is known as a *cusp of the second kind;* (3) if there are two distinct cusps at the same point, it is known as a *point of osculation.*

All points having two and only two tangents, whether real or imaginary, distinct or coincident, are called *double points* of the curve. Thus nodes and cusps of all kinds are double points. Triple points are such points on a curve for which there are three tangents; similarly for quadruple points, etc. An isolated point on a curve is also called a *conjugate point.* All points that are not ordinary points are known as *singular points.*

The conditions necessary for the existence of a singular point (x_1, y_1) on a curve $f(x, y) = 0$ are that

$$\frac{\partial f_1}{\partial x_1} = 0, \quad \text{and} \quad \frac{\partial f_1}{\partial y_1} = 0. \tag{1}$$

The reason for this will become clear when considering equation [1], §10—5, when applied to the point (x_1, y_1):

$$\frac{dy_1}{dx_1} = -\frac{\dfrac{\partial f_1}{\partial x_1}}{\dfrac{\partial f_1}{\partial y_1}}; \tag{1}$$

whenever $\dfrac{\partial f_1}{\partial x_1}$ and $\dfrac{\partial f_1}{\partial y_1}$ are *both* equal to zero, the value of $\dfrac{dy_1}{dx_1}$ becomes $-\dfrac{0}{0}$, which is indeterminate; hence nothing can be said about the tangent at such a point.

It should be noted, in passing, that if $\dfrac{\partial f_1}{\partial x_1} \neq 0$, but $\dfrac{\partial f_1}{\partial y_1} = 0$, then the value of $\dfrac{dy_1}{dx_1} = \dfrac{k}{0} = \pm \infty$; this means that the tangent to the curve at (x_1, y_1) is perpendicular to the X-axis, having the equation $x = x_1$.

10—10. Conditions for Singular Points. Without giving the proofs, we may state the conditions to be satisfied at singular points:

$$f(x, y) = 0; \quad \frac{\partial f}{\partial x} = 0; \quad \frac{\partial f}{\partial y} = 0;$$

and, in addition,

(I) for a *node:*

$$\left(\frac{\partial^2 f}{\partial x\, \partial y}\right)^2 - \frac{\partial^2 f}{\partial x^2}\, \frac{\partial^2 f}{\partial y^2} > 0;$$

(II) for a *cusp:*

$$\left(\frac{\partial^2 f}{\partial x\, \partial y}\right)^2 - \frac{\partial^2 f}{\partial x^2}\, \frac{\partial^2 f}{\partial y^2} = 0;$$

(III) for *isolated points:*

$$\left(\frac{\partial^2 f}{\partial x\, \partial y}\right)^2 - \frac{\partial^2 f}{\partial x^2}\, \frac{\partial^2 f}{\partial y^2} < 0.$$

EXAMPLE. Prove that the curve $y^2 = x^5 + x^4$ has a singular point at the origin.

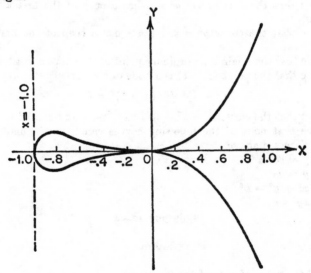

Solution.

$$f(x,y) = y^2 - x^5 - x^4 = 0.$$

At $(0,0)$, $\quad \dfrac{\partial f}{\partial x} = -5x^4 - 4x^3 = 0; \qquad \dfrac{\partial f}{\partial y} = 2y = 0;$

therefore the fundamental conditions for a singular point are satisfied.

Furthermore, at the point $(0,0)$, we have:

$$\frac{\partial^2 f}{\partial x^2} = 0, \qquad \frac{\partial^2 f}{\partial x\, \partial y} = 0, \qquad \frac{\partial^2 f}{\partial y^2} = 2;$$

therefore

$$\left(\frac{\partial^2 f}{\partial x\, \partial y}\right)^2 - \frac{\partial^2 f}{\partial x^2}\frac{\partial^2 f}{\partial y^2} = 0,$$

showing that the point $(0,0)$ is either a cusp or a point of osculation. The curve is symmetrical with respect to the X-axis; moreover, it extends to the left of the Y-axis as far as $x = -1$, as may be seen from the equation by solving for y: $y = \pm\sqrt{x^5 + x^4}$. Hence the origin is a point of osculation, since there is a distinct cusp on either side of the Y-axis.

EXERCISE 10—3

1. Prove that the curve $x^3 - y^2 = 0$ has a cusp of the first kind at the point $(0,0)$.

2. Prove that the curve $y^3 = 3x^2 + 4x^3$ has a cusp of the first kind at the origin.

3. Prove that the origin is an ordinary point of the curve $x = x^3 + y^3 + y$.

4. Prove that the point $(1, -1)$ is a node of the curve

$$(x - 1)^2 + y(y + 1)^2 = 0.$$

5. Prove that the curve $x^3 + y^3 = 3xy$ has a node at the origin.

6. Prove that none of the following curves can have singular points of the kind discussed above:

   (a) $y = \cos x$
   (b) $y = ae^x$
   (c) $x^2 + y^2 = k^2$
   (d) $xy = k$

EXERCISE 10—4

Review

1. Find the value of each of the following:

   (a) $\displaystyle\lim_{x \to \infty} \left(\frac{\log x}{x}\right)^{1/x}$

   (b) $\displaystyle\lim_{x \to \frac{\pi}{2}} \left(\frac{\tan x}{\tan 3x}\right)$

   (c) $\displaystyle\lim_{x \to 0} x^{1/x}$

2. Find the total differential:
  (a) $Z = ax^2 + by^3$
  (b) $Z = ax^2y^3$
  (c) $Z = x^y$

3. Find the rectangle of maximum area inscribed in a semicircle of radius r.

4. If the function $x \cdot f(x)$ is continuous, find the equation whose solution gives values of x which make this function a maximum or a minimum.

5. Examine the following for points of inflection:
  (a) $x^3 + y = ax^2$
  (b) $xy = x^3 + 1$
  (c) $y = (x - a)^3 - b$

6. Find the radius of curvature at any point on the curve:
  (a) $y^2 = 4px$
  (b) $e^x = \sin y$

7. What value of x will make the expression $\dfrac{\sin x}{1 + \tan x}$ a maximum?

Expansion of Functions

CHAPTER ELEVEN

INFINITE SERIES AND SIGMA NOTATION

11—1. Basic Ideas and Notation. The reader is already familiar, from his algebra, with the notion of arithmetic series and geometric series.

(A) $\qquad\qquad 1, -\frac{1}{3}, \frac{1}{9}, -\frac{1}{27}, \cdots .$

(B) $\qquad\qquad x, x - 2a, x - 4a, x - 6a, \cdots .$

Thus, (A) is a geometric series (common ratio $= -\frac{1}{3}$), while (B) is an arithmetic series (common difference $= -2a$).

Any sequence of n terms in the form

$$u_1 + u_2 + u_3 + \cdots + u_{n-1} + u_n,$$

where the nth term is a given function of n, is called a *series*. For example:

(C) $\ 1^3 + 2^3 + 3^3 + \cdots + (n - 1)^3 + n^3;$

(D) $\dfrac{1}{1 \cdot 2} + \dfrac{1}{2 \cdot 3} + \dfrac{1}{3 \cdot 4} + \cdots + \dfrac{1}{(n - 1)(n)} + \dfrac{1}{n(n + 1)};$

(E) $\ 1 + 1 + \dfrac{1}{2!} + \dfrac{1}{3!} + \cdots + \dfrac{1}{(n - 1)!} + \dfrac{1}{n!}.$

In series (C), (D), and (E), the *general term*, or the nth term, is seen to be a function of n. The law of formation in each case, while perhaps not as simply stated as for an arithmetic or a geometric series, is nevertheless just as definite. Although the nth term is the last one expressed, it need not be the "last" term, literally; in fact, the nth term generalizes the law of formation, in accordance with which it is always possible to form a "next" term, no matter how many terms have preceded it. When the number of terms in a series is regarded as endless, we speak of the series as an *infinite series*. In such a series there is no last term, and the number of terms is said to be infinitely great. An infinite series is represented simply by

$$u_1 + u_2 + u_3 + \cdots,$$

where u_1 is the "first" term, and the three dots represent the indefinite continuation of successive terms, all following the same law of formation followed in writing the first three terms.

When only a few (three or four) of the first terms of an infinite series are expressed, it is to be assumed that the general term, or nth term, is that function of n which takes the value of the first term when $n = 1$, the value of the second term when $n = 2$, the value of the third term when $n = 3$, the value of the 10th term when $n = 10$, etc. For example:

in $\quad \dfrac{1}{1 \cdot 2} + \dfrac{1}{2 \cdot 3} + \dfrac{1}{3 \cdot 4} + \cdots, \quad$ the nth term is $\dfrac{1}{n(n+1)}$;

in $\quad \dfrac{1}{2^3} - \dfrac{1}{4^3} + \dfrac{1}{6^3} - \cdots, \quad$ the nth term is $\dfrac{(-1)^{n-1}}{(2n)^3}$;

in $\quad 3 + \dfrac{5}{2!} + \dfrac{7}{3!} + \dfrac{9}{4!} + \cdots, \quad$ the nth term is $\dfrac{2n+1}{n!}$.

11—2. Convergent Series. Consider an infinite series

$$u_1 + u_2 + u_3 + \cdots, \tag{1}$$

and suppose that the sum of the first n items is represented by S_n; thus

$$S_n = u_1 + u_2 + u_3 + u_4 + \cdots + u_n,$$

where n is a finite, or definite number, regardless of how large or small. Now if, as n increases indefinitely, the value of $\lim\limits_{n \to \infty} S_n$ has a definite, finite value, the series (1) is called a *convergent series*. In other words, in such a series, the greater the number of terms taken, the closer in value the sum of these terms becomes to some limiting value, or "sum," namely, $\lim_{n \to \infty} S_n$.

While we speak of this value to which the series converges as a "sum," it is not a sum in the strict sense of the word, since the number of terms taken, or the number of addends, is *indefinite*. In reality, the value of S_n is the limit of *a series of different sums*, each, in turn, successively closer to the numerical value of $\lim_{n \to \infty} S_n$. For example, if we take the series

$$3 + 1 + \frac{1}{3} + \frac{1}{3^2} + \cdots + \frac{1}{3^{n-2}}, \tag{2}$$

we see that it is a geometric series in which $a = 3$ and $r = \frac{1}{3}$; hence

$$S_n = \frac{a - ar^n}{1 - r} = \frac{3 - 3(\frac{1}{3})^n}{1 - \frac{1}{3}}.$$

Passing to the limit:

$$\lim_{n \to \infty} S_n = \frac{3 - 0}{1 - \frac{1}{3}} = \frac{9}{2} = 4\frac{1}{2}.$$

Therefore, since $\lim_{n \to \infty} S_n$ has a definite finite value, the series (2) is convergent.

11—3. Divergent Series. There are some series, however, for which $\lim_{n \to \infty} S_n$ does *not* have a definite finite value; such series are called *divergent series*. The following are examples of divergent series.

EXAMPLE 1. In the arithmetic series

$$1 + 5 + 9 + 13 + 17 + \cdots + (4n - 3),$$

it is clear that as $n \to \infty$, S_n increases without limit, and the series is divergent. In other words, it is always possible to find a value for n such that the value of S_n is greater than any preassigned value.

EXAMPLE 2. Another type of divergent series is

$$1 - 1 + 1 - 1 + \cdots + (-1)^{n-1} + \cdots,$$

which is known as an *oscillating series*, since the values of S_n oscillate between 0 and 1, according as n is even or odd. The $\lim_{n \to \infty} S_n$ in such a series is meaningless.

EXAMPLE 3. Some series which upon superficial inspection would appear to be convergent are actually divergent. Such a series is the *harmonic series*.

(A) $$\frac{1}{2} + \frac{1}{3} + \frac{1}{4} + \cdots + \frac{1}{n} + \cdots.$$

Let us group the terms as follows:

(B) $$\left[\frac{1}{2}\right] + \left[\frac{1}{3} + \frac{1}{4}\right] + \left[\frac{1}{5} + \frac{1}{6} + \frac{1}{7} + \frac{1}{8}\right] +$$
$$+ \left[\frac{1}{9} + \frac{1}{10} + \cdots + \frac{1}{16}\right] + \cdots;$$

now compare the value of each of these groups with the corresponding group of the series

(C) $$\left[\frac{1}{2}\right] + \left[\frac{1}{4} + \frac{1}{4}\right] + \left[\frac{1}{8} + \frac{1}{8} + \frac{1}{8} + \frac{1}{8}\right] +$$
$$+ \left[\frac{1}{16} + \frac{1}{16} + \cdots + \frac{1}{16}\right] + \cdots.$$

It is easily seen that the value of each group in (B) is greater than the value of the corresponding group in (C); but each group in (C) is equal to $\frac{1}{2}$, so that the sum of these groups can be made as large as we please, simply by taking a sufficient number of terms. The same is therefore true of (B); hence (B), and therefore (A), is a divergent series. It should be noted that in the harmonic series, the *successive terms* approach zero as a limit, and it is difficult, at first sight, to believe that the *successive sums* of this series add up to a value greater than any number which we can assign.

11—4. Sigma Notation. We frequently designate an infinite series in the so-called *sigma notation* by writing

$$\sum_{\infty} u_n, \quad \text{read "summation } u_n,\text{"}$$

or $\displaystyle\sum_{n=1}^{\infty} u_n,$ read "summation u_n from $n = 1$ to $n = \infty$."

Thus

$$\sum_{n=1}^{\infty} u_n = u_1 + u_2 + u_3 + \cdots + u_n + \cdots.$$

The subscript n is the same as the ordinal number of the term.

Sometimes a series is written as

$$\sum_{n=0}^{\infty} u_n = u_0 + u_1 + u_2 + \cdots + u_{n-1} + u_n + \cdots ;$$

in this case the nth term is u_{n-1}, the subscript being one less than the number of the term; u_n becomes the $(n + 1)$st term of the series, and

$$S_n = u_0 + u_1 + u_2 + \cdots + u_{n-1}.$$

EXERCISE 11—1

Write the first five terms of each of the following series:

1. $\displaystyle\sum_{n=1}^{\infty} \frac{n}{2^n}$

2. $\displaystyle\sum_{n=1}^{\infty} (-1)^{n+1} n^2$

3. $\displaystyle\sum_{n=0}^{\infty} \frac{1}{3^n}$

4. $\displaystyle\sum_{n=0}^{\infty} (2k)^n$

5. $\displaystyle\sum_{n=1}^{\infty} n(2n - 1)$

6. $\displaystyle\sum_{n=1}^{\infty} (-1)^n \frac{1}{3^n - 1}$

7. $\displaystyle\sum_{n=1}^{\infty} \frac{1}{n!}$

8. $\displaystyle\sum_{n=0}^{\infty} \frac{2n + 1}{2^n}$

9. $\displaystyle\sum_{n=0}^{\infty} (-1)^n \frac{x^n}{2^n}$

10. $\displaystyle\sum_{n=0}^{\infty} (-1)^{n+1} \left(\frac{k}{2}\right)^n$

11. $\displaystyle\sum_{n=1}^{\infty} \frac{1}{(n + 2)(n + 3)}$

12. $\displaystyle\sum_{n=1}^{\infty} \frac{\sqrt{n}}{n!}$

13. $\displaystyle\sum_{n=1}^{\infty} \frac{2^{n+1}}{3^{2n+1}}$

14. $\displaystyle\sum_{n=0}^{\infty} \frac{1}{(2n + 1)^n}$

15. $\displaystyle\sum_{n=1}^{\infty} (-1)^{n+1} \frac{n}{(n + 1)^n}$

Write the general, or nth term, for each of the following:

16. $1 + \dfrac{1}{3} + \dfrac{1}{5} + \dfrac{1}{7} + \cdots .$

17. $\dfrac{1}{1 \cdot 2} - \dfrac{1}{2 \cdot 3} + \dfrac{1}{3 \cdot 4} - \dfrac{1}{4 \cdot 5} + \cdots .$

18. $\dfrac{3}{1 \cdot 2} + \dfrac{5}{2 \cdot 3} + \dfrac{7}{3 \cdot 4} + \dfrac{9}{4 \cdot 5} + \cdots .$

19. $\dfrac{1}{2!} + \dfrac{1}{4!} + \dfrac{1}{6!} + \dfrac{1}{8!} + \cdots$.

20. $1 + \dfrac{3}{5} + \dfrac{4}{10} + \dfrac{5}{17} + \dfrac{6}{26} + \cdots$.

21. $\dfrac{1}{2} - \dfrac{1}{4} + \dfrac{1}{6} - \dfrac{1}{8} + \dfrac{1}{10} - \cdots$.

22. $\dfrac{1}{3} + \dfrac{1}{5^2} + \dfrac{1}{7^3} + \dfrac{1}{9^4} + \cdots$.

23. $\dfrac{3}{2} + \dfrac{5}{4} + \dfrac{9}{8} + \dfrac{17}{16} + \cdots$.

24. $\dfrac{1}{3} + \dfrac{2}{3^2} + \dfrac{3}{3^3} + \dfrac{4}{3^4} + \cdots$.

25. $\dfrac{1}{2} + \dfrac{2}{3^2} + \dfrac{3}{4^3} + \dfrac{4}{5^4} + \cdots$.

TESTS FOR CONVERGENCE AND DIVERGENCE

11—5. Fundamental Conditions for Convergence. It can be shown that a series

$$u_1 + u_2 + u_3 + \cdots + u_n + \cdots$$

cannot converge unless

$$\lim_{n \to \infty} u_n = 0. \tag{1}$$

This is only a *necessary*, but *not a sufficient* condition for convergence. Thus, some series for which $\lim_{n \to \infty} u_n = 0$ are divergent. Hence this principle is useful only in establishing the fact that a given series *is not convergent*. We may say, therefore, without exception, that *in a series* $u_1 + u_2 + u_3 + \cdots + u_n + \cdots$, *if* $\lim_{n \to \infty} u_n$ *is a number different from zero, then the series is divergent.* [2]

EXAMPLES. Each of the following series is divergent, since in each case the $\lim_{n \to \infty} u_n \neq 0$:

(1) $\quad 1 + 3 + 5 + 7 + 9 + \cdots + (2n - 1) + \cdots ; \lim_{n \to \infty} u_n = \infty.$

(2) $\quad 1 \cdot 2 + 2 \cdot 3 + 3 \cdot 4 + \cdots + n(n - 1) + \cdots ; \lim_{n \to \infty} u_n = \infty.$

(3) $\dfrac{1}{2} + \dfrac{2}{3} + \dfrac{3}{4} + \dfrac{4}{5} + \cdots + \dfrac{n}{n+1} + \cdots ; \lim\limits_{n \to \infty} u_n = 1.$

(4) $\dfrac{1}{3} + \dfrac{2}{5} + \dfrac{3}{7} + \dfrac{4}{9} + \cdots + \dfrac{n}{2n+1} + \cdots ; \lim\limits_{n \to \infty} u_n = \dfrac{1}{2}.$

11—6. Testing for Convergence and Divergence. The reader should realize that the problem of determining whether a given series converges or not is a very different problem from finding the exact value to which it converges. In fact, the problem of summation, or finding the value of S_n, is, in general, complex and difficult, and must be left to more advanced treatises. There are, however, a number of tests available for finding out whether a series is convergent or divergent. Some of the most common of these tests will now be discussed.

11—7. Comparison Tests for Convergence and Divergence. This test is based on the following principle, which is here stated without proof.

RULE I. *A series of positive terms converges if each term is less than, or at most, equal to, the corresponding term of another series of positive terms that is known to be convergent.*

RULE II. *A series of positive terms diverges if each term is equal to or greater than the corresponding term of another series of positive terms that is known to be divergent.*

The known series, with which the series to be tested is compared, is called the *comparison series.* For purposes of such comparisons, we often use (1) the *geometric series;* (2) the so-called *p-series;* (3) the *telescopic series;* and (4) the *arithmetic series.*

(1) *The Geometric Series:*

$$a + ar + ar^2 + ar^3 + \cdots + ar^n + \cdots.$$

From algebra, it will be recalled that this series is convergent when $r < 1$, and divergent when $r \geqq 1$.

(2) *The p-Series:*

$$1 + \dfrac{1}{2^p} + \dfrac{1}{3^p} + \dfrac{1}{4^p} + \cdots + \dfrac{1}{n^p} + \cdots.$$

(a) When $p > 1$, the p-series converges, as is shown in standard texts in algebra.

(b) When $p = 1$, the p-series becomes the harmonic series, $1 + \dfrac{1}{2} + \dfrac{1}{3} + \dfrac{1}{4} + \cdots + \dfrac{1}{n} + \cdots$, which has already been shown to be divergent (§11—3, Example 3).

(c) When $p < 1$, the p-series is also divergent, since the terms, after the first, are greater than the corresponding terms of the harmonic series.

(3) *The Telescopic Series:*

$$\frac{1}{k(k+1)} + \frac{1}{(k+1)(k+2)} +$$
$$+ \frac{1}{(k+2)(k+3)} + \cdots + \frac{1}{(k+n-1)(k+n)} + \cdots$$

where k is a constant > 0. It can be shown that in this series,

$$\lim_{n \to \infty} S_n = \frac{1}{k};$$

hence, by §11—2, this series is convergent when $k > 0$.

(4) *The Arithmetic Series:* The series

$$a + (a + d) + (a + 2d) + \cdots + [a + (n-1)d] + \cdots$$

obviously diverges for all values of a and d, since, by §11—5, [2], $\lim\limits_{n \to \infty} [a + (n-1)d] = \infty$.

11—8. Summary of Series Used for Comparisons.

I. *Convergent Series:*

[1] $a + ar + ar^2 + \cdots + ar^n + \cdots$ (when $r < 1$).

[2] $1 + \dfrac{1}{2^p} + \dfrac{1}{3^p} + \dfrac{1}{4^p} + \cdots + \dfrac{1}{n^p} + \cdots$ (when $p > 1$).

[3] $\dfrac{1}{k(k+1)} + \dfrac{1}{(k+1)(k+2)} + \cdots$
$$+ \frac{1}{(k+n-1)(k+n)} + \cdots \quad \text{(where } k > 0\text{).}$$

II. *Divergent Series:*

[4] $a + ar + ar^2 + \cdots + ar^n + \cdots$ (when $r \geqq 1$).

[5] $1 + \dfrac{1}{2} + \dfrac{1}{3} + \cdots + \dfrac{1}{n} + \cdots$.

[6] $a + (a + d) + (a + 2d) + \cdots + [a + (n - 1)d] + \cdots$;

EXAMPLE 1. Show that the series

$$\frac{3}{1 \cdot 2} + \frac{4}{2 \cdot 3} + \frac{5}{3 \cdot 4} + \cdots \quad \text{is divergent.}$$

Solution. We write the given series as

$$\frac{3}{1}\left(\frac{1}{2}\right) + \frac{4}{2}\left(\frac{1}{3}\right) + \frac{5}{3}\left(\frac{1}{4}\right) + \cdots ; \tag{1}$$

Comparing (1) with the harmonic series, §11—8, [5], omitting the first term,

$$\frac{1}{2} + \frac{1}{3} + \frac{1}{4} + \cdots + \frac{1}{n} \tag{2}$$

we see that each term of (1) is greater than the corresponding term of (2). But (2) is known to be divergent; hence (1) is also divergent.

NOTE. The convergence or divergence of a series is not affected by dropping off (or adding on) a finite number of terms; this merely affects the *value* of S_n, not its *existence*.

EXAMPLE 2. Show that the series

$$1 + \frac{1}{3^2} + \frac{1}{5^2} + \frac{1}{7^2} + \cdots \quad \text{is convergent.} \tag{1}$$

Solution. Compare the given series with the p-series (§11—8, [2]) when $p = 2$:

$$1 + \frac{1}{4} + \frac{1}{9} + \frac{1}{16} + \cdots . \tag{2}$$

Each term of (1) is equal to or less than the corresponding term of (2); hence (1) is convergent.

EXAMPLE 3. Test for convergence or divergence:

$$1 + \frac{\sqrt{3}}{3} + \frac{\sqrt{5}}{5} + \frac{\sqrt{7}}{7} + \cdots . \tag{1}$$

Solution. Compare (1) with the harmonic series (§11—8, [5]):

$$1 + \frac{1}{2} + \frac{1}{3} + \frac{1}{4} + \cdots . \tag{2}$$

Each term of (1) is equal to or greater than the corresponding term of (2), since $\frac{\sqrt{3}}{3} > \frac{1}{2}$, $\frac{\sqrt{5}}{5} > \frac{1}{3}$, etc.; hence (2) is divergent.

EXAMPLE 4. Test for convergence:

$$1 + \frac{1}{2!} + \frac{1}{3!} + \cdots . \tag{1}$$

Solution. Compare (1) with the geometric series

$$1 + \frac{1}{2} + \frac{1}{4} + \frac{1}{8} + \cdots \quad (r < 1). \tag{2}$$

Series (1) may be written

$$1 + \frac{1}{2} + \frac{1}{6} + \frac{1}{24} + \cdots . \tag{3}$$

Since each term of (3) is equal to or less than the corresponding term of (2), then, by §11—8, [1], series (1) is convergent.

EXERCISE 11—2

Test the following series for convergence or divergence by using the comparison test:

1. $1 + 2^2 + 3^2 + 4^2 + \cdots .$

2. $\frac{1}{2^2} + \frac{1}{4^2} + \frac{1}{6^2} + \frac{1}{8^2} + \cdots .$

3. $1 \cdot 2 + 2 \cdot 3 + 3 \cdot 4 + 4 \cdot 5 + \cdots .$

4. $1 + \frac{1}{3} + \frac{1}{5} + \frac{1}{7} + \cdots .$

5. $\frac{1}{1 \cdot 2} + \frac{1}{3 \cdot 4} + \frac{1}{4 \cdot 5} + \frac{1}{5 \cdot 6} + \cdots .$

6. $1 + 2! + 3! + 4! + 5! + \cdots$.

7. $1 + \dfrac{\sqrt{2}}{2} + \dfrac{\sqrt{3}}{3} + \dfrac{\sqrt{4}}{4} + \cdots$.

8. $1 + \dfrac{1}{2^2} + \dfrac{1}{3^2} + \dfrac{1}{4^2} + \dfrac{1}{5^2} + \cdots$.

9. $1 + \dfrac{1}{2!} + \dfrac{1}{4!} + \dfrac{1}{6!} + \cdots$.

10. $1 + \dfrac{1}{2^2} + \dfrac{1}{3^3} + \dfrac{1}{4^4} + \dfrac{1}{5^5} + \cdots$.

11—9. Ratio Tests for Convergence and Divergence. In addition to the comparison tests discussed in §11—8, a number of so-called *ratio tests* can be used to determine convergence and divergence. The commonest of these is the ratio of the $(n + 1)$st term to the nth term, or the test ratio u_{n+1}/u_n. It can be shown that, if the absolute value of the limit of this ratio as $n \to \infty$ exists, and is designated by R, then:

(A) If $R < 1$, the series converges;
(B) If $R > 1$, the series diverges;
(C) If $R = 1$, the test fails;

where
$$R = \lim_{n \to \infty} \left| \frac{u_{n+1}}{u_n} \right|.$$

Using the absolute value of the ratio enables us to employ the test ratio for series which have negative as well as positive terms.

EXAMPLE 1. Using the ratio test, prove the convergence of
$$\frac{1}{3} + \frac{2}{3^2} + \frac{3}{3^3} + \cdots.$$

Solution. Here the general term u_n equals $\dfrac{n}{3^n}$; hence
$$u_{n+1} = \frac{n + 1}{3^{n+1}}.$$

Therefore:
$$\frac{u_{n+1}}{u_n} = \frac{n + 1}{3^{n+1}} \div \frac{n}{3^n} = \frac{n + 1}{3^{n+1}} \cdot \frac{3^n}{n} = \frac{n + 1}{3n};$$
$$\lim_{n \to \infty} \left| \frac{u_{n+1}}{u_n} \right| = \lim_{n \to \infty} \frac{n + 1}{3n} = \frac{1}{3}.$$

Hence the series is convergent, since $R < 1$.

EXAMPLE 2. Test for convergence or divergence:

$$\frac{2}{1\cdot 2} + \frac{2^2}{2\cdot 3} + \frac{2^3}{3\cdot 4} + \cdots + \frac{2^n}{n(n+1)} + \cdots.$$

Solution. Here

$$\frac{u_{n+1}}{u_n} = \frac{\dfrac{2^{n+1}}{(n+1)(n+2)}}{\dfrac{2^n}{n(n+1)}} = \frac{2n}{n+2};$$

$$\lim_{n\to\infty}\left|\frac{u_{n+1}}{u_n}\right| = \lim_{n\to\infty}\frac{2n}{n+2} = 2.$$

Hence $R > 1$, and the series is divergent.

EXAMPLE 3. Test for convergence or divergence:

$$2 + \frac{2}{3} + \frac{2}{5} + \frac{2}{7} + \cdots.$$

Solution. Here $u_n = \dfrac{2}{2n-1}$.

Hence
$$\frac{u_{n+1}}{u_n} = \frac{\dfrac{2}{2n+1}}{\dfrac{2}{2n-1}} = \frac{2n-1}{2n+1};$$

$$\lim_{n\to\infty}\frac{u_{n+1}}{u_n} = \lim_{n\to\infty}\frac{2n-1}{2n+1} = 1.$$

Thus the test fails in this case. Some other method must be used. Comparing the given series with the harmonic series

$$1 + \frac{1}{2} + \frac{1}{3} + \frac{1}{4} + \cdots$$

shows that the original series is divergent.

EXERCISE 11—3

By using the ratio test, determine the convergence or divergence of the following series:

1. $\dfrac{1}{2} + \dfrac{2}{2^2} + \dfrac{3}{2^3} + \dfrac{4}{2^4} + \cdots.$

2. $\dfrac{1}{3} + \dfrac{3}{5} + \dfrac{5}{7} + \dfrac{7}{9} + \cdots$.

3. $1 + \dfrac{2}{3^2} + \dfrac{3}{3^3} + \dfrac{4}{3^4} + \cdots$.

4. $1 + \dfrac{2}{2} + \dfrac{3}{2^2} + \dfrac{4}{2^3} + \cdots$.

5. $1 + \dfrac{2^2}{2!} + \dfrac{3^2}{3!} + \dfrac{4^2}{4!} + \cdots$.

6. $\dfrac{3}{1\cdot 2} + \dfrac{4}{2\cdot 3} + \dfrac{5}{3\cdot 4} + \dfrac{6}{4\cdot 5} + \cdots$.

7. $\dfrac{2}{1\cdot 2} + \dfrac{2^2}{2\cdot 3} + \dfrac{2^3}{3\cdot 4} + \cdots$.

8. $\dfrac{1}{2!} + \dfrac{1}{4!} + \dfrac{1}{6!} + \dfrac{1}{8!} + \cdots$.

9. $2 + \dfrac{2^2}{2!} + \dfrac{2^3}{3!} + \dfrac{2^4}{4!} + \cdots$.

10. $1 + \dfrac{2!}{3} + \dfrac{3!}{3^2} + \dfrac{4!}{3^3} + \cdots$.

11—10. Alternating Series. Two basic principles, given here without proof, are useful in discussing *alternating series*, that is, series whose terms are alternating positive and negative.

I. *An alternating series converges if the absolute value of each term is less than that of the preceding term, and if the nth term approaches zero as n becomes infinite.*

EXAMPLE. $1 - \tfrac{1}{2} + \tfrac{1}{3} - \tfrac{1}{4} + \cdots$ is convergent.

II. *A series of real terms converges if the series formed from the absolute values of the terms converges; that is, a series $u_1 + u_2 + u_3 + \cdots$ converges if the series $|u_1| + |u_2| + |u_3| + \cdots$ converges.*

EXAMPLE. The series

$$1 - \frac{1}{2^2} + \frac{1}{3^2} - \frac{1}{4^2} + \cdots \quad \text{is convergent}$$

because the series

$$1 + \frac{1}{2^2} + \frac{1}{3^2} + \frac{1}{4^2} + \cdots \quad \text{converges.}$$

POWER SERIES

11—11. Convergence of Power Series. Any series of the form

$$a_0 + a_1 x + a_2 x^2 + a_3 x^3 + \cdots + a_{n-1} x^{n-1} + a_n x^n + \cdots,$$

where $a_0, a_1, a_2, \cdots$ are constants, is known as a *power series* in x. Such series are important in higher mathematics, for they are often used in finding the values of a given function. In fact, we shall develop special methods for expressing any ordinary function in terms of a series.

Whether a power series converges or diverges depends upon the particular value assigned to x. Thus every power series converges when $x = 0$. A power series may converge for all values of x, or for no values of x other than zero. In general, however, a power series converges for some values of x besides $x = 0$, and diverges for other values of x.

To determine the values for which a power series converges, we use the following ratio test. Thus in a power series, if $\lim\limits_{n \to \infty} \left| \dfrac{a_{n-1}}{a_n} \right| = L$, then the series

    (a)   converges for all values of x such that $|x| < L$;

    (b)   diverges for all values of x such that $|x| > L$;

    (c)   no test if $|x| = L$.

11—12. Interval of Convergence of a Power Series. The values of x for which a power series converges are said to constitute the *interval of convergence* for the series. It can be proved that this interval, when plotted, will always have zero as the center. A series may or may not converge for the value of x at either end point of its interval of convergence. For all other values of x, the series is divergent.

EXAMPLE 1. Find the values of x for which the series is convergent:

$$x + \frac{x^3}{3} + \frac{x^5}{5} + \frac{x^7}{7} + \cdots.$$

Solution. Here $a_0 = 1$, $a_1 = \frac{1}{3}$, $a_2 = \frac{1}{5}$, etc.

$$a_n = \frac{1}{2n + 1}; \qquad a_{n-1} = \frac{1}{2n - 1}.$$

Hence
$$\frac{a_{n-1}}{a_n} = \frac{2n+1}{2n-1} = \frac{2 + \dfrac{1}{n}}{2 - \dfrac{1}{n}},$$

and
$$\lim_{n \to \infty} \left| \frac{a_{n-1}}{a_n} \right| = 1.$$

Hence the series converges for $|x| < 1$, that is, for $-1 < x < 1$; it diverges for $|x| > 1$, that is, for $-1 > x > 1$. To test the end points:

when $x = 1$, we have

$$1 + \frac{1}{3} + \frac{1}{5} + \frac{1}{7} + \cdots, \quad \text{which is divergent;}$$

when $x = -1$, we have

$$-1 - \frac{1}{3} - \frac{1}{5} - \frac{1}{7} - \cdots, \quad \text{which is also divergent.}$$

Hence the end points are not included in the interval of convergence.

EXAMPLE 2. Find the values of x for which the series is convergent:

$$x - \frac{x^2}{2} + \frac{x^3}{3} - \frac{x^4}{4} + \cdots.$$

Solution. Here $a_0 = 1$, $a_1 = -\frac{1}{2}$, $a_2 = \frac{1}{3}$, etc.

$$|a_n| = \frac{1}{n+1}; \qquad |a_{n-1}| = \frac{1}{n}.$$

Hence
$$\left| \frac{a_{n-1}}{a_n} \right| = \frac{n+1}{n}; \qquad \lim_{n \to \infty} \left| \frac{a_{n-1}}{a_n} \right| = 1.$$

Thus the series converges for $|x| < 1$, or for $-1 < x < 1$. To test the end points:

when $x = 1$, we have

$$1 - \frac{1}{2} + \frac{1}{3} - \frac{1}{4} + \cdots, \quad \text{which is convergent;}$$

when $x = -1$, we have

$$-1 - \frac{1}{2} - \frac{1}{3} - \frac{1}{4} - \cdots, \quad \text{which is divergent.}$$

Hence for the original power series, the end point $x = +1$ is included in the interval of convergence, but the other end point, -1, is not included.

EXAMPLE 3. Find the interval of convergence of

$$1 + x + \frac{x^2}{2!} + \frac{x^3}{3!} + \cdots.$$

Solution. Here $a_0 = 1$, $a_1 = 1$, $a_2 = \frac{1}{2!}$, etc.

$$a_n = \frac{1}{n!}, \qquad a_{n-1} = \frac{1}{(n-1)!}.$$

Hence $\dfrac{a_{n-1}}{a_n} = \dfrac{n!}{(n-1)!} = n;$ $\qquad \lim\limits_{n \to \infty} \left| \dfrac{a_{n-1}}{a_n} \right| = \infty.$

Therefore the series converges for $|x| < \infty$, that is, for $-\infty < x < \infty$, or for all positive and negative values of x.

EXERCISE 11—4

Determine the values of x for which the following series are convergent:

1. $1 - x + x^2 - x^3 + \cdots.$
2. $1 + 2x + 3x^2 + 4x^3 + \cdots.$
3. $x - \dfrac{x^3}{3} + \dfrac{x^5}{5} - \dfrac{x^7}{7} + \cdots.$
4. $1 + \dfrac{x}{2} + \dfrac{x^2}{3} + \dfrac{x^3}{4} + \cdots.$
5. $1 + \dfrac{x}{2} + \dfrac{x^2}{4} + \dfrac{x^3}{8} + \dfrac{x^4}{16} + \cdots.$
6. $1 + \dfrac{x^2}{2} + \dfrac{x^3}{3} + \dfrac{x^4}{4} + \cdots.$
7. $x + \dfrac{x^2}{\sqrt{2}} + \dfrac{x^3}{\sqrt{3}} + \cdots.$
8. $x - \dfrac{x^3}{3!} + \dfrac{x^5}{5!} - \dfrac{x^7}{7!} + \cdots.$

EXPANSION OF FUNCTIONS

11—13. Transforming a Function into a Power Series. It will be recalled from algebra that a function of the type $(1 + x)^k$ can be transformed into a series of terms which are powers of x. Thus if k is a positive integer, a finite series is obtained—the familiar binomial expansion of $(k + 1)$ terms:

$$(1 + x)^k = 1 + kx + \frac{k(k - 1)}{2!} x^2 + \frac{k(k - 1)(k - 2)}{3!} x^3 + \cdots . \quad (1)$$

If, on the other hand, k is a negative integer or a fraction, the use of the binomial expansion leads to an infinite series, but not of the same type as (1), since x appears with negative or fractional exponents. By actual division to as many terms as desired, we obtain, for example:

$$(1 + x)^{-k} = \frac{1}{(1 + x)^k}$$

$$= 1 - x + x^2 - x^3 + \cdots + (-1)^{n-1}x^{n-1} + \frac{(-1)^n x^n}{1 + x} . \quad (2)$$

By this method of expansion there will always be a remainder,

$$R = \frac{(-1)^n x^n}{1 + x} ;$$

this remainder will vanish only if $R \to 0$ as $n \to \infty$. The only condition under which $R \to 0$ as $n \to \infty$ is when $|x| < 1$. Hence series (2) can be used to find the approximate value of a function of the form $(1 + x)^{-k}$ only for values of x such that $-1 < x < 1$; in this case, $R \to 0$, and the series can be shown to be convergent for $-1 < x < 1$; the greater the number of terms taken, the closer the approximation will be.

No analogous method is available for the convenient expansion of $(1 + x)^k$ when k is a fraction.

11—14. Maclaurin's Series. In order to arrive at a more general method of expanding a function into a power series, let us assume that, for a given function, $f(x)$, there exists a power series equivalent to the function in question, and that this series has the form

$$f(x) = a_0 + a_1x + a_2x^2 + \cdots + a_{n-1}x^{n-1} + \cdots . \quad (1)$$

To determine the value of the coefficients (the a's) in (1), we may proceed as follows. Let $f(x) = 0$; then $f(0) = a_0$. Let us assume that derivatives of all orders exist for $f(x)$ at $x = 0$; let us also assume that $f(x)$ can be differentiated term by term indefinitely often. To be sure, not all functions meet these requirements, but many functions do, and the present discussion is limited to such functions. Then, by differentiation, we have:

$$f'(x) = a_1 + 2a_2x + 3a_3x^2 + 4a_4x^3 + \cdots; \quad f'(0) = 1 \cdot a_1.$$

$$f''(x) = 2a_2 + 3 \cdot 2a_3x + 4 \cdot 3a_4x^2 + \cdots; \quad f''(0) = 2 \cdot 1a_2.$$

$$f'''(x) = 3 \cdot 2a_3 + 4 \cdot 3 \cdot 2a_4x + \cdots; \quad f'''(0) = 3 \cdot 2 \cdot 1a_3.$$

$$f^{n-1}(x) = (n-1)!\, a_{n-1} + n!\, a_n x + \cdots; \quad f^{n-1}(0) = (n-1)!\, a_{n-1}.$$

Substituting these values for the coefficients a_0, a_1, a_2, $\cdots$ in equation (1), we obtain:

$$f(x) = f(0) + f'(0)x + \frac{f''(0)}{2!} x^2 + \frac{f'''(0)}{3!} x^3 + \cdots$$

$$+ \frac{f^{n-1}(0)}{(n-1)!} x^{n-1} + \cdots . \qquad [1]$$

The expansion of $f(x)$ given in [1] is known as *Maclaurin's series*. When a function has been expanded in this way in a Maclaurin's series, it is necessary (1) to bear in mind the assumptions made above, and (2) to determine the interval of convergence of the series.

EXAMPLE 1. Expand in a Maclaurin's series the function $f(x) = e^x$.

Solution.

$$f(x) = e^x \qquad f(0) = 1$$

$$f'(x) = e^x \qquad f'(0) = 1$$

$$f''(x) = e^x \qquad f''(0) = 1$$

$$\cdots\cdots\cdots \qquad \cdots\cdots\cdots$$

Hence

$$e^x = 1 + x + \frac{x^2}{2!} + \frac{x^3}{3!} + \cdots \frac{x^{n-1}}{(n-1)!} + \cdots .$$

Since

$$\lim_{n \to \infty} \left| \frac{u_{n+1}}{u_n} \right| = \lim_{n \to \infty} \left| \frac{x}{n} \right| = 0,$$

the series is convergent for all values of x.

EXAMPLE 2. Expand $\log (1 + x)$ in a Maclaurin's series.

Solution.

$$f(x) = \log (1 + x) \qquad\qquad f(0) = 0$$

$$f'(x) = \frac{1}{1 + x} \qquad\qquad f'(0) = 1$$

$$f''(x) = -\frac{1}{(1 + x)^2} \qquad\qquad f''(0) = -1$$

$$f'''(x) = +\frac{2}{(1 + x)^3} \qquad\qquad f'''(0) = +2$$

$$f^{iv}(x) = -\frac{2\cdot 3}{(1 + x)^4} \qquad\qquad f^{iv}(0) = -2\cdot 3$$

Hence

$$\log (1 + x) = x - \frac{x^2}{2} + \frac{x^3}{3} - \frac{x^4}{4} + \cdots + \frac{(-1)^{n-1}}{n} x^n + \cdots ;$$

Since

$$\lim_{n \to \infty} \left| \frac{u_{n+1}}{u_n} \right| = \lim_{n \to \infty} \left| \frac{x}{1 + \dfrac{1}{n}} \right| = |x|,$$

therefore the series is convergent for $|x| < 1$, or for $-1 < x < 1$. To test the end points, we note that when $x = +1$, the series is also convergent; but when $x = -1$, the series becomes, in effect, equivalent to the harmonic series, which is divergent.

11—15. The Binomial Expansion. In a similar manner, we can investigate, by using the Maclaurin's series, the expansion of $(1 + x)^m$. Thus,

$$f(x) = (1 + x)^m \qquad\qquad f(0) = 1$$

$$f'(x) = m(1 + x)^{m-1} \qquad\qquad f'(0) = m$$

$$f''(x) = m(m - 1)(1 + x)^{m-2} \qquad\qquad f''(0) = m(m - 1)$$

$$\cdots\cdots\cdots\cdots\cdots\cdots\cdots\cdots\cdots \qquad\qquad \cdots\cdots\cdots\cdots\cdots\cdots$$

Hence:

$$(1 + x)^m = 1 + mx + \frac{m(m - 1)}{2!} x^2 + \frac{m(m - 1)(m - 2)}{3!} x^3 + \cdots$$

$$+ \frac{m(m - 1) \cdots (m - n + 2)}{(n - 1)!} x^{n-1} + \cdots .$$

In other words, the familiar binomial theorem of algebra is valid not only when m is a positive integer, but also, *for certain values of x*, when m is a negative integer or a fraction. Applying the ratio test, we have:

$$\frac{u_{n+1}}{u_n} = \frac{m(m-1)\cdots(m-n+1)x^n}{n!} \cdot \frac{(n-1)!}{m(m-1)\cdots(m-n+2)x^{n-1}},$$

which, when simplified, becomes

$$\frac{(m-n+1)x}{n}, \quad \text{or} \quad \left(\frac{m+1}{n}-1\right)x.$$

Hence, $\lim\limits_{n\to\infty}\left|\dfrac{u_{n+1}}{u_n}\right| = |x|$; therefore the series is convergent for $|x| < 1$, or for $-1 < x < 1$. Consequently the binomial expansion is valid for any value of m for values of x between -1 and $+1$. Whether it is valid for the values ± 1, the end points of the interval of convergence, depends upon the particular value of m; a discussion of this problem is beyond the scope of this book.

11—16. The Exponential Series. Let us consider the expansion of e^x once more. In §11—14, Example 1, we learned that

$$e^x = 1 + x + \frac{x^2}{2!} + \frac{x^3}{3!} + \cdots + \frac{x^{n-1}}{(n-1)!} + \cdots. \tag{1}$$

Putting $x = 1$ in both sides of (1), we have:

$$e = 1 + 1 + \frac{1}{2!} + \frac{1}{3!} + \frac{1}{4!} + \cdots.$$

We may evaluate numerically as follows:

```
1st  term =  1.00000
2nd  term =  1.00000
3rd  term =   .50000
4th  term =   .16667 ···   (by dividing 3rd term by 3)
5th  term =   .04167 ···   (by dividing 4th term by 4)
6th  term =   .00833 ···   (by dividing 5th term by 5)
7th  term =   .00139 ···   (by dividing 6th term by 6)
8th  term =   .00019 ···   (by dividing 7th term by 7)
            ―――――――――
             2.71825 ···
```

Hence, the value of e, which may also be defined as

$$e = \lim_{x \to \infty} \left(1 + \frac{1}{x}\right)^x,$$

is equal to 2.7183, correct to four decimal places. This value is of considerable importance in applied mathematics; to ten decimal places, $e = 2.71828\ 18285$.

EXERCISE 11—5

Verify each of the following expansions; determine for what values of the variable they are convergent:

1. $e^{-x} = 1 - x + \dfrac{x^2}{2!} - \dfrac{x^3}{3!} + \dfrac{x^4}{4!} - \cdots$.

2. $\sin x = x - \dfrac{x^3}{3!} + \dfrac{x^5}{5!} - \dfrac{x^7}{7!} + \cdots$.

3. $\cos x = 1 - \dfrac{x^2}{2!} + \dfrac{x^4}{4!} - \dfrac{x^6}{6!} + \cdots$.

4. $\log (1 - x) = -x - \dfrac{x^2}{2} - \dfrac{x^3}{3} - \dfrac{x^4}{4} - \cdots$.

5. $a^x = 1 + x \log a + \dfrac{x^2 \log^2 a}{2!} + \dfrac{x^3 \log^3 a}{3!} + \cdots$.

6. $e^{\sin x} = 1 + x + \dfrac{x^2}{2} - \dfrac{x^4}{8} + \cdots$.

11—17. Taylor's Series. It is often useful to develop a function as a power series in terms of ascending powers of $(x - a)$, where a is some given constant. This leads to a series of the form

$$f(x) = c_0 + c_1(x - a) + c_2(x - a)^2 + \\ + c_3(x - a)^3 + \cdots + c_{n-1}(x - a)^{n-1} + \cdots.$$

Now, making the same assumptions with respect to this series as were made in §11—14, and following the same method of reasoning to determine the coefficients $c_0, c_1, c_2, \cdots c_n$, we obtain the series

$$f(x) = f(a) + \frac{f'(a)}{1!}(x - a) + \frac{f''(a)}{2!}(x - a)^2 + \cdots \\ + \frac{f^{(n-1)}(a)}{(n-1)!}(x - a)^{n-1} + \cdots. \quad [1]$$

This series is known as *Taylor's series*. It is a more general form of Maclaurin's series; in other words, Maclaurin's series is a special case of Taylor's series, obtained by putting $a = 0$ in [1] above.

It should be pointed out that Maclaurin's series is useful in computing the value of $f(x)$ when x has values *in the neighborhood of zero*, for it is then that the series converges with greatest rapidity; for the same reason, Taylor's series is most useful for computing the value of $f(x)$ when x has values near to a, whatever the constant value of a may be.

11—18. Other Forms of Taylor's Series. A function of the sum of two quantities, for example, $f(x + h)$, may be expanded in powers of one of those quantities. This may be deduced from equation [1] above by taking $a = h$, and substituting $(x + h)$ for x; thus

$$f(x + h) = f(h) + \frac{f'(h)}{1!} x + \frac{f''(h)}{2!} x^2 +$$
$$+ \frac{f'''(h)}{3!} x^3 + \cdots + \frac{f^{n-1}(h)}{(n-1)!} x^{n-1} + \cdots. \quad [2]$$

Finally, if we interchange x and h in equation [2], observing that $f(h + x) \equiv f(x + h)$, we obtain an alternative form of Taylor's series for $f(x + h)$ in powers of the number h; thus

$$f(x + h) = f(x) + \frac{f'(x)}{1!} h + \frac{f''(x)}{2!} h^2 +$$
$$+ \frac{f'''(x)}{3!} h^3 + \cdots + \frac{f^{n-1}(x)}{(n-1)!} h^{n-1} + \cdots. \quad [3]$$

EXAMPLE 1. Expand e^x in terms of powers of $(x - 2)$.

Solution. Here $f(x) = e^x$; $x - a = x - 2$, or $a = 2$.

Hence
$$f(x) = e^x, \quad \text{and} \quad f(a) = f(2) = e^2;$$
$$f'(x) = e^x, \quad \text{and} \quad f'(a) = f'(2) = e^2;$$
$$f''(x) = e^x, \quad \text{and} \quad f''(a) = f''(2) = e^2; \quad \text{etc.}$$

Substituting in [1]:
$$e^x = e^2 + e^2(x - 2) + \frac{e^2(x - 2)^2}{2!} + \frac{e^2(x - 3)^3}{3!} + \cdots,$$

or $e^x = e^2 \left[1 + (x - 2) + \frac{1}{2!}(x - 2)^2 + \frac{1}{3!}(x - 2)^3 + \cdots \right].$

EXAMPLE 2. Expand $\sin x$ in powers of $\left(x - \dfrac{\pi}{4}\right)$.

Solution. Here $f(x) = \sin x$; $x - a = x - \dfrac{\pi}{4}$, or $a = \dfrac{\pi}{4}$.

Hence

$$f(x) = \sin x, \quad \text{and} \quad f(a) = f\left(\frac{\pi}{4}\right) = \frac{\sqrt{2}}{2};$$

$$f'(x) = \cos x, \quad \text{and} \quad f'(a) = f'\left(\frac{\pi}{4}\right) = \frac{\sqrt{2}}{2};$$

$$f''(x) = -\sin x, \quad \text{and} \quad f''(a) = f''\left(\frac{\pi}{4}\right) = -\frac{\sqrt{2}}{2};$$

$$f'''(x) = -\cos x, \quad \text{and} \quad f'''(a) = f'''\left(\frac{\pi}{4}\right) = -\frac{\sqrt{2}}{2}; \quad \text{etc.}$$

Therefore, substituting in [1]:

$$\sin x = \frac{\sqrt{2}}{2} + \frac{\sqrt{2}}{2}\left(x - \frac{\pi}{4}\right) - \frac{\sqrt{2}}{2}\left(\frac{1}{2!}\right)\left(x - \frac{\pi}{4}\right)^2 -$$

$$- \frac{\sqrt{2}}{2}\left(\frac{1}{3!}\right)\left(x - \frac{\pi}{4}\right)^3 + \cdots,$$

or

$$\sin x = \frac{\sqrt{2}}{2}\left[1 + \left(x - \frac{\pi}{4}\right) - \frac{1}{2!}\left(x - \frac{\pi}{4}\right)^2 - \frac{1}{3!}\left(x - \frac{\pi}{4}\right)^3 + \cdots\right].$$

EXAMPLE 3. Expand $\sin (x + h)$ in terms of powers of h.

Solution. Here $f(x + h) = \sin (x + h)$, and $f(x) = \sin x$.
Hence $f'(x) = \cos x$; $f''(x) = -\sin x$; $f'''(x) = -\cos x$; etc.
Substituting in [3]:

$$\sin (x + h) = \sin x + \frac{\cos x}{1!}\,h - \frac{\sin x}{2!}\,h^2 - \frac{\cos x}{3!}\,h^3 + \cdots.$$

EXAMPLE 4. Expand $\log (x + h)$ in terms of powers of h.

Solution. Here $f(x + h) = \log (x + h)$; $f(x) = \log x$.

Hence

$$f'(x) = \frac{1}{x}; \qquad f''(x) = -\frac{1}{x^2};$$

$$f'''(x) = \frac{2}{x^3}; \qquad f^{iv}(x) = -\frac{2\cdot 3}{x^4}; \quad \text{etc.}$$

Substituting in [3]:

$$\log (x + h) = \log x + \frac{h}{x} - \frac{h^2}{2x^2} + \frac{h^3}{3x^3} - \frac{h^4}{4x^4} + \cdots .$$

EXERCISE 11—6

1. Expand e^x in powers of $(x + 3)$.
2. Expand e^{-x} in powers of $(x - 1)$.
3. Expand $\sin x$ in powers of $(x - a)$.
4. Expand $\cos x$ in powers of $(x - a)$.
5. Expand $\log x$ in powers of $(x - a)$.
6. Expand $\log x$ in powers of $(x - 3)$.
7. Expand $\log (x + h)$ in powers of x.
8. Expand $\sin (x + h)$ in powers of x.
9. Expand $\cos (x - h)$ in powers of h.
10. Expand e^{x+h} in powers of h.

11—19. The Remainder in Taylor's Series. We may write Taylor's series with a remainder, as follows:

$$f(x) = f(a) + (x - a)f'(a) + \frac{(x - a)^2}{2!} f''(a) + \cdots$$

$$+ \frac{(x - a)^n}{n!} f^{(n)}(a) + R. \qquad (1)$$

In advanced treatises it is shown that

$$R = \frac{(x - a)^{n+1}}{(n + 1)!} f^{(n+1)}(x_1), \qquad (2)$$

where all that we know about x_1 is that it lies between a and x. If $R \to 0$ as $n \to \infty$, the Taylor's series converges and so represents the function in question. The remainder is a measure of the difference between the value of the function $f(x)$ and the sum of the first $(n + 1)$ terms in equation (1).

11—20. Relation of Taylor's Series to the Theorem of Mean Value. Consider equation (1) of §11—19, with the value of R, as given in equation (2) of the same paragraph, substituted for R in (1). We thus have

$$f(x) = f(a) + (x - a)f'(a) + \frac{(x - a)^2}{2!}f''(a) + \cdots$$

$$+ \frac{(x - a)^{n+1}}{(n + 1)!} f^{(n+1)}(x_1).$$

If in this equation we take $n = 1$, we get:

$$f(x) = f(a) + (x - a)f'(x_1). \tag{1}$$

Now write $x = a + h$ in (1):

$$f(a + h) = f(a) + hf'(x_1), \tag{2}$$

where x_1 lies between a and $(a + h)$.

Both equations (1) and (2) will be recognized as the law of the mean (§9—3), stated in slightly different form.

11—21. Operations with Power Series. In advanced texts it is shown that power series can be combined under certain conditions. We shall state two of these principles without proof.

I. *Two power series may be added for every value of x for which both series converge.*

For example:
$$e^x = 1 + x + \frac{x^2}{2!} + \frac{x^3}{3!} + \cdots,$$

and
$$e^{-x} = 1 - x + \frac{x^2}{2!} - \frac{x^3}{3!} + \cdots;$$

hence
$$e^x + e^{-x} = 2\left[1 + \frac{x^2}{2!} + \frac{x^4}{4!} + \cdots\right].$$

II. *A power series may be differentiated term by term for every value of x within the interval of convergence, but not for end-point values of x.*

For example:
$$\sin x = x - \frac{x^3}{3!} + \frac{x^5}{5!} - \frac{x^7}{7!} + \cdots;$$

differentiating both sides:
$$\cos x = 1 - \frac{x^2}{2!} + \frac{x^4}{4!} - \frac{x^6}{6!} + \cdots.$$

THE VALUE OF π; EULER'S FORMULAS

11—22. Calculating Tables of Functions. Power series such as the expansion of sin x and cos x in §11—21, (II), are frequently used when computing tables of values of functions such as logarithms, trigonometric functions, exponential functions, etc. The two series in question are convergent for all finite values of x, where x is expressed in radian measure.

To compute the numerical value of π, any one of several series may be employed. For example, Gregory's series,

$$\theta = \tan \theta - \frac{1}{3} \tan^3 \theta + \frac{1}{5} \tan^5 \theta - \frac{1}{7} \tan^7 \theta + \cdots,$$

where θ is an angle such that $\tan \theta \geqq 1$, can be so used, since it holds for values of θ between $+\frac{\pi}{4}$ and $-\frac{\pi}{4}$, inclusive. If we set $\tan \theta = x$, where $-1 < x < +1$, we get:

$$\tan^{-1} x = x - \frac{x^3}{3} + \frac{x^5}{5} - \frac{x^7}{7} + \cdots; \qquad [1]$$

setting $x = 1$, we have:

$$\tan^{-1} 1 = \frac{\pi}{4} = 1 - \frac{1}{3} + \frac{1}{5} - \frac{1}{7} + \frac{1}{9} - \cdots. \qquad [2]$$

Unfortunately, since this series converges very slowly, it is not practical, for a great many terms would have to be added to obtain the value of π to any great degree of accuracy.

Another series, known as Euler's series, which converges somewhat more rapidly than series [2], may be derived from series [1] by utilizing the relation

$$\tan^{-1} \frac{1}{2} + \tan^{-1} \frac{1}{3} = \frac{\pi}{4}, \qquad [3]$$

and setting $x = \frac{1}{2}$ and $x = \frac{1}{3}$, in succession, in series [1]:

$$\frac{\pi}{4} = \tan^{-1} \frac{1}{2} + \tan^{-1} \frac{1}{3}$$

$$= \frac{1}{2} - \frac{1}{3} \cdot \frac{1}{2^3} + \frac{1}{5} \cdot \frac{1}{2^5} - \frac{1}{7} \cdot \frac{1}{2^7} + \cdots$$

$$+ \frac{1}{3} - \frac{1}{3} \cdot \frac{1}{3^3} + \frac{1}{5} \cdot \frac{1}{3^5} - \frac{1}{7} \cdot \frac{1}{3^5} + \cdots. \qquad [4]$$

Still other series are available which converge even more rapidly than [4].

11—23. Euler's Analytic Formulas for the Sine and Cosine. Consider the exponential series

$$e^x = 1 + x + \frac{x^2}{2!} + \frac{x^3}{3!} + \frac{x^4}{4!} + \cdots, \qquad [1]$$

and set $x = i\theta$, where $i = \sqrt{-1}$:

$$e^{i\theta} = 1 + i\theta - \frac{\theta^2}{2!} - \frac{i\theta^3}{3!} + \frac{\theta^4}{4!} + \frac{i\theta^5}{5!} - \cdots; \qquad [2]$$

Grouping the real and imaginary terms, we have

$$e^{i\theta} = \left(1 - \frac{\theta^2}{2!} + \frac{\theta^4}{4!} - \cdots\right) + i\left(\theta - \frac{\theta^3}{3!} + \frac{\theta^5}{5!} - \cdots\right) \qquad [3]$$

Comparing [3] with §11—21, (II), we note:

$$e^{i\theta} = \cos\theta + i\sin\theta, \qquad [4]$$

and, writing $-i$ for i,

$$e^{-i\theta} = \cos\theta - i\sin\theta. \qquad [5]$$

Solving [4] and [5] simultaneously:

$$\cos\theta = \frac{e^{i\theta} + e^{-i\theta}}{2}, \qquad [6]$$

$$\sin\theta = \frac{e^{i\theta} - e^{-i\theta}}{2i}. \qquad [7]$$

Equations [6] and [7] are Euler's analytic definitions for the sine and cosine functions; they have many uses in higher mathematics.

EXAMPLE 1.   Using Euler's formulas, prove that $\sin 2\theta = 2\sin\theta\cos\theta$.

Solution.   By multiplication, from [6] and [7],

$$\sin\theta\cos\theta = \left(\frac{e^{i\theta} - e^{-i\theta}}{2i}\right)\left(\frac{e^{i\theta} + e^{i\theta}}{2}\right) = \frac{e^{2i\theta} - e^{-2i\theta}}{4i}; \qquad (1)$$

or $\qquad\qquad 2\sin\theta\cos\theta = \left(\frac{e^{2i\theta} - e^{-2i\theta}}{2i}\right); \qquad (2)$

but the right-hand member of (2) is equal (by substitution in [7]) to $\sin 2\theta$. Hence $2\sin\theta\cos\theta = \sin 2\theta$.

EXAMPLE 2. Prove that $e^{i\pi} + 1 = 0$.

Solution. In the formula $e^{i\theta} = \cos\theta + i\sin\theta$, set $\theta = \pi$; then

$$e^{i\pi} = \cos\pi + i\sin\pi$$
$$e^{i\pi} = -1 + i\cdot 0$$
$$e^{i\pi} = -1.$$

EXERCISE 11—7

1. Prove, by Euler's formulas, that $\sin^2\theta + \cos^2\theta = 1$.

2. Prove, by Euler's formulas, that $\cos 2\theta = \cos^2\theta - \sin^2\theta$.

3. Prove that $e^{2i\pi} = 1$.

4. Find the value of $e^{-i\pi}$; is this consistent with the value of $e^{i\pi}$ found in Illustrative Example 2 above?

5. Prove that $e^{2+i\pi} = -e^2$.

6. Prove that $e^{\pi/2} = \sqrt[i]{i}$.

EXERCISE 11—8

Review

1. Write the first five terms of each of the following infinite series with the given general term:

(a) 2^{n-2}

(b) $\dfrac{n^2}{\sqrt{n+1}}$

(c) $\dfrac{x^n}{na^n}$

(d) $\dfrac{(-1)^{n-1}e^{nx}}{\sin nx}$

2. Determine whether the following series are convergent or divergent:

(a) $1 + \dfrac{2}{2} + \dfrac{3}{2^2} + \dfrac{4}{2^3} + \cdots + \dfrac{n}{2^{n-1}} + \cdots$.

(b) $\dfrac{1}{1\cdot 2} + \dfrac{1}{2\cdot 2^2} + \dfrac{1}{3\cdot 2^3} + \dfrac{1}{4\cdot 2^4} + \cdots + \dfrac{1}{n(2^n)} + \cdots$.

3. Show that each of the following series is convergent, and determine in each case the interval of convergence:

(a) $1 + x + \dfrac{x^2}{2!} + \dfrac{x^3}{3!} + \cdots$.

(b) $1 - \dfrac{x^2}{2!} + \dfrac{x^4}{4!} - \dfrac{x^6}{6!} + \cdots$.

(c) $1 + \dfrac{1}{x} + \dfrac{1}{2x^2} + \dfrac{1}{3x^3} + \cdots$.

4. Find the limiting value of:

(a) $\lim\limits_{x \to 0} \sin x \log \cot x$.

(b) $\lim\limits_{\theta \to 0} (1 + \sin \theta)^{\theta}$.

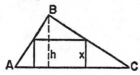

5. Find the altitude x of the rectangle of maximum area that can be inscribed in an acute-angled triangle of altitude h.

6. Test the equation $y^2 = x^2 - x^4$ for a singular point at the origin.

7. Find the total derivative:

(a) $u = 2axy + \log x$; $x = \sin y$.

(b) $u = y^2 + z^4 + zy$; $y = \sin x$, $z = \cos x$.

8. Find the values of x for which the following functions have a maximum or minimum value, and indicate which:

(a) $y = x^3 - 3x^2 - 24x + 15$.

(b) $y = \sqrt[x]{x}$.

(c) $y = (2 - \tan x) \cdot \tan x$.

9. Examine the following curves for multiple points:

(a) $y^2 = x^3 + 2x^2$.

(b) $y^2 = x \log (1 + x)$.

10. Examine the following curves for cusps:

(a) $x^3 = (y - x)^2$.

(b) $(x - y)^2 = (x - 1)^5$.

11. Prove that:

$$e^{x \sin x} = 1 + x^2 + \frac{x^4}{3} + \frac{x^6}{120} + \cdots .$$

12. Prove that:

$$e^x \cdot \log (1 + x) = x + \frac{x^2}{2!} + \frac{2x^3}{3!} + \frac{9x^5}{5!} + \cdots .$$

General Methods of Integration

CHAPTER TWELVE

INTEGRATION AS THE INVERSE
OF DIFFERENTIATION

12—1. The Integral as an Anti-derivative. The reader is already familiar with *inverse* operations and *inverse* functions. Consider the following:

$$y = x^3, \qquad x = \sqrt[3]{y};$$

$$y = x^n, \qquad x = (y)^{1/n};$$

$$y = a^x, \qquad x = \log_a y;$$

$$y = \tan \theta, \qquad \theta = \text{arc } \tan y.$$

Here the right-hand members in either column are the inverses of the right-hand members of the other column, and are obtained in each case by inverse operations. Similarly, the inverse operation of finding a derivative is known as *integration*.

In the Differential Calculus we answered the question: *given a function, what is its derivative?* Now we shall concern ourselves, in the Integral Calculus, with the inverse problem: *given a derivative, what is the function from which it was obtained?* For example, integral calculus is concerned with questions such as:

$3x^2$ is the derivative of what function?

or, in differential notation:

$3x^2 \, dx$ is the differential of what expression?

The answer, of course, can be seen by inspection; the anti-derivative of $3x^2$ is x^3, which may be checked at once by differentiating x^3.

The process of finding the function, given its derivative, is called *integration;* the result is called the *integral.* Hence an integral may be regarded as an *anti-derivative,* or as an anti-differential. The operations of the Differential Calculus may be symbolized as follows:

$$\frac{d}{dx} f(x) = f'(x),$$

or, in differential notation,

$$df(x) = f'(x) \, dx.$$

In the Integral Calculus we shall solve problems such as:

1. *Find a function $f(x)$ whose derivative $f'(x) = \phi(x)$ is given;*
2. *Given the differential of a function, find the function itself; i.e.,* $df(x) = f'(x) \, dx = \phi(x) \, dx.$

The function, $f(x)$, which we are seeking, is the *integral* of the given differential; we use the symbol $\int$ to denote the process of integration. We write, for example,

$$\int 3x^2 \, dx = x^3,$$

which is read: *"the integral of $3x^2 \, dx$ is x^3."* The symbol $\int$ is called the *integral sign;* the expression to be integrated ($3x^2$ in our illustration) is called the *integrand.* To illustrate integration from this point of view, namely, as the process of finding an anti-derivative, consider the following examples.

EXAMPLE 1. If $f(x) = 5x^2$, then $f'(x) \, dx = 10x \, dx$,

and $$\int 10x \, dx = 5x^2.$$

EXAMPLE 2.   If $f(x) = -\cos x$, then $f'(x)\, dx = \sin x\, dx$,

and
$$\int \sin x\, dx = -\cos x.$$

EXAMPLE 3.   If $f(x) = \log x$, then $f'(x)\, dx = \dfrac{1}{x}\, dx$,

and
$$\int \frac{1}{x}\, dx = \log x.$$

EXAMPLE 4.   If $f(x) = \sin 2x$, then $f'(x)\, dx = 2\cos 2x\, dx$,

and
$$\int 2\cos 2x\, dx = \sin 2x.$$

It should therefore be clear that the operation represented by the symbol $\dfrac{d}{dx}$ is the inverse of the operation denoted by the symbol $\int \cdots dx$. If we are using differential notation, then d and $\int$ are symbols of inverse operations.

12—2. The Constant of Integration.   Suppose we were to find the derivative of each of the following expressions:

$$x^4 + 10, \quad \text{and} \quad x^4 - 3.$$

The derivative in each case is $4x^3$. Now, if we were given the expression $4x^3$, and were asked to find its integral, it is obvious that we could not tell which constant to assign to the x^4; in fact, there are an infinite number of functions having the derivative $4x^3$, differing only in the constant term.   Hence we may write

$$\int 4x^3\, dx = x^4 + C,$$

where the symbol C represents an arbitrary constant and is called the *constant of integration.*   Thus the constant C may have any one of an infinite number of values, and the number of functions obtained from a given integration is infinite, unless some additional condition of the problem permits the determination of the value of C.   If we can find a definite value for C, then we refer to the result of the integration as a *particular integral;* if not, we speak of it as a *general integral,* or an *indefinite integral.*   We shall discuss the constant of integration again in a later chapter.   For the present, we may summarize the concept involved by stating, without formal proof, the two following general principles:

I. *If two functions differ by a constant, they have the same derivative.*
II. *If two functions have the same derivative, their difference is a constant.*

FUNDAMENTAL PRINCIPLES OF INTEGRATION

12—3. Basic Relationships. From what has already been said, it will be seen that many integrations can be performed "by inspection." If the reader has thoroughly mastered the fundamental formulas for derivatives, he will, with practice, be able to integrate many expressions at sight. However, we would wish to impress him with the need for having the fundamental formulas for differentiating, especially in differential notation, "at fingers' ends." Ready command of these formulas will go a long way toward success in integrating. Facility in integrating can be achieved only by diligent practice.

One of the basic principles for integrating is, of course, that

$$\int dx = x + C, \tag{1}$$

which follows at once from the definitions of the symbols and the meaning of the constant of integration.

Another principle is the following:

since
$$\frac{d}{dx}(av) = a\frac{dv}{dx},$$

or
$$d(av) = a\,dv;$$

therefore
$$\int a\,dv = a\int dv. \tag{2}$$

Or, expressing the role played by a constant multiplier in the integrand in another way, we have,

$$\int av\,dx = a\int v\,dx.$$

In words, *a constant factor may be removed from one side of the integration sign to the other without affecting the value of the integral.* This is a very useful procedure, but the reader should be warned that this can be done *only with a constant*, and never with a variable.   For example:

$$\int 2x^3 \, dx = 2 \int x^3 \, dx.$$

$$\int \pi a^2 x^2 \, dx = \pi a^2 \int x^2 \, dx.$$

$$\frac{3}{4} \int \sin x \, dx = \int \frac{3}{4} \sin x \, dx.$$

$$\frac{2}{3} \int \frac{3}{2} e^x \, dx = \int e^x \, dx.$$

A third basic principle is the following:

$$\int (du + dv - dw) = \int du + \int dv - \int dw, \qquad [3]$$

or, in different symbolism,

$$\int \{f(x) + \phi(x) - F(x)\} \, dx = \int f(x) \, dx + \int \phi(x) \, dx - \int F(x) \, dx.$$

In words, *the integral of the sum of any number of functions is equal to the sum of the integrals of the several functions.*   This follows, of course, from the fact that

$$\frac{d}{dx} (u + v - w) = \frac{du}{dx} + \frac{dv}{dx} - \frac{dw}{dx}.$$

Example 1.   Find $\int (ax^2 + bx + c) \, dx.$

Solution.

$$\int (ax^2 + bx + c) \, dx = \int ax^2 \, dx + \int bx \, dx + \int c \, dx$$

$$= a \int x^2 \, dx + b \int x \, dx + c \int dx$$

$$= a \left(\frac{x^3}{3}\right) + b \left(\frac{x^2}{2}\right) + cx + C$$

$$= \frac{1}{3} ax^3 + \frac{1}{2} bx^2 + cx + C.$$

EXAMPLE 2. Find $\int \left(2 \sin x - \dfrac{\pi}{4} x \right) dx$.

Solution.

$$\int \left(2 \sin x - \frac{\pi}{4} x \right) dx = \int 2 \sin x \, dx - \int \frac{\pi}{4} x \, dx$$

$$= 2 \int \sin x \, dx - \frac{\pi}{4} \int x \, dx$$

$$= 2(- \cos x) - \frac{\pi}{4} \left(\frac{x^2}{2} \right) + C$$

$$= -2 \cos x - \frac{\pi}{8} x^2 + C.$$

12—4. The Process of Integration. It is important to point out that the result of an integration does not always lead to a simple function. That is to say, performing an inverse operation in mathematics sometimes yields unusual results. Thus in arithmetic, when adding positive numbers only, the inverse operation of subtraction may lead to negative numbers, which differ in kind from the sort of numbers operated upon in arithmetic in the first place. Again, in algebra, when raising real numbers to the second power, the inverse operation of taking the square root may lead to imaginaries, a different kind of number. So in calculus: when we differentiate various so-called *elementary functions* (i.e., algebraic expressions resulting from the operations of addition, subtraction, multiplication, division, involution, evolution; exponential and logarithmic functions; and trigonometric functions and their inverses) the result of the differentiation contains only elementary functions; but the inverse operation of integration may yield a new kind of function, not an elementary function at all. However, in this book we shall limit ourselves to integrals which are elementary functions.

12—5. The Form $\int u^n \, du$. The integral of a power of a variable, when the power is not -1, is given by the following:

$$\int u^n \, du = \frac{u^{n+1}}{n + 1} + C, \quad \text{where } n \neq -1. \qquad [4]$$

This may be readily verified by reversing the procedure:

$$\frac{d}{du} \left(\frac{u^{n+1}}{n + 1} + C \right) = \frac{n + 1}{n + 1} u^n = u^n.$$

Thus, *to integrate a variable to a given power, we increase the exponent by that amount and also divide the expression by the amount of the new exponent.*

EXAMPLE 1. Find $\int x^6\, dx$.

Solution.

$$\int x^6\, dx = \frac{x^{6+1}}{6+1} + C = \frac{x^7}{7} + C.$$

EXAMPLE 2. Find $\int ax^{\frac{1}{3}}\, dx$.

Solution.

$$\int ax^{\frac{1}{3}}\, dx = a\int x^{\frac{1}{3}}\, dx = a\left(\frac{x^{\frac{1}{3}+1}}{\frac{1}{3}+1}\right) = a\,\frac{x^{\frac{4}{3}}}{\frac{4}{3}} = \frac{3ax^{\frac{4}{3}}}{4} + C.$$

EXAMPLE 3. Find $\int \frac{6}{x^3}\, dx$.

Solution.

$$\int \frac{6}{x^3}\, dx = 6\int x^{-3}\, dx = 6\left(\frac{x^{-3+1}}{-3+1}\right) = \frac{6x^{-2}}{-2} = -3x^{-2} = -\frac{3}{x^2} + C.$$

EXAMPLE 4. Find $\int x^{\frac{3}{4}}\, dx$.

Solution.

$$\int x^{\frac{3}{4}}\, dx = \frac{x^{\frac{3}{4}+1}}{\frac{3}{4}+1} = \frac{4}{7}x^{\frac{7}{4}} + C.$$

EXAMPLE 5. Find $\int \frac{dx}{\sqrt[3]{x}}$.

Solution.

$$\int \frac{dx}{\sqrt[3]{x}} = \int x^{-\frac{1}{3}}\, dx = \frac{x^{\frac{2}{3}}}{\frac{2}{3}} = \frac{3}{2}\,\sqrt[3]{x^2} + C.$$

EXAMPLE 6. Find $\int (2a - 3x)^2\, dx$.

Solution. Multiplying out first:

$$\int (2a - 3x)^2 \, dx = \int (4a^2 - 12ax + 9x^2) \, dx$$

$$= 4a^2 \int dx - 12a \int x \, dx + 9 \int x^2 \, dx$$

$$= 4a^2(x) - 12a\left(\frac{x^2}{2}\right) + 9\left(\frac{x^3}{3}\right)$$

$$= 4a^2x - 6ax^2 + 3x^3 + C.$$

EXERCISE 12—1

Integrate each of the following; check your result by differentiating:

1. $\int 8x^3 \, dx$

2. $\int 10ax^4 \, dx$

3. $\int y^4 \, dy$

4. $\int x^{3/4} \, dx$

5. $\int 2\sqrt{x} \, dx$

6. $\int \frac{2}{3} x^{2/3} \, dx$

7. $\int z^{3/2} \, dz$

8. $\int 5y^{2/5} \, dy$

9. $\int x^{m/n} \, dx$

10. $\int \frac{6dx}{x^4}$

11. $\int \sqrt{2px} \, dx$

12. $\int y^{p+q} \, dy$

13. $\frac{6}{5} \int \sqrt[5]{z} \, dz$

14. $\int \frac{10}{3} z^{2/3} \, dz$

15. $\int kz^{1/4} \, dz$

16. $\int 2x^{-2} \, dx$

17. $\int a\sqrt{x^3} \, dx$

18. $\int \frac{4dy}{\sqrt[3]{y^5}}$

19. $\int \left(\frac{x^2}{4} - \frac{2x}{3}\right) dx$

20. $\int \left(2x^3 - 8x^2 + \frac{x}{4}\right) dx$

21. $\int -\frac{2}{3} x^{-5/6} \, dx$

22. $\int \left(\frac{12}{x^2} - \frac{3}{x^4}\right) dx$

23. $\int (x + 3)^2 \, dx$

24. $\int (3a - x^2)^2 \, dx$

25. $\int 5(x + 2)^3 \, dx$

12—6. Variations of the Form $\int u^n\, dv.$ At this point it might be
well to suggest to the reader that there is no general or infallible method
of finding the integral of a given function, since it is impossible always
to retrace the steps of a differentiation. What we must do is this:
depending upon how familiar we are with the derivative formulas, we
try to recognize the given function as the derivative of some known
function, or to change it to a form which can so be recognized. Every
process of integration involves bringing the integrand to a form to
which some standard, known formula applies. Considerable ingenuity
is often required to do this.

EXAMPLE 1. Find $\int \sqrt{a + x}\, dx.$

Solution.

$$\int (\sqrt{a + x})\, dx = \int (a + x)^{\frac{1}{2}}\, dx.$$

Now, we may write $dx = d(a + x).$

Hence, the given integral may be written:

$$\int (a + x)^{\frac{1}{2}}\, dx = \int (a + x)^{\frac{1}{2}}\, d(a + x);$$

here, if we consider $u = a + x,$ and $n = \frac{1}{2},$ then our integral can be
found by means of equation [4]:

$$\int (u^n)\, du = \int (a + x)^{\frac{1}{2}}\, d(a + x)$$

$$= \frac{(a + x)^{\frac{3}{2}}}{\frac{3}{2}} + C = \frac{2}{3}\,(a + x)^{\frac{3}{2}} + C.$$

EXAMPLE 2. Find $\int \dfrac{x\, dx}{\sqrt{1 - x^2}}.$

Solution. Regard $\sqrt{1 - x^2} = (1 - x^2)^{\frac{1}{2}}$ as $u^n,$ where $u = (1 - x^2)$
and $n = \dfrac{1}{2}.$ Since $\dfrac{du}{dx} = \dfrac{1}{2}\,(1 - x^2)^{-\frac{1}{2}}(-2x),$

or, $du = -x(1 - x^2)^{-\frac{1}{2}}\, dx,$ then

$$\int \frac{x\, dx}{\sqrt{1 - x^2}} = -(1 - x^2)^{\frac{1}{2}} = -\sqrt{1 - x^2} + C.$$

EXAMPLE 3. Find $\int (b^2 - a^2x^2)^{1/2}x \, dx.$

Solution. We note that

$$d(b^2 - a^2x^2) = -2a^2x \, dx.$$

Multiplying inside the integral sign by $-2a^2$, and dividing outside by $-2a^2$, leaves the integral unchanged in value, but transforms it to the form $\int u^n \, du$, where $u = (b^2 - a^2x^2)$, and $n = \frac{1}{2}$;

hence,

$$\int u^n \, du = \frac{1}{-2a^2} \int (b^2 - a^2x^2)^{1/2}(-2a^2x \, dx)$$

$$= \frac{1}{-2a^2} \int (b^2 - a^2x^2)^{1/2} \, d(b^2 - a^2x^2)$$

$$= -\frac{1}{2a^2}\left[\frac{(b^2 - a^2x^2)^{3/2}}{\frac{3}{2}}\right] = -\frac{1}{3a^2}(b^2 - a^2x^2)^{3/2} + C.$$

EXAMPLE 4. Find $\int (x^3 + 1)^2 x^2 \, dx.$

Solution. We can first multiply out:

$$\int (x^3 + 1)^2 x^2 \, dx = \int (x^6 + 2x^3 + 1)x^2 \, dx$$

$$= \int (x^8 + 2x^5 + x^2) \, dx$$

$$= \frac{x^9}{9} + \frac{x^6}{3} + \frac{x^3}{3} + C. \tag{1}$$

Or, an alternative, more elegant solution. Since $d(x^3 + 1) = 3x^2 \, dx$, we may replace the "$x^2 \, dx$" in the original problem by "$\frac{1}{3}d(x^3 + 1)$"; thus

$$\int (x^3 + 1)^2 x^2 \, dx = \frac{1}{3} \int (x^3 + 1)^2 \, d(x^3 + 1)$$

$$= \frac{1}{3} \cdot \frac{(x^3 + 1)^3}{3} = \frac{(x^3 + 1)^3}{9} + C'. \tag{2}$$

We leave it to the reader to verify the fact that the right-hand members of (1) and (2), respectively, will both check when differentiated.

Integrate; check by differentiating:

1. $\int \dfrac{dx}{\sqrt{1-x}}$

8. $\int (\sqrt{m^2 - x^2})x\, dx$

2. $\int \dfrac{3dx}{(x+a)^2}$

9. $\int \dfrac{4a\, dx}{(b-x)^5}$

3. $\int \dfrac{1}{(x+a)^3}\, dx$

10. $\int (\sqrt{x^2 + a^2})x\, dx$

4. $\int \dfrac{dx}{2(ax-b)^3}$

11. $\int \dfrac{x\, dx}{\sqrt[3]{1-x^2}}$

5. $\int \sqrt{a+x}\, dx$

12. $\int \sqrt{z-2}\, dz$

6. $\int \sqrt{1-3x}\, dx$

13. $\int \sqrt{z^2 - 2}\, z\, dz$

7. $\int \dfrac{k}{(ax+b)^2}\, dx$

14. $\int (m+x)(a+x)\, dx$

15. Integrate $\int (x^4 - 1)^2 x^3\, dx$ in two ways, explain why the results are not identical, and show that both are correct.

12—7. The Form $\int \dfrac{du}{u}$. The integral of the reciprocal of a variable is the logarithm of the variable; or

$$\int \frac{du}{u} = \int \frac{1}{u}\, du = \log u + C. \qquad [5]$$

This, of course, is readily verified by differentiation; thus,

$$\frac{d}{du}\,(\log u) = \frac{1}{u}.$$

NOTE. Strictly speaking, we should write $\int \dfrac{du}{u} = \log |u| + C$, since, although $\dfrac{1}{u}$ is defined for all values of $u \neq 0$, $\log u$ is defined only for values of $u > 0$.

EXAMPLE 1. Find $\int \dfrac{3dx}{x}$.

Solution.

$$\int \frac{3dx}{x} = 3 \int \frac{dx}{x} = 3 \log x + C.$$

EXAMPLE 2. Find $\int \dfrac{dx}{x+1}$.

Solution. Since, from the definition of a differential, we may write, if we care to, $dx = d(x+1)$.

Hence $\quad \displaystyle\int \frac{dx}{x+1} = \int \frac{d(x+1)}{x+1} = \log (x+1) + C,$

where we are regarding u as equal to $x+1$, and

$$\frac{du}{dx} = 1, \quad \text{or} \quad du = 1(dx) = dx.$$

By differentiating the answer, the result is seen to check; thus

$$\frac{d}{dx} \log (x+1) + C = \frac{1}{x+1}.$$

In fact, *whenever the numerator of an integrand is (or can be made) the differential of the denominator, the integral is the logarithm of the denominator.*

EXAMPLE 3. Find $\int \dfrac{x\,dx}{x^2+1}$.

Solution. $d(x^2 + 1) = 2x.$

Rewrite the given example:

$$\int \frac{x\,dx}{x^2+1} = \frac{1}{2} \int \frac{2x\,dx}{x^2+1};$$

now, considering u as $x^2 + 1$, and $\dfrac{du}{dx}$ as $2x$, or $du = 2x\,dx$, we have:

$$\frac{1}{2} \int \frac{2x\,dx}{x^2+1} = \frac{1}{2} \log (x^2 + 1) + C.$$

EXAMPLE 4. Find $\int \dfrac{1-x}{x^2}\,dx$.

Solution. First rewrite the integral as follows:

$$\int \left(\frac{1-x}{x^2}\right) dx = \int \frac{1}{x^2}\,dx - \int \frac{1}{x}\,dx$$

$$= \int \frac{dx}{x^2} - \int \frac{dx}{x} = -\frac{1}{x} - \log x + C.$$

EXERCISE 12—3

Find the following integrals; check by differentiation:

1. $3\int \dfrac{dx}{x}$	**6.** $\int \dfrac{dz}{2-3z}$	**11.** $\int \dfrac{2y^2-2}{y}\,dy$
2. $\int \dfrac{4}{3}\dfrac{dx}{x}$	**7.** $\int \dfrac{2x\,dx}{x^2+2}$	**12.** $\int \dfrac{z^3\,dz}{z^2-1}$
3. $\int \dfrac{5}{x}\,dx$	**8.** $\int \dfrac{x\,dx}{x^2-4}$	**13.** $\int \dfrac{x^2\,dx}{x^3-1}$
4. $\int \dfrac{2dx}{x}$	**9.** $\int \dfrac{z^2\,dz}{z^3-1}$	**14.** $\int \dfrac{3x^5\,dx}{x^3+2}$
5. $\int \dfrac{dx}{2x+3}$	**10.** $\int \dfrac{z^2-1}{z^3}\,dz$	**15.** $\int \dfrac{3x^2-2}{x^3-2x+1}\,dx$

STANDARD ELEMENTARY INTEGRAL FORMS

12—8. Standard Integrals. Thus far we have discussed five so-called standard elementary integrals, which we restate below for convenience:

$$\int dx = x + C \qquad\qquad\qquad\qquad\qquad [1]$$

$$\int a\,dv = a\int dv \qquad\qquad\qquad\qquad\qquad [2]$$

$$\int (du + dv - dw) = \int du + \int dv - \int dw \qquad\qquad [3]$$

$$\int v^n \, dv = \frac{v^{n+1}}{n+1} + C, \text{ where } n \neq -1 \qquad [4]$$

$$\int \frac{dv}{v} = \log v + C \qquad [5]$$

We now give a list of the remaining standard integral forms for reference:

$$\int a^v \, dv = \frac{a^v}{\log a} + C \qquad [6]$$

$$\int e^v \, dv = e^v + C \qquad [7]$$

$$\int \sin v \, dv = -\cos v + C \qquad [8]$$

$$\int \cos v \, dv = \sin v + C \qquad [9]$$

$$\int \sec^2 v \, dv = \tan v + C \qquad [10]$$

$$\int \csc^2 v \, dv = -\cot v + C \qquad [11]$$

$$\int \sec v \tan v \, dv = \sec v + C \qquad [12]$$

$$\int \csc v \cot v \, dv = -\csc v + C \qquad [13]$$

$$\int \tan v \, dv = \log \sec v + C \qquad [14]$$

$$\int \cot v \, dv = \log \sin v + C \qquad [15]$$

$$\int \sec v \, dv = \log (\sec v + \tan v) + C \qquad [16]$$

$$\int \csc v \, dv = \log (\csc v - \cot v) + C \qquad [17]$$

$$\int \frac{dv}{v^2 + a^2} = \frac{1}{a} \arctan \frac{v}{a} + C \qquad [18]$$

$$\int \frac{dv}{v^2 - a^2} = \frac{1}{2a} \log \frac{v - a}{v + a} + C \qquad [19]$$

$$\int \frac{dv}{\sqrt{v^2 \pm a^2}} = \log (v + \sqrt{v^2 \pm a^2}) + C \qquad [20]$$

$$\int \frac{dv}{\sqrt{a^2 - v^2}} = \text{arc sin} \frac{v}{a} + C \qquad [21]$$

$$\int \frac{dv}{v\sqrt{v^2 - a^2}} = \frac{1}{a} \text{arc sec} \frac{v}{a} + C \qquad [22]$$

$$\int \frac{dv}{\sqrt{2av - v^2}} = \text{arc vers} \frac{v}{a} + C, \text{ where vers} \frac{v}{a} = 1 - \cos \frac{v}{a}. \qquad [23]$$

12—9. The Forms $\int a^v \, dv$ and $\int e^v \, dv$. These formulas are seen to be true from the following considerations. We know that

$$d(a^v) = \log a \cdot a^v \cdot dv;$$

integrating both sides, we get:

$$a^v = \int \log a \cdot a^v \cdot dv,$$

or

$$a^v = \log a \int a^v \, dv;$$

hence

$$\int a^v \, dv = \frac{a^v}{\log a} + C.$$

In a similar manner, since

$$d(e^v) = e^v \, dv,$$

it follows that

$$\int e^v \, dv = e^v + C.$$

The reader can readily verify these results by differentiation.

EXAMPLE 1. Find $\int e^{2x} \, dx$.

Solution. If we let $v = 2x$, then $dv = 2dx$. Now, in order to make the expression $\int e^{2x} \, dx$ conform to the standard form $\int e^v \, dv$, since

$dv = 2dx$ when $v = 2x$, we insert the factor 2 before the dx, and the factor $\frac{1}{2}$ before the integral sign; thus

$$\int e^{2x}\, dx = \frac{1}{2} \int e^{2x}\, 2dx;$$

but $\quad \int e^v\, dv = e^v + C, \quad$ hence $\quad \frac{1}{2} \int e^{2x}\, 2dx = \frac{1}{2} e^{2x} + C.$

EXAMPLE 2. Find $\int 2a^{3x}\, dx.$

Solution. Let $v = 3x$; then $dv = 3dx$. Insert 3 before the dx, and $\frac{1}{3}$ before the integral sign; then

$$\int 2a^{3x}\, dx = \frac{1}{3} \int 2a^{3x}\, 3dx = \frac{2}{3} \int a^{3x}\, 3dx = \frac{2}{3} \frac{a^{3x}}{\log a} + C.$$

EXAMPLE 3. Find $\int e^{\cos x}(\sin x)\, dx.$

Solution. Let $v = \cos x$; $dv = - \sin x\, dx$; then

$$\int e^{\cos x}(\sin x\, dx) = - \int e^{\cos x}(- \sin x\, dx) = -e^{\cos x} + C.$$

EXERCISE 12—4

Find the following integrals; check by differentiation:

1. $\int \dfrac{4}{5} e^{4x}\, dx$

2. $\int ka^{mx}\, dx$

3. $\int e^{\sin x}(\cos x)\, dx$

4. $\int e^{2 \sin x}(\cos x)\, dx$

5. $\int e^{-3x}\, dx$

6. $\int a^{x/n}\, dx$

7. $\int a^{2x-1}\, dx$

8. $\int e^{2x^3} \cdot x^2\, dx$

9. $\int e^x(e^x + 1)\, dx$

10. $\int \dfrac{8dx}{a^{2x}}$

11. $\int xk^{x^2}\, dx$

12. $\int (e^{x/2} + e^{-x/2})\, dx$

12—10. Standard Forms [8] through [13]. These formulas follow immediately from the corresponding formulas for differentiation, as given in §5—9, §5—10, §5—11, §5—12, equations [6]–[11]. They may be verified simply and directly by differentiation.

EXAMPLE 1. Find $\int \cos 3ax\, dx$.

Solution. Let $v = 3ax$; then $dv = 3a\, dx$.

Therefore
$$\int \cos 3ax\, dx = \frac{1}{3a} \int \cos 3ax \cdot 3a\, dx$$

$$= \frac{1}{3a} \int \cos 3ax \cdot d(3ax)$$

$$= \frac{1}{3a}\, (\sin 3ax) + C.$$

EXAMPLE 2. Find $\int \csc \frac{\theta}{m}\, d\theta$.

Solution. Let $v = \frac{\theta}{m}$; then $dv = \frac{d\theta}{m}$.

Hence,
$$\int \csc \frac{\theta}{m}\, d\theta = m \int \csc \frac{\theta}{m} \cdot \frac{1}{m}\, d\theta$$

$$= m \log \left(\csc \frac{\theta}{m} - \cot \frac{\theta}{m} \right) + C.$$

EXAMPLE 3. Find $\int \csc^2 x^2 \cdot x\, dx$.

Solution. Let $v = x^2$; then $dv = 2x\, dx$.

Hence,
$$\int \csc^2 x^2 \cdot x\, dx = \frac{1}{2} \int \csc^2 x^2 \cdot 2x\, dx$$

$$= \frac{1}{2}\, (- \cot x^2) + C.$$

EXAMPLE 4. Find $\int \dfrac{d\theta}{\sin^2 \theta}$.

Solution. Since $\sin^2 \theta = \dfrac{1}{\csc^2 \theta}$,

then $\displaystyle\int \frac{d\theta}{\sin^2 \theta} = \int \csc^2 \theta \, d\theta = -\cot \theta + C.$

EXERCISE 12—5

Find the following integrals; check by differentiation:

1. $\displaystyle\int \cos 4x \, dx$

2. $\displaystyle\int \sin \frac{3x}{5} \, dx$

3. $\displaystyle\int \tan \frac{\theta}{2} \, d\theta$

4. $\displaystyle\int \sec^2 a\theta \, d\theta$

5. $\displaystyle\int \cot \frac{\theta}{3} \, d\theta$

6. $\displaystyle\int \tan a\theta \, d\theta$

7. $\displaystyle\int \cot \frac{x^2}{m} x \, dx$

8. $\displaystyle\int \csc^2 m\theta \, d\theta$

9. $\displaystyle\int \cot e^\theta \cdot e^\theta \, d\theta$

10. $\displaystyle\int \cos (\log x) \frac{dx}{x}$

11. $\displaystyle\int \sec^2 x^4 \cdot x^3 \, dx$

12. $\displaystyle\int \sec \frac{\theta}{m} \, d\theta$

13. $\displaystyle\int (\tan x + \cot x)^2 \, dx$

14. $\displaystyle\int (1 + \tan^2 x) \, dx$

Verify the following:

15. $\displaystyle\int \frac{d\theta}{\cos^2 \theta} = \tan \theta + C$

16. $\displaystyle\int \cot \frac{\theta}{m} \, d\theta = m \log \sin \frac{\theta}{m} + C$

17. $\displaystyle\int \sec 3\theta \, d\theta = \frac{1}{3} \log (\sec 3\theta + \tan 3\theta) + C$

18. $\displaystyle\int \csc^2 \theta^3 \cdot \theta^2 \, d\theta = -\frac{1}{3} \cot \theta^3 + C$

19. $\displaystyle\int \tan 3\theta \, d\theta = \frac{1}{3} \log \sec 3\theta + C$

20. $\displaystyle\int (\sin \theta + \cos \theta)^2 \, d\theta = \theta + \sin^2 \theta + C$

12—11. Standard Forms [14]–[17]. These four formulas may be proved by transforming the integrand in each case so that we may apply equation [5]; the proofs follow.

Form [14]:

$$\int \tan v \, dv = \int \frac{\sin v \, dv}{\cos v} = -\int \frac{-\sin v \, dv}{\cos v}$$

$$= -\int \frac{d(\cos v)}{\cos v} = -\log \cos v + C.$$

But $- \log \cos v = - \log\left(\frac{1}{\sec v}\right) = -\log 1 + \log \sec v = \log \sec v$;

hence, $\qquad \int \tan v \, dv = \log \sec v.$

Form [15]:   In a similar manner,

$$\int \cot v \, dv = \int \frac{\cos v \, dv}{\sin v} = \int \frac{d(\sin v)}{\sin v} = \log \sin v + C.$$

Form [16]:   To transform the integrand so that it will be in the form $\int \frac{dv}{v}$, we write:

$$\sec v = \sec v \, \frac{\sec v + \tan v}{\sec v + \tan v} = \frac{\sec v \tan v + \sec^2 v}{\sec v + \tan v}$$

Hence $\displaystyle \int \sec v \, dv = \int \frac{\sec v \tan v + \sec^2 v}{\sec v + \tan v} \, dv$

$$= \int \frac{d(\sec v + \tan v)}{\sec v + \tan v} = \log (\sec v + \tan v) + C.$$

Form [17]:   This may be derived in a manner similar to the proof for [16]. We leave it as an exercise for the reader; multiply the integrand $\csc v$ by the fraction $\dfrac{\csc v - \cot v}{\csc v - \cot v}$.

12—12. Forms [18]–[23]. The formulas for the standard forms [18]–[23] may be derived by suitable transformations or substitutions. Formulas [22] and [23] follow at once from the corresponding formulas for differentiation.

EXAMPLE 1.   Find $\int \dfrac{dx}{9x^2 + 4}$.

Solution.   Let $v^2 = 9x^2$, and $a^2 = 4$; then $v = 3x$, $dv = 3dx$, and $a = 2$.
From formula [18], we get:

$$\int \frac{dx}{9x^2 + 4} = \frac{1}{3} \int \frac{d(3x)}{(3x)^2 + (2)^2} = \frac{1}{3} \cdot \frac{1}{2} \text{ arc tan } \frac{3x}{2} = \frac{1}{6} \text{ arc tan } \frac{3x}{2} + C.$$

EXAMPLE 2.   Find $\int \dfrac{dx}{4x^2 - 25}$.

Solution.   Let $v^2 = 4x^2$, and $a^2 = 25$; then $v = 2x$, $dv = 2dx$, and
$a = 5$.
From formula [19], we get:

$$\int \frac{dx}{4x^2 - 25} = \frac{1}{2} \int \frac{2dx}{(2x)^2 - (5)^2} = \frac{1}{2} \int \frac{d(2x)}{(2x)^2 - (5)^2}$$

$$= \frac{1}{2} \cdot \frac{1}{(2)(5)} \log \frac{2x - 5}{2x + 5}$$

$$= \frac{1}{20} \log \frac{2x - 5}{2x + 5} + C.$$

EXAMPLE 3.   Find $\int \dfrac{dx}{\sqrt{4x^2 + 9}}$.

Solution.   Let $v^2 = 4x^2$, and $a^2 = 9$; then $v = 2x$, $dv = d(2x)$, and
$a = 3$.
From formula [20]:

$$\int \frac{dx}{\sqrt{4x^2 + 9}} = \frac{1}{2} \int \frac{d(2x)}{\sqrt{(2x)^2 + (3)^2}}$$

$$= \frac{1}{2} \log \left(2x + \sqrt{4x^2 + 9}\right) + C.$$

EXAMPLE 4.   Find $\int \dfrac{dx}{5x^2 - 3}$.

Solution.   Let $v^2 = 5x^2$, and $a^2 = 3$; then $v = \sqrt{5}\,x$, $dv = \sqrt{5}\,dx$,
and $a = \sqrt{3}$.

From formula [19]:

$$\int \frac{dx}{5x^2 - 3} \; \frac{1}{\sqrt{5}} \int \frac{\sqrt{5}\,dx}{(\sqrt{5}\,x)^2 - (\sqrt{3})^2} = \frac{1}{\sqrt{5}} \cdot \frac{1}{2\sqrt{3}} \log \frac{\sqrt{5}\,x - \sqrt{3}}{\sqrt{5}\,x + \sqrt{3}} + C.$$

EXAMPLE 5. Find $\displaystyle\int \frac{dx}{x\sqrt{4x^2 - 25}}$.

Solution. Rewrite as follows:

$$\int \frac{dx}{x\sqrt{4x^2 - 25}} = \int \frac{2dx}{2x\sqrt{(2x)^2 - (5)^2}}.$$

Here $v = 2x$, $a = 5$, $dv = 2dx$.
From formula [22]:

$$\int \frac{dx}{x\sqrt{4x^2 - 25}} = \frac{1}{5} \text{ arc sec } \frac{2x}{5} + C.$$

EXAMPLE 6. Find $\displaystyle\int \frac{x\,dx}{\sqrt{x^4 + 36}}$.

Solution. Let $v^2 = x^4$, and $a^2 = 36$; then $v = x^2$, $dv = 2x\,dx$, and
$a = 6$.
From formula [20]:

$$\int \frac{x\,dx}{\sqrt{x^4 + 36}} = \frac{1}{2} \int \frac{2x\,dx}{\sqrt{(x^2)^2 + (6)^2}}$$
$$= \frac{1}{2} \log\,(x^2 + \sqrt{x^4 + 36}) + C.$$

EXAMPLE 7. Find $\displaystyle\int \frac{dx}{x^2 + 4x + 20}$.

Solution. Rewrite as follows, completing the square in the denominator:

$$\int \frac{dx}{(x^2 + 4x + 4) + 16} = \int \frac{dx}{(x + 2)^2 + (4)^2}.$$

Now apply formula [18], where $v = x + 2$, $a = 4$.

$$\int \frac{dx}{(x + 2)^2 + (4)^2} = \frac{1}{4} \text{ arc tan } \frac{x + 2}{4} + C.$$

EXERCISE 12—6

Find the following integrals; check by differentiation:

1. $\displaystyle\int \frac{dx}{25x^2 - 16}$

2. $\displaystyle\int \frac{3dx}{x^2 + 4b^2}$

3. $\displaystyle\int \frac{dx}{\sqrt{x^2 + a^2b^2}}$

4. $\displaystyle\int \frac{dx}{\sqrt{m^2x^2 - 25}}$

5. $\displaystyle\int \frac{dx}{4x^2 + 9}$

6. $\displaystyle\int \frac{dx}{\sqrt{16 - 9x^2}}$

7. $\displaystyle\int \frac{dx}{\sqrt{9x^2 + 16}}$

8. $\displaystyle\int \frac{dx}{a^2x^2 - b^2}$

9. $\displaystyle\int \frac{dx}{x\sqrt{9x^2 - 5}}$

10. $\displaystyle\int \frac{dx}{3x^2 + 7}$

11. $\displaystyle\int \frac{x \, dx}{m^4 + x^4}$

12. $\displaystyle\int \frac{dx}{x\sqrt{3x^2 - 4}}$

Verify the following:

13. $\displaystyle\int \frac{x \, dx}{5x^4 + 3} = \frac{1}{2\sqrt{15}} \arctan \frac{\sqrt{5} \, x^2}{\sqrt{3}} + C$

14. $\displaystyle\int \frac{x \, dx}{\sqrt{x^4 - m^4}} = \frac{1}{2} \log (x^2 + \sqrt{x^4 - m^4}) + C$

15. $\displaystyle\int \frac{dx}{3x\sqrt{2x^2 - 7}} = \frac{1}{3\sqrt{7}} \operatorname{arc\,sec} \frac{\sqrt{2} \, x}{\sqrt{7}} + C$

16. $\displaystyle\int \frac{dx}{x^2 + 4x - 8} = \frac{1}{4\sqrt{3}} \log \frac{(x + 2) - 2\sqrt{3}}{(x + 2) + 2\sqrt{3}} + C$

17. $\displaystyle\int \frac{dx}{\sqrt{2 - 3x^2}} = \frac{\sqrt{3}}{3} \arcsin \frac{\sqrt{6} \, x}{2} + C$

18. $\displaystyle\int \frac{dx}{x\sqrt{3x^2 - 4}} = \frac{1}{2} \operatorname{arc\,sec} \frac{\sqrt{3} \, x}{2} + C$

19. $\displaystyle\int \frac{dx}{x\sqrt{a^2 - x^2}} = -\frac{1}{a} \log \left(\frac{a + \sqrt{a^2 - x^2}}{x} \right) + C$

20. $\displaystyle\int \frac{dx}{\sqrt{ax^2 + bx + c}} = \left(\frac{1}{\sqrt{a}} \right) \log (2ax + b + 2\sqrt{a^2x^2 + abx + ac}) + C$

EXERCISE 12—7

Review

Find the following integrals:

1. $\displaystyle\int \sqrt{t}\, dt$

2. $\displaystyle\int a^{3x}\, dx$

3. $\displaystyle\int \sqrt{2x + 3}\, dx$

4. $\displaystyle\int e^{-x^2}x\, dx$

5. $\displaystyle\int (\log x)^4\, \frac{dx}{x}$

6. $\displaystyle\int \frac{dx}{x + 5}$

7. $\displaystyle\int \frac{z\, dz}{z^2 - 1}$

8. $\displaystyle\int \frac{x^3 - 8}{x}\, dx$

9. $\displaystyle\int \frac{x^2\, dx}{3x + 2}$

10. $\displaystyle\int (mx + b)^2\, dx$

11. $\displaystyle\int \frac{dx}{9x^2 + 16}$

12. $\displaystyle\int \frac{dx}{9x^2 - 16}$

13. $\displaystyle\int e^{\cos x}\cdot\sin x\, dx$

14. The slope of a certain curve is given by $\dfrac{dy}{dx} = 6x^2 - 10x + 8$. If the curve passes through the point (2,10), find its equation.

15. If $\dfrac{d^2y}{dx^2} = 12x + 6$, find y in terms of x if it is known that when $x = 2$, $\dfrac{dy}{dx} = 28$, and when $x = -3$, $y = -1$.

Special Methods of Integration

CHAPTER THIRTEEN

INTEGRATION BY PARTS

13—1. Need for Special Methods. The standard forms discussed in the preceding chapter enable us to integrate many expressions that arise. However, other types of expressions may be encountered for which these standard forms will not suffice, and so certain special methods must be employed. One of the most common of these is the method of *integration by parts*.

13—2. Integration by Parts. Consider u and v, which represent functions of x. We have:

$$\frac{d}{dx}(uv) = v\frac{du}{dx} + u\frac{dv}{dx},$$

which may also be written in the differential form

$$d(uv) = v\left(\frac{du}{dx}\right)dx + u\left(\frac{dv}{dx}\right)dx,$$

or
$$d(uv) = v\,du + u\,dv, \tag{1}$$

where
$$du = \frac{du}{dx}dx, \quad \text{and} \quad dv = \frac{dv}{dx}dx.$$

Now, by transposing, equation (1) may be written

$$u \, dv = d(uv) - v \, du. \tag{2}$$

By integrating both sides of (2), we obtain

$$\int u \, dv = uv - \int v \, du. \tag{[1]}$$

It will be seen that we may make use of equation [1] whenever it is possible to find the integral of $v \, du$. The method of "integrating by parts" thus amounts to this: if we cannot integrate $f(x) \, dx$ directly, we try to break the expression $f(x) \, dx$ into *two* factors or "parts," say u and dv, such that the integrals of both dv and $v \, du$ are readily found. The method of procedure is suggested by the following:

$$\int f(x) \, dx = \int u \, dv; \qquad \int u \, dv = uv - \int v \, du. \tag{[2]}$$

Frankly, no general rule can be given to indicate the best way of selecting the factors u and dv in all cases; each example must be studied individually. However, with practice the reader will doubtless develop facility in using this method, which is one of the most useful of all special methods of integration. It may help to remember that we try to select u, and the corresponding dv, in such a way that not only can we find v from dv, but also that $\int v \, du$ is easier to evaluate than the original integral, $\int f(x) \, dx$. In some cases it may be necessary to repeat the process one or more times. The following examples will illustrate the application of the method of integration by parts.

EXAMPLE 1. Find $\int x \sin x \, dx$.

Solution. Let $u = x$, and $dv = \sin x \, dx$; then $du = dx$, and $v = \int \sin x \, dx = -\cos x$.

Substituting in equation [2] above:

$$\int x \sin x \, dx = -x \cos x - \int -\cos x \, dx$$
$$= -x \cos x + \sin x + C.$$

EXAMPLE 2. Find $\int x \log x \, dx$.

Solution. Let $u = \log x$, and $dv = x \, dx$; then $du = \dfrac{dx}{x}$, and $v = \int x \, dx = \dfrac{x^2}{2}$.

Substituting in [2]:

$$\int x \log x \, dx = \frac{1}{2} x^2 \log x - \int \frac{x \, dx}{2}$$

$$= \frac{1}{2} x^2 \log x - \frac{1}{2} \int x \, dx$$

$$= \frac{1}{2} x^2 \log x - \frac{x^2}{4} + C.$$

EXAMPLE 3. Find $\int x e^x \, dx$.

Solution. Let $u = x$, and $dv = e^x \, dx$; then $du = dx$, and $v = e^x$.

Hence: $\int x e^x \, dx = x e^x - \int e^x \, dx$

$$= x e^x - e^x + C = e^x (x - 1) + C.$$

EXAMPLE 4. Find $\int \arcsin x \, dx$.

Solution. Let $u = \arcsin x$, and $dv = dx$; then $du = \dfrac{dx}{\sqrt{1 - x^2}}$, and $v = x$.

Hence, $\int \arcsin x \, dx = x \arcsin x - \int \dfrac{x \, dx}{\sqrt{1 - x^2}}$,

But, by §12—6, Example 2, we see that

$$\int \frac{x \, dx}{\sqrt{1 - x^2}} = -\sqrt{1 - x^2};$$

hence $\int \arcsin x \, dx = x \arcsin x + \sqrt{1 - x^2} + C.$

EXAMPLE 5.   Find $\int x^2 \sin x \, dx$.

Solution. Let $u = x^2$, and $dv = \sin x \, dx$; then $du = 2x \, dx$, and $v = \int \sin x \, dx = -\cos x$.

Hence: $\int x^2 \sin x \, dx = -x^2 \cos x + 2 \int x \cos x \, dx$.   (1)

But, to find the integral $\int x \cos x \, dx$ in equation (1), we must apply the method of integration by parts again; thus

let $\qquad\qquad u = x$, and $dv = \cos x \, dx$;

$\qquad\qquad\qquad du = dx$, and $v = \sin x$.

Hence, $\qquad \int x \cos x \, dx = x \sin x - \int \sin x \, dx$

$\qquad\qquad\qquad\qquad = x \sin x + \cos x + C$.   (2)

Therefore, substituting (2) in (1):

$$\int x^2 \sin x \, dx = -x^2 \cos x + 2(x \sin x + \cos x) + C.$$
$$= -x^2 \cos x + 2x \sin x + 2 \cos x + C'.$$

EXAMPLE 6.   Find $\int x^2 \cos 2x \, dx$.

Solution. Let $u = x^2$, and $dv = \cos 2x \, dx$; then $du = 2x \, dx$, and $v = \int \cos 2x \, dx = \frac{1}{2} \sin 2x$.

Hence: $\int x^2 \cos 2x \, dx = \frac{1}{2} x^2 \sin 2x - \int x \sin 2x \, dx$.   (1)

But, to find the integral $\int x \sin 2x \, dx$, we use the method once more:

let $\qquad\qquad u = x$, and $dv = \sin 2x \, dx$;

$\qquad\qquad\qquad du = dx$, and $v = \dfrac{-\cos 2x}{2}$.

Hence, $\displaystyle\int x \sin 2x \, dx = \frac{-x \cos 2x}{2} + \frac{1}{2}\int \cos 2x \, dx$

$$= \frac{-x \cos 2x}{2} + \frac{\sin 2x}{4}. \tag{2}$$

Therefore, substituting (2) in (1):

$$\int x^2 \cos 2x \, dx = \frac{1}{2}x^2 \sin 2x - \frac{1}{2}x \cos 2x + \frac{1}{4}\sin 2x + C.$$

EXERCISE 13—1

Find the following integrals, using the method of integration by parts:

1. $\displaystyle\int x \cos x \, dx$

2. $\displaystyle\int \log x \, dx$

3. $\displaystyle\int xe^{2x} \, dx$

4. $\displaystyle\int x^2 e^x \, dx$

5. $\displaystyle\int e^x \cos x \, dx$

6. $\displaystyle\int \log^2 x \, dx$

7. $\displaystyle\int x^2 \cos x \, dx$

8. $\displaystyle\int x^3 \log x \, dx$

9. $\displaystyle\int \theta \sec^2 \theta \, d\theta$

10. $\displaystyle\int \arctan x \, dx$

11. $\displaystyle\int x^2 e^{-x} \, dx$

12. $\displaystyle\int x^2 \log x \, dx$

13. $\displaystyle\int \theta \tan^2 \theta \, d\theta$

14. $\displaystyle\int z \sec^2 z \, dz$

15. $\displaystyle\int e^x \cos 2x \, dx$

16. $\displaystyle\int \cos x \log \sin x \, dx$

17. $\displaystyle\int x^2 \sin 2x \, dx$

18. $\displaystyle\int x \cos 2x \, dx$

TRIGONOMETRIC INTEGRALS

13—3. Trigonometric Reduction. Many differentials containing trigonometric functions can be reduced to standard forms for integration by first making appropriate trigonometric transformations.

EXAMPLE 1. Find $\int \sin^2 x \, dx$.

Solution. By trigonometry, $\sin^2 x = \dfrac{1}{2}(1 - \cos 2x)$.

Hence,
$$\int \sin^2 x \, dx = \frac{1}{2} \int (1 - \cos 2x) \, dx$$

$$= \frac{1}{2} \int dx - \frac{1}{2} \int \cos 2x \, dx$$

$$= \frac{1}{2} x - \frac{1}{4} \int \cos 2x \, 2dx;$$

therefore,
$$\int \sin^2 x \, dx = \frac{1}{2} x - \frac{1}{4} \sin 2x.$$

NOTE: This may also be written, by trigonometry, as

$$\int \sin^2 x \, dx = \frac{1}{2} x - \frac{1}{2} \sin x \cos x,$$

since $\sin 2x = 2 \sin x \cos x$.

EXAMPLE 2. Find $\int \sin^3 \theta \, d\theta$.

Solution. By trigonometry,

$$\int \sin^3 \theta \, d\theta = \int \sin^2 \theta \cdot \sin \theta \, d\theta = \int (1 - \cos^2 \theta) \sin \theta \, d\theta$$

$$= \int \sin \theta \, d\theta - \int \cos^2 \theta \sin \theta \, d\theta$$

$$= -\cos \theta + \frac{\cos^3 \theta}{3} + C.$$

EXAMPLE 3. Find $\int \sin^5 \theta \cos^2 \theta \, d\theta$.

Solution.

$$\int \sin^5 \theta \cos^2 \theta \, d\theta = \int \sin \theta \cdot \sin^4 \theta \cos^2 \theta \, d\theta$$

$$= \int \cos^2 \theta (1 - \cos^2 \theta)^2 \sin \theta \, d\theta$$

$$= \int (\cos^2 \theta - 2 \cos^4 \theta + \cos^6 \theta) \sin \theta \, d\theta$$

$$= \int \cos^2 \theta \sin \theta \, d\theta - \int 2 \cos^4 \theta \sin \theta \, d\theta$$

$$+ \int \cos^6 \theta \sin \theta \, d\theta$$

$$= \frac{-\cos^3 \theta}{3} + \frac{2 \cos^5 \theta}{5} - \frac{\cos^7 \theta}{7} + C.$$

EXAMPLE 4. Find $\int \cot^4 x \, dx$.

Solution.

$$\int \cot^4 x \, dx = \int \cot^2 x (\csc^2 x - 1) \, dx$$

$$= \int \cot^2 x \csc^2 x \, dx - \int \cot^2 x \, dx$$

$$= -\int (\cot x)^2 \, d(\cot x) - \int (\csc^2 x - 1) \, dx$$

$$= -\frac{\cot^3 x}{3} + \cot x + x + C.$$

EXAMPLE 5. Find $\int \sin^2 \theta \cos^4 \theta \, d\theta$.

Solution. See §13—4, II, below, for trigonometric substitutions:

$$\int \sin^2 \theta \cos^4 \theta \, d\theta = \int (\sin \theta \cos \theta)^2 \cos^2 \theta \, d\theta$$

$$= \int \left(\frac{1}{2} \sin 2\theta\right)^2 \cdot \left(\frac{1}{2} + \frac{1}{2} \cos 2\theta\right) d\theta$$

$$= \int \frac{1}{4} \sin^2 2\theta \left(\frac{1}{2} + \frac{1}{2} \cos 2\theta\right) d\theta$$

$$= \int \left(\frac{1}{8} \sin^2 2\theta + \frac{1}{8} \sin^2 2\theta \cos 2\theta\right) d\theta$$

$$= \frac{1}{8} \int \left(\frac{1}{2} - \frac{1}{2} \cos 4\theta\right) d\theta + \frac{1}{8} \int \sin^2 2\theta \cos 2\theta \, d\theta$$

$$= \frac{\theta}{16} - \frac{\sin 4\theta}{64} + \frac{\sin^3 2\theta}{48} + C.$$

13—4. Summary of Trigonometric Reductions.

I. *To find* $\int \sin^m x \cos^n x \, dx$, *when either m or n is a positive, odd whole number*, we transform the expression to be integrated by means of the relation $\sin^2 \theta = 1 - \cos^2 \theta$, or $\cos^2 \theta = 1 - \sin^2 \theta$, in such a way that we may apply the standard form

$$\int v^n \, dv = \frac{v^{n+1}}{n+1}.$$

II. *To find* $\int \sin^m x \cos^n x \, dx$, *when both m and n are positive even whole numbers*, we make use of the trigonometric relations:

$$\sin \theta \cos \theta = \tfrac{1}{2} \sin 2\theta;$$

$$\sin^2 \theta = \tfrac{1}{2} - \tfrac{1}{2} \cos 2\theta;$$

$$\cos^2 \theta = \tfrac{1}{2} + \tfrac{1}{2} \cos 2\theta.$$

III. *To find* $\int \sin mx \cos nx \, dx$, $\int \sin mx \sin nx \, dx$, *or* $\int \cos mx \cos nx \, dx$, we use standard reduction formulas; the proofs of these formulas are not given here, but they are based upon the addition formulas of trigonometry, that is, $\sin X + \sin Y = 2 \sin \tfrac{1}{2}(X + Y) \cos \tfrac{1}{2}(X - Y)$.

$$\int \sin mx \cos nx \, dx = -\frac{\cos (m+n)x}{2(m+n)} - \frac{\cos (m-n)x}{2(m-n)} + C;$$

$$\int \sin mx \sin nx \, dx = -\frac{\sin (m+n)x}{2(m+n)} + \frac{\sin (m-n)x}{2(m-n)} + C;$$

$$\int \cos mx \cos nx \, dx = \frac{\sin (m+n)x}{2(m+n)} + \frac{\sin (m-n)x}{2(m-n)} + C.$$

EXERCISE 13—2

Find the following:

1. $\int \cos^2 \theta \, d\theta$

2. $\int \cos^3 \theta \, d\theta$

3. $\int \sin^5 \theta \, d\theta$

4. $\int \cos^2 \theta \sin^2 \theta \, d\theta$

5. $\int \sin^4 \theta \, d\theta$

6. $\int \cot^2 \theta \, d\theta$

7. $\int \sin^3 \theta \cos \theta \, d\theta$

8. $\int \sin^2 \theta \cos \theta \, d\theta$

9. $\int \cos^3 \theta \sin \theta \, d\theta$

10. $\int \tan^3 \theta \, d\theta$

11. $\int \sin^3 \theta \cos^2 \theta \, d\theta$

12. $\int (\tan 2\theta)^2 \, d\theta$

INTEGRATION BY SUBSTITUTION; CHANGE OF VARIABLE

13—5. Algebraic Substitution. Frequently an expression to be integrated can be transformed, by the suitable substitution of a new variable, into one of the fundamental standard forms. Some of the simpler kinds of such substitutions will now be illustrated.

EXAMPLE 1. Find $\int x\sqrt{3x-2} \, dx$.

Solution.

Let $\sqrt{3x-2} = z;$

then $\quad 3x - 2 = z^2,\quad$ and $\quad x = \dfrac{z^2 + 2}{3};\quad dx = \dfrac{2}{3} z\, dz.$

Hence, $\displaystyle \int x\sqrt{3x - 2}\, dx = \int \dfrac{z^2 + 2}{3} \cdot z \cdot \dfrac{2}{3} z\, dz$

$$= \dfrac{2}{9} \int (z^4 + 2z^2)\, dz$$

$$= \dfrac{2}{9} \int z^4\, dz + \dfrac{4}{9} \int z^2\, dz$$

$$= \dfrac{2}{45} z^5 + \dfrac{4}{27} z^3 = \dfrac{2}{9} z^3 \left(\dfrac{z^2}{5} + \dfrac{2}{3} \right).$$

Substituting $(3x - 2)^{\frac{1}{2}}$ for z:

$$\int x\sqrt{3x - 2}\, dx = \dfrac{2}{9} (3x - 2)^{\frac{3}{2}} \cdot \left(\dfrac{3x - 2}{5} + \dfrac{2}{3} \right) + C.$$

EXAMPLE 2. Find $\displaystyle \int \dfrac{x\, dx}{(3x + 1)^{\frac{2}{3}}}$.

Solution.

Let $\qquad\qquad\qquad z = (3x + 1)^{\frac{1}{3}};$

then $\quad x = \dfrac{z^3 - 1}{3},\, dx = z^2\, dz,\quad$ and $\quad (3x + 1)^{\frac{2}{3}} = z^2.$

Hence, $\displaystyle \int \dfrac{x\, dx}{(3x + 1)^{\frac{2}{3}}} = \dfrac{1}{3} \int \dfrac{(z^3 - 1)z^2\, dz}{z^2} = \dfrac{1}{3} \int (z^3 - 1)\, dz$

$$= \dfrac{1}{3} \int z^3\, dz - \dfrac{1}{3} \int dz$$

$$= \dfrac{1}{12} z^4 - \dfrac{1}{3} z = \dfrac{1}{12} z(z^3 - 4).$$

Substituting $(3x + 1)^{\frac{1}{3}} = z$:

$$\int \dfrac{x\, dx}{(3x + 1)^{\frac{2}{3}}} = \dfrac{1}{12} \sqrt[3]{3x + 1}\, (3x + 1 - 4)$$

$$= \dfrac{1}{4} \sqrt[3]{3x + 1}\, (x - 1) + C.$$

EXAMPLE 3. Find $\int \dfrac{dx}{\sqrt[3]{x} + \sqrt{x}}$.

Solution. Let $x^{1/6} = z$; then $x = z^6$, $dx = 6z^5\, dz$, $x^{1/2} = z^3$, and $x^{1/3} = z^2$.

Hence, $\displaystyle\int \frac{dx}{x^{1/3} + x^{1/2}} = 6\int \frac{z^5\, dz}{z^2 + z^3} = 6\int \frac{z^3\, dz}{1 + z}$

$$= 6\int \left(z^2 - z + 1 - \frac{1}{z + 1} \right) dz$$

$$= 2z^3 - 3z^2 + 6z - 6\log(z + 1).$$

Substituting for z, z^2, and z^3:

$$\int \frac{dx}{x^{1/3} + x^{1/2}} = 2\sqrt{x} - 3\sqrt[3]{x} + 6\sqrt[6]{x} - \log(\sqrt[6]{x} + 1) + C.$$

EXAMPLE 4. Find $\int \dfrac{dx}{x - 3 + \sqrt{x - 3}}$.

Solution. Let $z = \sqrt{x - 3}$; then $x = z^2 + 3$, and $dx = 2z\, dz$.

Hence, $\displaystyle\int \frac{dx}{x - 3 + \sqrt{x - 3}} = 2\int \frac{z\, dz}{z^2 + z} = 2\int \frac{dz}{z + 1}$

$$= 2\log(z + 1).$$

Substituting $\sqrt{x - 3} = z$:

$$\int \frac{dx}{x - 3 + \sqrt{x - 3}} = 2\log(\sqrt{x - 3} + 1) + C.$$

EXERCISE 13—3

Find the following:

1. $\int \dfrac{x\, dx}{\sqrt{3 + 5x}}$; let $z = \sqrt{3 + 5x}$.

2. $\int \dfrac{dx}{x\sqrt{x + 1}}$; let $z = \sqrt{x + 1}$.

3. $\int \dfrac{dx}{x^{1/4} + x^{3/4}}$; let $z^4 = x$.

4. $\int \dfrac{x\, dx}{\sqrt[3]{x - 1}}$; let $z^3 = x - 1$.

5. $\int \dfrac{dx}{\sqrt{x} + 1}$

6. $\int \dfrac{x\, dx}{\sqrt{x - 1}}$

7. $\int t\sqrt{t + 1}\, dt$

8. $\int \dfrac{dx}{x^2\sqrt{x^2 + 1}}$

Verify the following:

9. $\int x^2 \sqrt{x-1}\, dx = 2\left[\dfrac{(x-1)^{5/2}}{5} + \dfrac{(x-1)^2}{2} + \dfrac{(x-1)^{3/2}}{3}\right] + C.$

10. $\int \dfrac{t^2\, dt}{\sqrt{(1-t^2)^3}} = \dfrac{t}{\sqrt{1-t^2}} - \text{arc sin } t + C.$

11. $\int x\sqrt{2x+5}\, dx = \dfrac{1}{10}(2x+5)^{5/2} - \dfrac{5}{6}(2x+5)^{3/2} + C.$

12. $\int \dfrac{d\theta}{\theta\sqrt{1-\theta^2}} = -\log \dfrac{1+\sqrt{1-\theta^2}}{\theta} + C.$

13—6. Trigonometric Substitutions. When the integrand contains expressions such as $\sqrt{a^2 - v^2}$ or $\sqrt{v^2 \pm a^2}$, the integration may be performed by using the following trigonometric substitutions:

I.   **For $\sqrt{a^2 - v^2}$, we put $v = a \sin \theta$;** the expression then becomes:
$$\sqrt{a^2 - a^2 \sin^2 \theta} = \sqrt{a^2(1 - \sin^2 \theta)} = a \cos \theta.$$

II. **For $\sqrt{v^2 + a^2}$, we put $v = a \tan \theta$;** the expression then becomes:
$$\sqrt{a^2 \tan^2 \theta + a^2} = \sqrt{a^2(\tan^2 \theta + 1)} = a \sec \theta.$$

III. **For $\sqrt{v^2 - a^2}$, we put $v = a \sec \theta$;** the expression then becomes:
$$\sqrt{a^2 \sec^2 \theta - a^2} = \sqrt{a^2(\sec^2 \theta - 1)} = a \tan \theta.$$

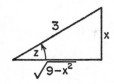

EXAMPLE 1.   Find $\int \sqrt{9 - x^2}\, dx.$

Solution. Let $a^2 = 9$, $a = 3$; $x = 3 \sin z$,
$dx = 3 \cos z\, dz$; $\sqrt{9 - x^2} = 3 \cos z$. Therefore,

$$\int \sqrt{9 - x^2}\, dx = \int (3 \cos z)(3 \cos z\, dz)|$$

$$= \int \frac{9}{2}(2 \cos^2 z)\, dz$$

$$= \frac{9}{2}\int (1 + \cos 2z)\, dz$$

$$= \frac{9}{2}\int dz + \frac{9}{2}\int \cos 2z\, dz$$

$$= \frac{9}{2}\left(z + \frac{1}{2}\sin 2z\right) = \frac{9}{2}(z + \sin z \cos z).$$

But, $x = 3 \sin z$; from the triangle of reference we have $z = \arc \sin \dfrac{x}{3}$, and $\cos z = \dfrac{1}{3} \sqrt{9 - x^2}$; hence

$$\int \sqrt{9 - x^2}\, dx = \frac{9}{2} \left(\arc \sin \frac{x}{3} + \frac{x}{9} \sqrt{9 - x^2} \right)$$

$$= \frac{9}{2} \arc \sin \frac{x}{3} + \frac{x}{2} \sqrt{9 - x^2} + C.$$

EXAMPLE 2. Find $\displaystyle\int \frac{dx}{\sqrt{x^2 + 4}}$.

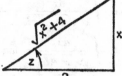

Solution. Let $x = 2 \tan z$; $dx = 2 \sec^2 z \, dz$;
$\sqrt{x^2 + 4} = 2 \sec z$.

Hence, $\displaystyle\int \frac{dx}{\sqrt{x^2 + 4}} = \int \frac{2 \sec^2 z \, dz}{2 \sec z} = \int \sec z \, dz$

$$= \log (\sec z + \tan z).$$

Therefore, from the triangle of reference,

$$\int \frac{dx}{\sqrt{x^2 + 4}} = \log \left(\frac{\sqrt{x^2 + 4} + x}{2} \right) + C.$$

EXAMPLE 3. Find $\displaystyle\int \frac{\sqrt{x^2 - a^2}}{x}\, dx$.

Solution. Let $x = a \sec z$; $dx = a \sec z \tan z \, dz$; $\sqrt{x^2 - a^2} = a \tan z$.
Therefore,

$$\int \frac{\sqrt{x^2 - a^2}}{x}\, dx = \int \frac{(a \tan z) \cdot (a \sec z \tan z \, dz)}{a \sec z}$$

$$= a \int \tan^2 z \, dz = a \int (\sec^2 z - 1) \, dz$$

$$= a \int \sec^2 z \, dz - a \int dz$$

$$= a \tan z - az;$$

But $\sec z = \dfrac{x}{a}$; hence, from the triangle:

$$\int \frac{\sqrt{x^2 - a^2}}{x}\, dx = \sqrt{x^2 - a^2} - a \arc \cos \frac{a}{x} + C.$$

EXERCISE 13—4

Verify the following:

1. $\displaystyle\int \frac{dx}{\sqrt{x^2 - 9}} = \log (x + \sqrt{x^2 - 9}) + C$

2. $\displaystyle\int \sqrt{a^2 - x^2}\, dx = \frac{x}{2} \sqrt{a^2 - x^2} + \frac{a^2}{2} \arc\sin \frac{x}{a} + C$

3. $\displaystyle\int \frac{dx}{x^2\sqrt{a^2 + x^2}} = -\frac{\sqrt{x^2 + a^2}}{a^2 x} + C$

4. $\displaystyle\int \frac{dx}{x\sqrt{x^2 + 1}} = \log \frac{\sqrt{x^2 + 1} - 1}{x} + C$

5. $\displaystyle\int \frac{\sqrt{a^2 - x^2}}{x^2}\, dx = -\frac{\sqrt{a^2 - x^2}}{x} - \arc\sin \frac{x}{a} + C$

6. $\displaystyle\int x\sqrt{x^2 - a^2}\, dx = \frac{1}{3} (x^2 - a^2)^{3/2} + C$

TABLES OF INTEGRALS

13—7. Use of Integral Tables. The reader will find a list of integrals given on pages 371–378. Such a collection of formulas should not be regarded as a substitute for a good working knowledge of methods of integration. It is far more important for the reader to cultivate facility in recognizing forms, and to become proficient in transforming a given expression to a standard form, than it is to rely upon a table of integrals. No such table, however elaborate, is ever complete. The chief value of a table is that its use saves time.

When using a table of integrals, we note that it is arranged systematically, according to the form of the integrand. It must not be supposed that any given integrand will always be found in the table. If, upon searching the table until we find an integrand which is *exactly* of the form given, the integral can be written at once by substituting in the formula taken from the table. On the other hand, it may be necessary to transform a given integrand algebraically or trigonometrically in order that it will assume the *exact form* of some integrand given in the table before the formula in the table can be applied. The following examples will show how this is done.

EXAMPLE 1. Find $\int \dfrac{x\,dx}{5+3x^2}\,dx$.

Solution. From the table, page 374, formula [39] gives

$$\int \frac{x\,dx}{a+bx^2} = \frac{1}{2b}\log\left(x^2+\frac{a}{b}\right),$$

which is exactly in the form of the given integrand. Here, then, $a = 5$ and $b = 3$; hence

$$\int \frac{x\,dx}{5+3x^2}\,dx = \frac{1}{6}\log\left(x^2+\frac{5}{3}\right) + C.$$

EXAMPLE 2. Find $\int \dfrac{5\,dx}{3-7x^2}$.

Solution. From the table, page 373, formula [36] gives

$$\int \frac{dv}{a^2-v^2} = \frac{1}{2a}\log\frac{a+v}{a-v},$$

where x has been replaced by v. The given integrand, however, must first be rewritten; let $a = \sqrt{3}$, and $v^2 = 7x^2$, or $v = \sqrt{7}\,x$, and $dv = d(\sqrt{7}\,x)$. Substituting in the formula, we obtain:

$$\int \frac{5dx}{3-7x^2} = 5\int \frac{d(\sqrt{7}\,x)}{(\sqrt{3})^2-(\sqrt{7}\,x)^2}$$

$$= \frac{5}{\sqrt{7}}\left(\frac{1}{2\sqrt{3}}\right)\log\frac{\sqrt{3}+\sqrt{7}\,x}{\sqrt{3}-\sqrt{7}\,x}$$

$$= \frac{5\sqrt{21}}{42}\log\frac{\sqrt{3}+\sqrt{7}\,x}{\sqrt{3}-\sqrt{7}\,x} + C.$$

EXAMPLE 3. Find $\int \dfrac{dx}{9x^2-12x+10}$.

Solution. Completing the square in the denominator:

$$\int \frac{dx}{(9x^2-12x+4)+6} = \int \frac{dx}{(3x-2)^2+(\sqrt{6})^2}.$$

Using formula [18], page 372, we have:

$$v = (3x-2), \qquad a = \sqrt{6}, \qquad dv = 3dx.$$

Hence

$$\frac{1}{3} \int \frac{3dx}{(3x-2)^2 + (\sqrt{6})^2} = \frac{1}{3}\left(\frac{1}{\sqrt{6}}\right) \text{arc tan } \frac{3x-2}{\sqrt{6}}$$

$$= \frac{\sqrt{6}}{18} \text{ arc tan } \frac{3x-2}{\sqrt{6}} + C.$$

EXERCISE 13—5

Integrate the following, using the tables at the back of the book, pages 371–378; for practice, verify your result by differentiation:

1. $\displaystyle\int \frac{dx}{x^2(3+5x)}$

2. $\displaystyle\int \frac{x\,dx}{(2x^2-9)^{\frac{1}{2}}}$

3. $\displaystyle\int e^{3x} \sin 2x\, dx$

4. $\displaystyle\int \frac{x\,dx}{\sqrt{9-4x^2}}$

5. $\displaystyle\int \frac{dx}{3-2x^2}$

6. $\displaystyle\int \frac{dx}{(2x^2+1)^{\frac{3}{2}}}$

7. $\displaystyle\int \frac{x^2\,dx}{\sqrt{x^2-5}}$

8. $\displaystyle\int \frac{dx}{x\sqrt{9x^2-16}}$

9. $\displaystyle\int \frac{dx}{(3x^2+5)^{\frac{1}{2}}}$

10. $\displaystyle\int \frac{dx}{x\sqrt{3x-2}}$

11. $\displaystyle\int \frac{dx}{x+7x^3}$

12. $\displaystyle\int \frac{(2x^2-9)^{\frac{1}{2}}\,dx}{x}$

13. $\displaystyle\int \frac{dx}{x\sqrt{4+3x}}$

14. $\displaystyle\int \frac{dx}{x^2\sqrt{3-x^2}}$

15. $\displaystyle\int \sin 6x \cos 3x\, dx$

INTEGRATION OF RATIONAL FRACTIONS

13—8. Rational Fractions. As used here, this term refers to a fraction in x whose numerator and denominator contain x to positive integral powers only—in other words, a fraction whose numerator and denominator are polynomials in x. For example,

(a) $\dfrac{x+1}{x^2-3x+2}$, (b) $\dfrac{2x^2+1}{(x^2-1)(x+3)(x-2)}$, (c) $\dfrac{x^4+2x^3-8}{x^3-1}$, are rational fractions. If the degree in x of the numerator is greater than that of the denominator, as in (c) above, the fraction can be changed, by division, to the form below; thus

$$\frac{x^4 + 2x^3 - 8}{x^3 - 1} = x + 2 + \frac{x - 6}{x^3 - 1}.$$

We shall now show how to integrate fractions of the form (a) and (b); fractions such as (c) are first reduced by division, and the resulting quotient and remainder may then be integrated.

13—9. Partial Fractions. To integrate a rational fraction, we first resolve it, by a well-known method of algebra, into partial fractions. For our purposes, we shall consider only three types, limiting the discussion to the simpler situations; other types may be dealt with by analogous methods, but are more cumbersome and occur less frequently in practice.

13—10. Case I: *When the Denominator May Be Factored into Real, Linear Factors, None of Which Are Repeated.* The following examples illustrate both the method of resolving the integrand into partial fractions, as well as the actual integration; the latter is, of course, a relatively simple matter.

EXAMPLE 1. Find $\displaystyle\int \frac{7x + 6}{x(x + 3)}\, dx.$

Solution. By algebra, we may write:

$$\frac{7x + 6}{x(x + 3)} = \frac{A}{x} + \frac{B}{x + 3}.$$

To find the values of A and B, we clear of fractions, obtaining:

$$7x + 6 = A(x + 3) + B(x) \tag{1}$$

In equation (1), let $x = -3$; then $7(-3) + 6 = B(-3)$, or $B = +5$. Similarly, let $x = 0$; then $6 = 3A$, or $A = 2$.

Hence: $$\frac{7x + 6}{x(x + 3)} = \frac{2}{x} + \frac{5}{x + 3}.$$

Therefore, $\displaystyle\int \frac{7x + 6}{x(x + 3)}\, dx = \int \frac{2}{x}\, dx + \int \frac{5}{x + 3}\, dx$

$$= 2 \log x + 5 \log (x + 3) + C.$$

EXAMPLE 2. Find $\int \dfrac{2x - 34}{3x^2 - 11x - 4}\, dx$.

Solution. Factoring the denominator:

$$3x^2 - 11x - 4 = (3x + 1)(x - 4).$$

Resolving into partial fractions:

$$\frac{2x - 34}{(3x + 1)(x - 4)} = \frac{A}{3x + 1} + \frac{B}{x - 4}.$$

Clearing of fractions:

$$2x - 34 = A(x - 4) + B(3x + 1).$$

Letting $x = +4$, $B = -2$; when $x = -\frac{1}{3}$, $A = 8$.

Hence: $\dfrac{2x - 34}{(3x + 1)(x - 4)} = \dfrac{8}{3x + 1} - \dfrac{2}{x - 4}.$

Therefore,

$$\int \frac{2x - 34}{(3x + 1)(x - 4)}\, dx = \int \frac{8}{3x + 1}\, dx - \int \frac{2}{x - 4}\, dx$$

$$= 8 \int \frac{1}{3x + 1}\, dx - 2 \int \frac{1}{x - 4}\, dx$$

$$= \frac{8}{3} \int \frac{3}{3x + 1}\, dx - 2 \int \frac{1}{x - 4}\, dx$$

$$= \frac{8}{3} \log (3x + 1) - 2 \log (x - 4) + C.$$

EXAMPLE 3. Find $\int \dfrac{6x^2 + 12x - 30}{x(x - 3)(x + 2)}\, dx$.

Solution. By the method of partial fractions:

$$\frac{6x^2 + 12x - 30}{x(x - 3)(x + 2)} = \frac{A}{x} + \frac{B}{x - 3} + \frac{C}{x + 2},$$

or, $6x^2 + 12x - 30 = A(x - 3)(x + 2) + Bx(x + 2) + Cx(x - 3).$

Letting $x = 0$, $A = 5$; letting $x = 3$, $B = 4$; letting $x = -2$, $C = -3$.

Hence, $\dfrac{6x^2 + 12x - 30}{x(x - 3)(x + 2)} = \dfrac{5}{x} + \dfrac{4}{x - 3} - \dfrac{3}{x + 2}.$

Therefore:

$$\int \frac{6x^2 + 12x - 30}{x(x-3)(x+2)}\, dx = \int \frac{5}{x}\, dx + \int \frac{4}{x-3}\, dx - \int \frac{3}{x+2}\, dx$$
$$= 5 \log x + 4 \log (x-3) - 3 \log (x+2) + C.$$

NOTE. The result may be put into a more compact form by writing $5 \log x + 4 \log (x-3) - 3 \log (x+2)$ in the form

$$\log x^5 + \log (x-3)^4 - \log (x+2)^3,$$

or $$\log \frac{x^5(x-3)^4}{(x+2)^3} + C.$$

EXERCISE 13—6

Find the following:

1. $\displaystyle\int \frac{12x + 16}{x(x+2)}\, dx$

2. $\displaystyle\int \frac{3x - 6}{x^2 - 3x}\, dx$

3. $\displaystyle\int \frac{7(x-2)}{(x-4)(x+3)}\, dx$

4. $\displaystyle\int \frac{x + 35}{x^2 - 25}\, dx$

5. $\displaystyle\int \frac{-x - 1}{(x-1)(x-2)}\, dx$

6. $\displaystyle\int \frac{6x^2 + 13x - 4}{x(x+4)(x-1)}\, dx$

7. $\displaystyle\int \frac{x - 31}{(2x+1)(x-3)}\, dx$

8. $\displaystyle\int \frac{25x^2 + 6x - 12}{x(x+4)(3x-1)}\, dx$

9. $\displaystyle\int \frac{x(p-q)}{(p+x)(q+x)}\, dx$

10. $\displaystyle\int \frac{x^2 + ab}{x(x-a)(x+b)}\, dx$

13—11. Case II: *When the Factors of the Denominator Are All of the First Degree and Some Are Repeated.* This type of rational fraction will now be integrated.

EXAMPLE 1. Find $\displaystyle\int \frac{5x^2 - 3x - 2}{x^2(x-2)}\, dx$.

Solution. Let $\dfrac{5x^2 - 3x - 2}{x^2(x-2)} = \dfrac{A}{x} + \dfrac{B}{x^2} + \dfrac{C}{x-2}$.

Then, clearing of fractions:

$$5x^2 - 3x - 2 = Ax(x-2) + B(x-2) + Cx^2. \qquad (1)$$

Now, letting $x = 2$, we find $C = 3$; letting $x = 0$, $B = 1$; substituting the values of B and C so found from (1), we get $A = 2$.

NOTE.   An alternative method for finding the values of A, B, and C is to use the method of undetermined coefficients.   Accordingly, we rewrite (1) as follows:

$$5x^2 - 3x - 2 = Ax^2 - 2Ax + Bx - 2B + Cx^2,$$

or, $$5x^2 - 3x - 2 = (A + C)x^2 + (B - 2A)x - 2B. \tag{2}$$

Equating coefficients of like powers of x, we obtain:

$$A + C = 5; \qquad B - 2A = -3; \qquad -2B = -2.$$

Solving these three simultaneous equations by the methods of elementary algebra yields $A = 2$, $B = 1$, $C = 3$, as before.

Hence: $$\frac{5x^2 - 3x - 2}{x^2(x - 2)} = \frac{2}{x} + \frac{1}{x^2} + \frac{3}{x - 2};$$

therefore,

$$\int \frac{5x^2 - 3x - 2}{x^2(x - 2)} \, dx = \int \frac{2}{x} \, dx + \int \frac{1}{x^2} \, dx + \int \frac{3}{x - 2} \, dx$$

$$= 2 \log x - \frac{1}{x} + 3 \log (x - 2) + C.$$

EXAMPLE 2.   Find $\displaystyle \int \frac{-2x^2 - x - 9}{(x + 2)^2(2x + 1)} \, dx.$

Solution.   Let $\displaystyle \frac{-2x^2 - x - 9}{(x + 2)^2(2x + 1)} = \frac{A}{x + 2} + \frac{B}{(x + 2)^2} + \frac{C}{2x + 1}.$

Then

$$-2x^2 - x - 9 = A(x + 2)(2x + 1) + B(2x + 1) + C(x + 2)^2$$
$$= (2A + C)x^2 + (5A + 2B + 4C)x + (2A + B + 4C).$$

Hence, equating coefficients of like powers of x:

$$\left. \begin{array}{l} 2A + C = -2 \\ 5A + 2B + 4C = -1 \\ 2A + B + 4C = -9 \end{array} \right\};$$

or $$A = 1; \quad B = 5; \quad C = -4.$$

Therefore,

$$\int \frac{-2x^2 - x - 9}{(x+2)^2(2x+1)} \, dx = \int \frac{1}{x+2} \, dx + \int \frac{5}{(x+2)^2} \, dx - \int \frac{4}{2x+1} \, dx$$

$$= \log (x+2) - \frac{5}{x+2} - 2 \log (2x+1) + C.$$

EXAMPLE 3. Find $\int \dfrac{-6x^2 + 9x - 2}{x(x-1)^3} \, dx$.

Solution. Let $\dfrac{-6x^2 + 9x - 2}{x(x-1)^3} = \dfrac{A}{x} + \dfrac{B}{x-1} + \dfrac{C}{(x-1)^2} + \dfrac{D}{(x-1)^3}$.

Clearing of fractions:

$$-6x^2 + 9x - 2 = A(x-1)^3 + Bx(x-1)^2 + Cx(x-1) + Dx, \quad (1)$$

or

$$-6x^2 + 9x - 2 = (A+B)x^3 + (-3A - 2B + C)x^2$$
$$+ (3A + B - C + D)x - A. \quad (2)$$

From (1), if $x = 0$, $A = 2$; if $x = 1$, $D = 1$; substituting these values of A and B in (2), we obtain:

$$-6x^2 + 9x - 2 = (2+B)x^3 + (-6 - 2B + C)x^2$$
$$+ (6 + B - C + 1)x - 2. \quad (3)$$

Equating coefficients of like powers of x in (3):

$$0 = 2 + B, \quad \text{or} \quad B = -2;$$

$$-6 = -6 - 2B + C, \quad \text{or} \quad C = 2B = -4.$$

Hence:

$$\int \frac{-6x^2 + 9x - 2}{x(x-1)^3} \, dx$$

$$= \int \frac{2}{x} \, dx - \int \frac{2}{x-1} \, dx - \int \frac{4}{(x-1)^2} \, dx + \int \frac{1}{(x-1)^3} \, dx$$

$$= 2 \log x - 2 \log (x-1) + \frac{4}{x-1} - \frac{1}{2(x-1)^2} + C.$$

Find the following:

1. $\int \dfrac{7x^2 + x - 3}{x^2(5x + 3)}\, dx$

2. $\int \dfrac{6x^2 + 41x + 50}{x(x + 5)^2}\, dx$

3. $\int \dfrac{12x^2 - 25x + 15}{x^2(2x - 3)}\, dx$

4. $\int \dfrac{3x^2 + 11x + 8}{(x + 2)^3}\, dx$

5. $\int \dfrac{4x^2 + x - 8}{(x - 2)^2(x + 3)}\, dx$

6. $\int \dfrac{6x^2 + 22x + 12}{x^3 + 2x^2}\, dx$

7. $\int \dfrac{5x^2 + 13x + 2}{(x^2 - 4)(x + 2)}\, dx$

8. $\int \dfrac{9x^3 - 3x^2 - 5x - 3}{x^3(x + 1)}\, dx$

13—12. Case III: *When the Denominator Contains One or More Factors of the Second Degree, but None of Them Repeated.* The following examples show how such a fraction may be decomposed into partial fractions; note that to each non-repeated second-degree factor in the denominator of the given fraction there corresponds a partial fraction of the form

$$\frac{Ax + B}{px^2 + qx + r},$$

where A, B, p, q, r = constants.

EXAMPLE 1. Find $\int \dfrac{13x + 36}{(2x - 3)(x^2 + 7)}\, dx$.

Solution. Let $\dfrac{13x + 36}{(2x - 3)(x^2 + 7)} = \dfrac{A}{2x - 3} + \dfrac{Bx + C}{x^2 + 7}$.

Clear of fractions:

$$13x + 36 = Ax^2 + 7A + 2Bx^2 - 3Bx + 2Cx - 3C$$
$$= (A + 2B)x^2 + (-3B + 2C)x + (7A - 3C).$$

Hence,

$$A + 2B = 0; \quad -3B + 2C = 13; \quad 7A - 3C = 36.$$

Solving, $A = 6$, $B = -3$,   and $C = 2$;

and $\dfrac{13x + 36}{(2x - 3)(x^2 + 7)} = \dfrac{6}{2x - 3} + \dfrac{-3x + 2}{x^2 + 7}$.

Therefore:

$$\int \frac{13x + 36}{(2x - 3)(x^2 + 7)} \, dx$$

$$= \int \frac{6}{2x - 3} + \int \frac{-3x + 2}{x^2 + 7}$$

$$= \int \frac{6}{2x - 3} \, dx - \int \frac{3x}{x^2 + 7} \, dx + \int \frac{2}{x^2 + 7} \, dx$$

$$= 3 \log (2x - 3) - \frac{3}{2} \log (x^2 + 7) + \frac{2}{\sqrt{7}} \arctan \frac{x}{\sqrt{7}} + C.$$

EXERCISE 13—8

Find the following integrals:

1. $\int \frac{9x^2 + 2x - 12}{x(x^2 - 3)} \, dx$

2. $\int \frac{6x + 13}{(x^2 + 1)(2x + 1)} \, dx$

3. $\int \frac{-2x^2 + 3x - 30}{2x^3 + 5x} \, dx$

4. $\int \frac{9x^2 - 5x + 3}{x + x^3} \, dx$

5. $\int \frac{-2x^2 + 18x + 9}{(3x - 1)(x^2 + 2)} \, dx$

6. $\int \frac{5x^3 + 7x^2 - x - 8}{x^4 - x^2 - 2} \, dx$

7. $\int \frac{2x + 4}{(1 + x^2)(1 - 2x)} \, dx$

8. $\int \frac{7x^3 + 2x^2 + 5x - 2}{x^4 - 1} \, dx$

EXERCISE 13—9

Review

Verify the following integrations without the use of tables:

1. $\int \frac{2x \, dx}{(x^2 + 1)^2} = -\frac{1}{(x^2 + 1)} + C$

2. $\int \sqrt{3 + x} \, dx = \frac{2}{3} (3 + x)^{3/2} + C$

3. $\int \frac{dx}{x \log x} = \log (\log x) + C$

4. $\int \dfrac{3dx}{\sqrt{x^2 - 5}} = 3 \log (x + \sqrt{x^2 - 5}) + C$

5. $\int \dfrac{x^2\, dx}{a^3 - x^3} = -\dfrac{1}{3} \log (a^3 - x^3) + C$

6. $\int \dfrac{5x + 1}{(x + 2)(x - 1)}\, dx = 2 \log (x - 1) + 3 \log (x + 2) + C$

7. $\int x^3 \log x\, dx = \dfrac{x^4 \log x}{4} - \dfrac{x^4}{16} + C$

8. $\int \cos (x^2)\, x\, dx = \dfrac{1}{2} \sin (x^2) + C$

9. $\int x \sin 3x\, dx = -\dfrac{1}{3} x \cos 3x + \dfrac{1}{9} \sin 3x + C$

10. $\int \dfrac{\log (ax)}{x}\, dx = \log ax + \dfrac{1}{2} (\log x)^2 + C$

11. $\int \dfrac{3dx}{9x^2 + 4} = \dfrac{1}{2} \arctan \dfrac{3x}{2} + C$

12. $\int \dfrac{8x^3\, dx}{a + 2x^4} = \log (a + 2x^4) + C$

13. $\int \dfrac{3x^2 - 1}{x(x^2 - 1)}\, dx = \log (x^3 - x) + C$

14. $\int \dfrac{1}{a^2 - x^2}\, dx = \dfrac{1}{2a} \log \dfrac{a + x}{a - x} + C$

15. $\int \arcsin x\, dx = x \arcsin x + \sqrt{1 - x^2} + C$

The Definite Integral

CHAPTER FOURTEEN

INTEGRATION BETWEEN LIMITS

14—1. Generating an Area. Consider the shaded area "under" the curve $y = \phi(x)$, that is, the area $RSMP$, bounded by the ordinates SR and MP, by part of the X-axis, SM, and by the arc RP of the curve AB,

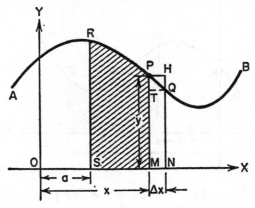

or $y = \phi(x)$. This area may be considered as having been generated by a variable ordinate moving from an initial position of SR to a terminal position MP. If we represent this *variable area* by u, then it is clear that u is a function of x, and that $u = 0$ when $x = a$.

Now if we let x take on a small increment Δx, the area u takes on an increment Δu, or the area $PMNQ$. It is clear that, in the rectangles concerned,

$$\text{area } MNQT < \text{area } PMNQ < \text{area } MNHP$$

or,
$$NQ \cdot \Delta x < \Delta u < MP \cdot \Delta x;$$

dividing by Δx:
$$NQ < \frac{\Delta u}{\Delta x} < MP.$$

As $\Delta x \to 0$, MP remains constant and $NQ \to MP$ as a limit. Hence

$$\frac{du}{dx} = MP = y,$$

or, in differential notation,

$$du = y \, dx. \tag{1}$$

This equation may be stated as a general principle:

The differential of the area bounded by any curve, the X-axis, and two ordinates, is equal to the product of the ordinate which terminates the area and the differential of the corresponding abscissa.

14—2. The Definite Integral and Its Meaning. From the preceding paragraph, we have

$$y = \phi(x),$$

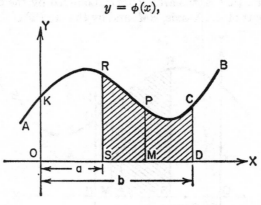

and $du = y \, dx$; hence $du = \phi(x) \, dx$, where du represents the differential of the area between the curve $y = \phi(x)$, the X-axis, and any two ordinates such as SR and DC. Integrating the last equation, we have

$$u = \int \phi(x) \, dx \tag{1}$$

Since the expression $\int \phi(x)\,dx$ is a function of x, we may designate it as $f(x) + C$; thus

$$u = f(x) + C. \tag{2}$$

It is possible to determine the value of C if we know the value of u for some value of x, as follows.

Let us arbitrarily begin to measure the area from the Y-axis.

Then:

when $x = a$, $u = $ area $OSRK$;

when $x = b$, $u = $ area $ODCK$;

hence, if $x = 0$, $u = 0$. Substituting in (2):

$$u = f(0) + C, \quad \text{or} \quad C = -f(0). \tag{3}$$

Then equation (2) yields, by substituting (3) in (2):

$$u = f(x) - f(0). \tag{4}$$

But equation (4) is a general expression for the area from the Y-axis to *any* ordinate, such as MP. So, to find the area between *two particular ordinates*, say SR and DC, we proceed as follows, substituting the appropriate abscissa for x in (4):

$$\text{Area } OSRK = f(a) - f(0) \tag{5}$$

$$\text{Area } ODCK = f(b) - f(0) \tag{6}$$

Subtracting equation (5) from (6):

$$\text{Area } SDCR = f(b) - f(a). \tag{7}$$

14—3. Limits of Integration. The difference of the values of $\int y\,dx$ for $x = a$ and for $x = b$, as found in equation (7) of the preceding paragraph, is called "the integral from a to b of $y\,dx$." It is represented by the symbol

$$\int_a^b y\,dx, \quad \text{or} \quad \int_a^b \phi(x)\,dx.$$

It is called a *definite integral*, since its value is always definite, depending, of course, upon the values of a and b, as well as upon the nature of the function $\int y\,dx$. The process of finding the value of a definite integral is called "integrating between limits"; the value a is referred to as the *lower limit*, and b as the *upper limit*.

The reasoning of the previous paragraph may now be summarized symbolically as follows:

if $$\int \phi(x)\, dx = f(x) + C,$$

then $$\int_a^b \phi(x)\, dx = [f(x) + C]_a^b$$

$$= [f(b) + C] - [f(a) + C],$$

or $$\int_a^b \phi(x)\, dx = f(b) - f(a).$$

Although the expression $\int_a^b \phi(x)\, dx$, as we have just seen, represents a definite area, the numerical value of the definite integral, namely, $f(b) - f(a)$ may represent some quantity other than an area, or just a number, depending upon the problem from which the integral arises.

14—4. Calculating the Value of a Definite Integral. To find the numerical value of a definite integral, therefore, we first find the indefinite integral of the given differential; then, in this indefinite integral, we simply substitute for the variable first the upper limit and then the lower limit, and subtract the latter from the former. The procedure is illustrated in the following examples.

EXAMPLE 1. Find $\int_2^6 x^3\, dx.$

Solution. $\int_2^6 x^3\, dx = \left[\dfrac{x^4}{4}\right]_2^6 = 324 - 4 = 320.$

EXAMPLE 2. Find $\int_0^{2\pi} \cos x\, dx.$

Solution. $\int_0^{2\pi} \cos x\, dx = [\sin x]_0^{2\pi} = 0 - 0 = 0.$

EXAMPLE 3. Find $\int_{-a}^{+a} \dfrac{dx}{\sqrt{a^2 - x^2}}\,.$

Solution.

$$\int_{-a}^{+a} \frac{dx}{\sqrt{a^2 - x^2}} = \left[\arcsin \frac{x}{a}\right]_{-a}^{+a} = \arcsin 1 - \arcsin(-1)$$

$$= \frac{\pi}{2} - \frac{3\pi}{2} = -\pi.$$

EXERCISE 14—1

Find the value of each of the following definite integrals:

1. $\displaystyle\int_3^4 9x^2\,dx$

9. $\displaystyle\int_0^8 2\sqrt[3]{x}\,dx$

2. $\displaystyle\int_1^2 4x^3\,dx$

10. $\displaystyle\int_{-1}^{+1} (x^2 + 2x - 3)\,dx$

3. $\displaystyle\int_0^4 ax^{3/2}\,dx$

11. $\displaystyle\int_0^3 4e^x\,dx$

4. $\displaystyle\int_{2a}^{8a} \sqrt{2ax}\,dx$

12. $\displaystyle\int_0^{\pi/4} \cos 2x\,dx$

5. $\displaystyle\int_0^4 \frac{2\,dx}{\sqrt{1 + 2x}}$

13. $\displaystyle\int_1^2 \frac{dx}{a^{2x}}$

6. $\displaystyle\int_1^{2e} \frac{dx}{x}$

14. $\displaystyle\int_0^{\pi} e^{\cos x} \sin x\,dx$

7. $\displaystyle\int_0^k \frac{k\,dx}{\sqrt{k^2 - x^2}}$

15. $\displaystyle\int_0^{-3} \frac{dx}{\sqrt{1 - x}}$

8. $\displaystyle\int_1^3 \frac{x\,dx}{x^2 + 1}$

16. $\displaystyle\int_1^e \left(6x^2 - \frac{3}{x}\right) dx$

AREA UNDER A CURVE

14—5. Calculation of Areas. The discussion in §14—1 and §14—2 suggests a convenient method for calculating the area under a specific portion of a curve whose equation is known.

EXAMPLE 1. Find the area bounded by the parabola $y = x^2$, the X-axis, and the ordinates $x = 1$ and $x = 3$.

Solution.

$$A = \int_1^3 x^2\,dx = \left[\frac{x^3}{3}\right]_1^3$$

$$= \frac{27}{3} - \frac{1}{3} = \frac{26}{3} = 8\frac{2}{3}.$$

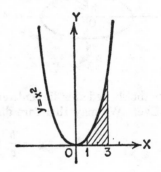

EXAMPLE 2.   Determine the area under the curve $y^2 = x$ and the line $x = 9$.

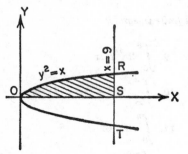

Solution. $y^2 = x$, or $y = x^{\frac{1}{2}}$.

$$\text{Area } ORS = \int_0^9 x^{\frac{1}{2}} \, dx = \frac{2}{3} [x^{\frac{3}{2}}]_0^9$$

$$= \frac{2}{3} (27 - 0) = 18.$$

Hence

$$\text{area } ORT = (2)(18) = 36.$$

EXAMPLE 3.   Find the area under the curve $y = \cos x$ from $x = -\pi/2$ to $x = \pi/2$.

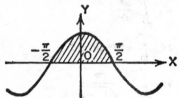

Solution.

$$A = \int_{-\pi/2}^{\pi/2} \cos x \, dx$$

$$= [\sin x]_{-\pi/2}^{\pi/2} = (1) - (-1) = 2.$$

EXAMPLE 4.   Find the area of the ellipse

$$\frac{x^2}{a^2} + \frac{y^2}{b^2} = 1.$$

Solution. We first find the area under this curve lying in the first quadrant.

$$A = \int_0^a y \, dx.$$

From the given equation,

$$y = \pm \frac{b}{a} \sqrt{a^2 - x^2}.$$

Since the shaded area lies entirely in the first quadrant, both a and b are positive. We may therefore disregard the $\pm$ sign. Hence

$$A = \frac{b}{a} \int_0^a \sqrt{a^2 - x^2} \, dx.$$

By formula [68], page 376:

$$A = \frac{b}{a}\left[\frac{x}{2}\sqrt{a^2 - x^2} + \frac{a^2}{2}\arcsin\frac{x}{a}\right]_0^a,$$

$$A = \frac{b}{a}\left(\frac{\pi a^2}{4} - 0\right) = \frac{\pi ab}{4}.$$

Hence, area of entire ellipse equals $4A$, or πab.

The reader should note carefully that this method of determining areas is a perfectly general method only if $y = f(x)$ is a continuous, single-valued function for values of x between $x = a$ and $x = b$, and if the limits a and b are finite. If either of these conditions does not hold, the method may or may not yield correct results, and special methods must be used.

EXERCISE 14—2

1. Find the area bounded by the parabola $y = 6x^2$, the X-axis, and the ordinates at $x = 2$ and $x = 4$.

2. Find the area bounded by the parabola $y^2 = 4x$, the Y-axis, and the line $y = 4$. *Hint:* First find the area A (the whole square); then find area B by difference.

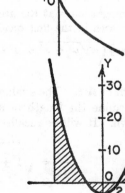

3. Find the shaded area, given that the equation of the curve is $y = x^2 - 4$.

4. Find the area under one arch of the curve $y = \sin x$, from $x = 0$ to $x = \pi$.

5. Find by integration the area included by the X-axis and the lines $y = x + 4$ and $2x + y = 10$; verify the result by elementary geometry.

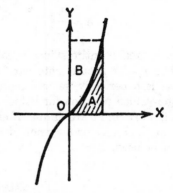

6. Find the area A bounded by the curve $y = x^3$, the X-axis, and the line $x = 2$. Find also the un-shaded area B.

7. Find the area of the figure bounded by the curve $y = x^2 + 2x - 6$, the X-axis, and the ordinates at $x = 2$ and $x = 6$.

8. Find the area included between the hyperbola $xy = a^2$, the X-axis, and the ordinates at $x = a$ and $x = b$.

9. Find the area bounded by the parabola $x^{1/2} + y^{1/2} = a^{1/2}$ and the coordinate axes.

10. Determine the area lying between the two parabolas $y^2 = 8x$ and $x^2 = 8y$.

11. Prove that the area bounded by a parabola and any chord drawn perpendicular to its axis is equal to two-thirds of the rectangle in which this area is inscribed.

12. Show by integration that the area of the circle $x^2 + y^2 = a^2$ equals πa^2; first find the area in the first quadrant, using the integration formula

$$\int \sqrt{a^2 - x^2}\, dx = \tfrac{1}{2}\left(x\sqrt{a^2 - x^2} + a^2 \arcsin \frac{x}{a} \right).$$

14—6. Length of Arc. The evaluation of a definite integral can also be used to determine the length of a particular arc of a curve whose equation is known. It will be recalled that

$$\frac{ds}{dx} = \sqrt{1 + \left(\frac{dy}{dx}\right)^2}.\qquad (1)$$

Now, consider the point P_0 as a fixed point on the curve $y = f(x)$, and regard the length along the curve of an arc to be generated by a moving,

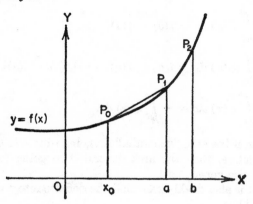

variable point (x,y). Reasoning as in §14—2, and integrating equation (1) above, we have:

$$\text{length of arc} = s = \int \sqrt{1 + \left(\frac{dy}{dx}\right)^2}\, dx,$$

remembering that the constant of integration is to be chosen so that $s = 0$ when $x = x_0$.

Thus,

$$\text{length of arc } P_0P_1 = \left[\int \sqrt{1 + \left(\frac{dy}{dx}\right)^2}\, dx\right]_{x=a},$$

$$\text{length of arc } P_0P_2 = \left[\int \sqrt{1 + \left(\frac{dy}{dx}\right)^2}\, dx\right]_{x=b}.$$

Hence, by difference,

$$\text{length of arc } P_1P_2 = \int_a^b \sqrt{1 + \left(\frac{dy}{dx}\right)^2}\, dx.$$

We shall discuss the length of arc further in Chapter Fifteen.

THE DEFINITE INTEGRAL AND ITS LIMITS

14—7. The Limits of Integration. It is easily shown that interchanging the limits of a definite integral is equivalent to changing the sign of the definite integral.

Since $\displaystyle\int_a^b \phi(x)\,dx = f(b) - f(a),$

and $\displaystyle\int_b^a \phi(x)\,dx = f(a) - f(b) = -[f(b) - f(a)],$

then $\displaystyle\int_a^b \phi(x)\,dx = -\int_b^a \phi(x)\,dx.$

In other words, if the area "generated" in going from a to b is arbitrarily considered positive, then the area generated in going from b to a is negative, and vice versa.

Moreover, it is also readily seen that the definite integral is a function of its limits. Thus,

if $\displaystyle\int_a^b \phi(x)\,dx = f(b) - f(a),$

then $\displaystyle\int_a^b \phi(z)\,dz = f(b) - f(a).$

In other words, the value of the definite integral depends upon its limits as well as upon the nature of the function which comprises the integrand.

14—8. Breaking-up or Combining Intervals of Integration. From the figure, it will be seen that

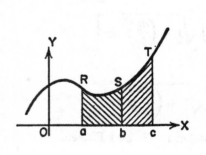

$$\int_a^b \phi(x)\,dx = f(b) - f(a), \quad \text{and}$$

$$\int_b^c \phi(x)\,dx = f(c) - f(b); \text{ hence,}$$

by addition, we have:

$$\int_a^b \phi(x)\,dx + \int_b^c \phi(x)\,dx$$
$$= f(c) - f(a). \quad [1]$$

But $\displaystyle\int_a^c \phi(x)\,dx = f(c) - f(a);$

then, by comparing the right-hand members of the last two equalities, it is clear that

$$\int_a^c \phi(x)\,dx = \int_a^b \phi(x)\,dx + \int_b^c \phi(x)\,dx. \qquad [2]$$

The relationship is quite general. A definite integral can be broken up
into any number of separate definite integrals in this manner; likewise,
any number of separate definite integrals can be combined, provided the
integrands are identical, by combining the limits. In fact, in the illustration, the limit b could just as well have been outside the interval
from a to c; by giving attention to the *signs* of the areas as well as to
their magnitudes, the relationship is still valid.

14—9. Improper Integrals. In the discussion thus far, the limits of
integration have been *finite* quantities. If, however, either the upper or
the lower limit, or both, should be *infinite*, the integral may or may not
exist. For example, suppose that in the definite integral

$$\int_a^b \phi(x)\, dx, \tag{1}$$

the upper limit $b \to +\infty$, that is, b increases indefinitely, while the lower
limit a remains constant. The succession of values taken by the integral
may then approach some definite, particular limiting value; if it does,
this limiting value is taken as the "integral of $\phi(x)$ from $x = a$ to
$x = +\infty$," and is designated by the expression

$$\int_a^\infty \phi(x)\, dx. \tag{2}$$

However, the succession of values assumed by the integral as $b \to \infty$
may *not* approach a limit; in this case, the expression given in equation
(2) has no meaning, and we say that the integral does not exist.

We may then set forth the following relations by way of definition:

$$\int_a^{+\infty} \phi(x)\, dx = \lim_{b\to\infty} \int_a^b \phi(x)\, dx, \tag{3}$$

$$\int_{-\infty}^b \phi(x)\, dx = \lim_{a\to-\infty} \int_a^b \phi(x)\, dx. \tag{4}$$

Here, in equations [3] and [4], the value of the expression on the left
side is said to exist only if the limit on the right side exists.

$$\int_{-\infty}^{+\infty} \phi(x)\, dx = \int_{-\infty}^0 \phi(x)\, dx + \int_0^{+\infty} \phi(x)\, dx. \tag{5}$$

Here, in equation [5], the integral on the left exists only if *both* integrals
on the right exist.

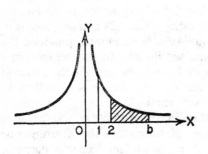

EXAMPLE 1. Find $\displaystyle\int_2^{+\infty} \frac{dx}{x^2}$.

Solution.

$$\int_2^{+\infty} \frac{dx}{x^2} = \lim_{b\to\infty} \int_2^b \frac{dx}{x^2}$$

$$= \lim_{b\to\infty} \left[-\frac{1}{x} \right]_2^b$$

$$= \lim_{b\to\infty} \left[-\frac{1}{b} + \frac{1}{2} \right] = \frac{1}{2}.$$

EXAMPLE 2. Find $\displaystyle\int_1^{+\infty} \frac{dx}{x}$.

Solution. $\displaystyle\int_1^{+\infty} \frac{dx}{x} = \lim_{b\to\infty} \int_1^b \frac{dx}{x} = \lim_{b\to\infty} [\log x]_1^b$

$$= \lim_{b\to\infty} [\log b - \log 1] = \lim_{b\to\infty} [\log b - 0] = +\infty.$$

In other words, in this case, the

$$\lim_{b\to\infty} \int_1^b \frac{dx}{x}$$

is infinite, or does not exist; hence the definite integral $\displaystyle\int_1^{\infty} \frac{dx}{x}$ does not exist.

EXAMPLE 3. Find $\displaystyle\int_{-\infty}^{+\infty} \frac{dx}{x^2}$.

Solution. $\displaystyle\int_{-\infty}^{+\infty} \frac{dx}{x^2} = \int_{-\infty}^0 \frac{dx}{x^2} + \int_0^{+\infty} \frac{dx}{x^2}$

$$= \left[-\frac{1}{x} \right]_{-\infty}^0 + \left[-\frac{1}{x} \right]_0^{+\infty}$$

$$= [(-\infty) - 0] + [(-0) - (-\infty)]$$

$$= -\infty + \infty = 0.$$

This result being absurd, it is clear that the integral $\displaystyle\int_{-\infty}^{+\infty} \frac{dx}{x^2}$ has no meaning and does not exist.

EXAMPLE 4. Find $\displaystyle\int_0^{\infty} \frac{dx}{x+1}$.

Solution. $\displaystyle\int_0^\infty \frac{dx}{x+1} = [\log (x+1)]_0^\infty$

$$= \log \infty - \log 0 = \infty - (-\infty) = \infty.$$

Thus the integral $\displaystyle\int_0^\infty \frac{dx}{x+1}$ is meaningless.

EXAMPLE 5. Find the total area under the curve $y = \dfrac{a^3}{x^2 + a^2}$.

Solution.

$$\text{Area } ORPS = \int_0^x y \, dx$$

$$= \int_0^x \frac{a^3 \, dx}{x^2 + a^2}$$

$$= \left[a^2 \arctan \frac{x}{a} \right]_0^x$$

$$= a^2 \arctan \frac{x}{a}.$$

Since the X-axis is an asymptote, the limits of integration will of necessity be $+\infty$ and $-\infty$. As $OS = x \to \infty$, $\arctan \dfrac{x}{a} \to \dfrac{\pi}{2}$ as a limit; hence $\displaystyle\int_0^\infty \frac{a^3 \, dx}{x^2 + a^2} = a^2 \left(\frac{\pi}{2}\right)$. The entire area under the curve, since the Y-axis is an axis of symmetry, is therefore twice the area to the right of the Y-axis, or $2\left(\dfrac{a^2 \pi}{2}\right) = \pi a^2$.

EXERCISE 14—3

Evaluate each of the following, if the integral exists:

1. $\displaystyle\int_1^\infty \frac{dx}{x^2}$

2. $\displaystyle\int_2^\infty \frac{dx}{x^3}$

3. $\displaystyle\int_0^\infty \frac{dx}{x^2 + a^2}$

4. $\displaystyle\int_0^\infty e^{-x} \, dx$

5. $\displaystyle\int_{-\infty}^1 e^x \, dx$

6. $\displaystyle\int_0^\infty \cos x \, dx$

7. $\displaystyle\int_0^\infty \frac{dx}{m^2 + x^2}$

8. $\displaystyle\int_{-\infty}^{+\infty} \frac{dx}{x^2 + 1}$

14—10. Interpretation of an Integral. We have seen that a definite integral may be taken to represent an area. It does not follow, however, that every integral can only represent an area; it may represent the length of an arc along a curve. Or it may be possible to interpret an integral in terms of a *physical quantity* rather than a geometric magnitude. The meaning of an integral thus depends upon the nature of the quantities which are represented by the variables x and y. If for example, the variable x represents *time* and the variable y represents *velocity*, then the integral represents *distance traveled*. In fact, many physical quantities may be represented by integrals—for example, *force, work, fluid pressure, center of gravity, moments of inertia,* etc.

It is true, however, that even when an integral stands for a physical quantity, it may be *represented geometrically* as an area. In this con-

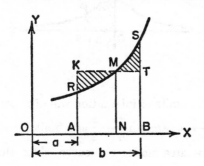

nection, it may be interesting to note how the integral is useful in defining the mean value of a function. Thus, if $y = f(x)$ is a given function, then the mean value of $f(x)$ for the interval from $x = a$ to $x = b$ is given by

$$\frac{\int_a^b f(x)\,dx}{b - a}.$$

For, as the figure shows,

$$\text{area } ARSB = \int_a^b f(x)\,dx;$$

furthermore, if the area of rectangle $AKTB$, on AB, or $(b - a)$, as a base, is equal in area to the area $ARSB$, then the

$$\text{mean value} = \frac{\text{area } AKTB}{b - a} = \frac{(AB)(MN)}{AB} = MN.$$

In other words, if M is so located that area RKM = area MTS, then MN is the mean value of $f(x)$ for the interval from $x = a$ to $x = b$.

14—11. Change of Limits when Changing the Variable. When evaluating a definite integral by the method of substitution, it is not necessary to go back to the original variable after the substitution has been made and the integration has been performed; it is only necessary to change the limits of integration to correspond to the change in the variable, as shown below.

EXAMPLE 1. Evaluate $\int_8^{15} \dfrac{dx}{x\sqrt{x+1}}$.

Solution. Let $z = \sqrt{x+1}$; then $x = z^2 - 1$, and $dx = 2z\,dz$. Also, when $x = 15$, $z = 4$, and when $x = 8$, $z = 3$.

Hence, $\int_8^{15} \dfrac{dx}{x\sqrt{x+1}} = \int_3^4 \dfrac{2z\,dz}{(z^2-1)z} = 2\int_3^4 \dfrac{dz}{z^2-1}$;

using formula [19], page 372, we have

$$2\int_3^4 \frac{dz}{z^2-1} = 2\left[\frac{1}{2}\log\frac{z-1}{z+1}\right]_3^4 = \log\frac{6}{5}.$$

EXAMPLE 2. Evaluate $\int_0^{a/2} \sqrt{a^2 - x^2}\,dx$.

Solution. Let $x = a\sin\theta$; then $\sqrt{a^2 - x^2} = a\sqrt{1 - \sin^2\theta} = a\cos\theta$; also, $dx = a\cos\theta\,d\theta$. Now when $x = \dfrac{a}{2}$, $\dfrac{1}{2}a = a\sin\theta$, or $\sin\theta = \dfrac{1}{2}$; therefore $\theta = \pi/6$. Again, when $x = 0$, $a\sin\theta = 0$, and $\theta = 0$.

Hence, $\int_0^{a/2} \sqrt{a^2 - x^2}\,dx = \int_0^{\pi/6} a^2\cos^2\theta\,d\theta$;

using formula [90], page 377, we have

$$a^2\int_0^{\pi/6} \cos^2\theta\,d\theta = a^2\left[\frac{\theta}{2} + \frac{1}{4}\sin 2\theta\right]_0^{\pi/6} = a^2\left(\frac{\pi}{12} + \frac{\sqrt{3}}{8}\right).$$

DERIVED CURVES AND INTEGRAL CURVES

14—12. Derived Curves. As we have already learned, for a curve whose equation is $y = f(x)$, the *slope* of the curve at any point on the curve is given by $\dfrac{dy}{dx}$, that is, the *differential coefficient of its ordinate with*

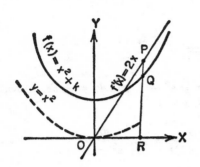

respect to its abscissa. If now on the same set of axes, we draw the curve whose equation is $y = f'(x)$,

where $f'(x)$ represents $\dfrac{d}{dx} f(x)$,

then at any point on this curve, the numerical measure of the ordinate is the same as that of the slope of the first curve at the point having the same abscissa. For example, if

$$f(x) = x^2 + k$$

and $f'(x) = 2x$,

then, for any point R on the X-axis, the ordinate RQ represents the function $f(x)$ for $x = OR$. The ordinate RP (where P lies on the derived curve for $x = OR$) is numerically equal to the slope of the original curve at Q; but the ordinate RP also represents the rate of change of the function $f(x)$ when $x = OR$.

Furthermore, since

$$\text{area } OPR = \int_0^x 2x\, dx = x^2,$$

the function $f(x)$ for $x = OR$ is represented by the area $OPR + k$.

In other words: for a function $f(x)$, if we draw the curve

$$y = f(x), \tag{1}$$

and also its first derived curve,

$$y = f'(x), \tag{2}$$

then the rate of change of the function for any value of x is represented not only by the slope of the first curve, but also by the ordinate of the derived curve for that value of x; and secondly, the function itself for any value of x is represented not only by the corresponding ordinate of the primary curve, but also by the area of the derived curve increased by the constant amount $f(0)$.

The derived curve (equation [2]) is known as the *curve of slopes* of the first curve (equation [1]). Two such related curves are shown here. The horizontal scale is the same for both curves, but the ordinates on the primary curve represent lengths, while the ordinates on the derived curve represent tangents of angles. Thus at any point at which the primary curve has a maximum or minimum ordinate, the slope equals zero, and therefore the corresponding ordinate on the derived curve equals zero. Conversely, wherever the derived curve intersects the X-axis, the corresponding ordinate of the primary curve is a maximum or a minimum.

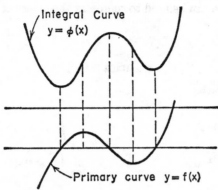

14—13. Integral Curves. Consider the graph of the curve

$$y = f(x). \tag{1}$$

Let the anti-derivative of $f(x)$ be designated by $\phi(x)$, and draw the graph of the curve

$$y = \int_0^x f(x)\, dx, \tag{2}$$

or, that is, the graph of

$$y = \phi(x) - \phi(0) = F(x). \tag{3}$$

The curve $y = \phi(x) - \phi(0)$, or $y = F(x)$, is known as the *first integral curve* of the curve [1] above. It is easily seen that

$$\frac{dF(x)}{dx} = \frac{d\phi(x)}{dx} = f(x). \tag{4}$$

From these equations we see that the following generalizations hold:

    I.   For any given abscissa x, the numerical value that gives the *length* of the ordinate of the first integral curve is the same as the value that gives the *area* between the primary curve, the axes, and the ordinate for this abscissa. Hence the ordinates of the first integral curve can be used to represent the areas of the primary curve when bounded as just described.

II. For any given abscissa x, the numerical value which gives the slope of the first integral curve is the same as the value that gives the length of the ordinate of the primary curve. Hence the ordinates of the primary curve can be used to represent the slopes of the first integral curve.

EXERCISE 14—4

Evaluate the following:

1. $\int_0^2 (4 - 3x)^2 \, dx$

2. $\int_0^a (a^2x - x^3) \, dx$

3. $\int_0^{\pi/2} \sin^2 x \, dx$

4. $\int_2^3 \frac{x \, dx}{1 + x^2}$

5. $\int_1^e z \log z \, dz$

6. $\int_0^\infty e^{-ax} \, dx$

7. $\int_2^6 \frac{x \, dx}{x + 2}$

8. $\int_1^e \log z \, dz$

9. Find the area of the curve $y = x^2 - 9$ which lies below the X-axis.

10. Find the area between the equilateral hyperbola $xy = 1$, the ordinates at $x = a$, $x = b$, and the X-axis.

11. Find the area under one arch of the sine curve $y = \cos x$.

12. Find the area under the semicubical parabola $y^2 = ax^3$ from the point where $x = 0$ to $x = a$.

EXERCISE 14—5

Review

Integrate by parts:

1. $\int e^x \sin x \, dx$

2. $\int xe^x \, dx$

Integrate by the method of partial fractions:

3. $\int \dfrac{(x-1)\,dx}{(x-3)^2}$

4. $\int \dfrac{(x-1)\,dx}{x(x+1)^2}$

5. $\int \dfrac{x^2-3x+3}{x^2-3x+2}\,dx$

6. $\int \dfrac{(1+x^2)\,dx}{x(x+1)(x-1)}$

7. $\int \dfrac{(x^2+1)\,dx}{x(1-x^2)}$

Find by using the Table of Integrals, pages 371–378:

8. $\int \dfrac{x^2\,dx}{(x+2)^2}$

9. $\int \dfrac{dx}{(x^2+1)^2}$

10. $\int \dfrac{x^2\,dx}{(1+x)^2}$

Test for convergence or divergence, and if convergent, find the interval of convergence:

11. $\sqrt{10} + \sqrt[3]{10} + \sqrt[4]{10} + \cdots.$

12. $1 + \dfrac{2}{x^2} + \dfrac{3}{x^3} + \dfrac{4}{x^4} + \cdots.$

Integration as
a Process of Summation

CHAPTER FIFTEEN

A BASIC PRINCIPLE

15—1. Two Aspects of Integration. Thus far we have regarded integration as the inverse of the operation of differentiation, and the integral was thought of as an anti-derivative. It is possible, however, to consider integration from another point of view, namely, as a *process of summation*, or as the addition of many similar elements. Indeed, it is largely from this point of view that the Integral Calculus developed historically, growing out of early attempts to determine the area bounded by various curves. A given area was subdivided into many small parts, and these "infinitesimal parts" were then added. The integral sign $\left(\int \right)$ is in fact simply an elongated S, originally used as an abbreviation for "sum." The method of summation not only throws additional light on the nature of the process of integration, but it also affords a convenient and powerful tool for solving many practical problems in science and technology to which the Integral Calculus is applied.

15—2. Finding an Area by Summing Up Its Component Parts. We have already seen that the area bounded by the curve $y = \phi(x)$, the X-axis, and the ordinates at $x = a$ and $x = b$ is given by the definite integral

$$\int_a^b y\, dx, \quad \text{or} \quad \int_a^b \phi(x)\, dx;$$

but by §14—3, this may be written as

$$\int_a^b \phi(x)\, dx = f(b) - f(a), \tag{1}$$

where $\phi(x)$ is the derivative of $f(x)$. Let us suppose that the interval from $x = a$ to $x = b$ is divided into any number of equal subintervals,

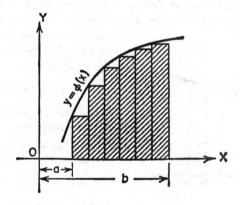

say n of them. If ordinates are erected at the points of division, and lines drawn through their extremities parallel to the X-axis, a series of rectangles will be obtained. The shaded area, consisting of the n rectangles so obtained, is an *approximation* to the area represented by [1] above. It will be seen that the greater the number of such subdivisions, the closer will be the approximation. If the number of rectangles is increased indefinitely, the *limit* of the sum of these rectangles will *equal* the area under the curve.

An alternative construction (which includes the above as a special case) would be the following. This time let us divide the interval from a to b into n subintervals, *not necessarily equal*, and erect ordinates at the points of division. Then, the sum of the rectangles constructed as shown in the figure will, as before, approximate the area under the curve; and the *limit of this sum*, as $n \to \infty$, and each subinterval approaches zero, will be exactly equal to the area in question. Thus the area under a curve may be regarded as the *limit of a sum*.

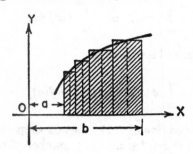

This can be expressed analytically as shown here. Let the lengths of the successive subintervals be designated by $\Delta x_1, \Delta x_2, \Delta x_3, \cdots, \Delta x_n$,

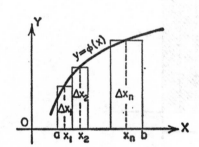

and the abscissas of the points chosen in the subintervals be designated by

$$x_1, x_2, x_3, \cdots, x_n.$$

Since the equation of the curve is $y = \phi(x)$, then the ordinates corresponding to $x_1, x_2, x_3, \cdots x_n$, are given by

$$\phi(x_1), \phi(x_2), \cdots \phi(x_n).$$

Hence the areas of these successive rectangles are given by

$$\phi(x_1)\,\Delta x_1, \; \phi(x_2)\,\Delta x_2, \; \phi(x_3)\,\Delta x_3, \; \cdots \; \phi(x_n)\,\Delta x_n.$$

Therefore, in view of what has been said above, the area under the curve is given by

$$\lim_{n \to \infty} [\phi(x_1)\,\Delta x_1 + \phi(x_2)\,\Delta x_2 + \phi(x_3)\,\Delta x_3 + \cdots + \phi(x_n)\,\Delta x_n].$$

Since this area is also expressed by

$$\int_a^b \phi(x)\,dx,$$

we have:

$$\int_a^b \phi(x)\,dx = \lim_{n \to \infty} [\phi(x_1)\,\Delta x_1 + \phi(x_2)\,\Delta x_2 + \cdots + \phi(x_n)\,\Delta x_n] \quad [2]$$

15—3. The Fundamental Theorem of the Integral Calculus.
Equation [2] of the preceding paragraph is, essentially, the fundamental
theorem of the Integral Calculus. We restate this basic principle com-
pletely, as follows:

If $\phi(x)$ is a continuous function in the interval from $x = a$ to $x = b$,
and if this interval is divided into subintervals of length $\Delta x_1, \Delta x_2, \cdots \Delta x_n$,
and if points are chosen, one in each subinterval, such that their abscissas
are $x_1, x_2, \cdots x_n$, we may form the sum

$$\phi(x_1) \, \Delta x_1 + \phi(x_2) \, \Delta x_2 + \cdots + \phi(x_n) \, \Delta x_n = \sum_{i=1}^{n} \phi(x_i) \, \Delta x_i; \qquad [3]$$

then, as n approaches infinity, and each subinterval approaches zero, the
limiting value of this sum is equal to the definite integral $\int_{a}^{b} \phi(x) \, dx$.

Equation [2] may be abbreviated as follows:

$$\int_{a}^{b} \phi(x) \, dx = \lim_{\substack{n \to \infty \\ \Delta x_i \to 0}} \sum_{i=1}^{n} \phi(x_i) \, \Delta x_i. \qquad [4]$$

It should be noted that each term in the sum in [3], that is, each of
the quantities $\phi(x_i) \, \Delta x_i$, is a *differential* expression, inasmuch as each of
the Δx_i factors approaches zero as a limit. Each of these $\phi(x_i) \, \Delta x_i$ terms
is called an *element* of the total quantity (whether area, volume, pressure,
etc.) whose sum is to be determined in this way.

15—4. Applying the Fundamental Theorem. The above discussion
leads to the following rule of procedure to be followed when using this
principle in practical problems:

I. *Let the required quantity be divided into similar parts, or elements,
in such a way that the limit of the sum of these parts will yield the quantity
desired;*

II. *Express the magnitudes of these parts in such a way as to yield a
sum of the form given by equation [3] above;*

III. *Select the appropriate limits $x = a$ and $x = b$, apply the formula*

$$\lim_{\substack{n \to \infty \\ \Delta x_i \to 0}} \sum_{i=1}^{n} \phi(x_i) \, \Delta x_i = \int_{a}^{b} \phi(x) \, dx,$$

and integrate.

AREAS OF PLANE CURVES

15—5. Curves in Rectangular Coordinates. We have already learned how an area may be determined by using a definite integral. We shall illustrate the use of the summation idea of integration to find areas, first of curves in rectangular coordinates, and then of curves in parametric equation form and in polar coordinate form.

As we have seen, the areas (I) and (II) below are given by the corresponding definite integrals, where the cross-hatched rectangular strip may be regarded as the type of similar strips which are summed up; the area of the strip shown being equal to $y \cdot dx$, or $x \cdot dy$, respectively.

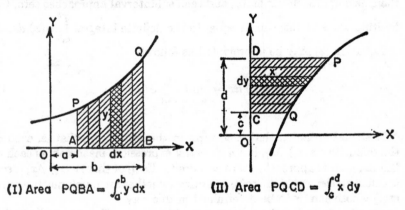

(I) Area $PQBA = \int_{a}^{b} y\, dx$ **(II) Area $PQCD = \int_{c}^{d} x\, dy$**

EXAMPLE 1. Find the area between the curve $x^2 = y^3$, the line $y = 4$, and the Y-axis.

Solution.

$$\text{Area } OCB = \lim_{n \to \infty} \sum_{i=1}^{n} \phi(y_i)\, \Delta y_i;$$

or $$\text{area } OCB = \int_{0}^{4} \phi(y)\, dy = \int_{0}^{4} x\, dy.$$

But $x = \sqrt{y^3}$; hence

$$\text{area } OCB = \int_{0}^{4} y^{3/2}\, dy = \left[\frac{2}{5} y^{5/2} \right]_{0}^{4} = \frac{64}{5} = 12\frac{4}{5}.$$

Hence, $$\text{total area } AOBCA = 2\left(12\frac{4}{5} \right) = 25\frac{3}{5}.$$

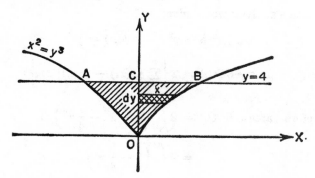

EXAMPLE 2. Find the area included between the curves $x^2 = 4y$ and $x^2 = 5 - y$.

Solution. The required area $ABCOA$ = area $DECBAD$ − area $DOECOAD$. Solving the given equations simultaneously, the points of

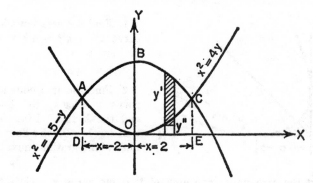

intersection A and C are given by $(2,1)$ and $(-2,1)$. The area $OBCE$ under the curve ABC is equal to:

$$\int_0^2 y \ dx = \int_0^2 (5 - x^2) \ dx = \left[5x - \frac{x^3}{3} \right]_0^2 = 10 - \frac{8}{3} = \frac{22}{3}.$$

Area OEC under curve AOC equals:

$$\int_0^2 y \ dx = \int_0^2 \frac{x^2}{4} dx = \frac{x^3}{12} \Big]_0^2 = \frac{2}{3}.$$

Hence,   area $OCBO$ equals $\frac{22}{3} - \frac{2}{3} = \frac{20}{3}$,

and   area $ABCOA$ equals $2 \left(\frac{20}{3} \right) = \frac{40}{3}$.

An alternative analysis would give:

$$\phi(x) = 5 - x^2 \qquad \psi(x) = \tfrac{1}{4}x^2$$

Then,   area $ABCOA = 2\left\{ \sum_{i=1}^{n} [\phi(x_i) - \psi(x_i)] \, \Delta x_i \right\}$,

therefore   area $ABCOA = 2\int_0^2 \left(5 - x^2 - \frac{1}{4}x^2 \right) dx$

$$= 2\int_0^2 \left(5 - \frac{5}{4}x^2 \right) dx$$

$$= 2\left[5x - \frac{5x^3}{12} \right]_0^2 = \frac{40}{3}.$$

EXERCISE 15—1

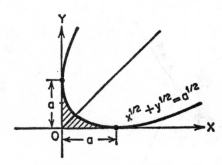

1. Find the area bounded by the curve $x^{\frac{1}{2}} + y^{\frac{1}{2}} = a^{\frac{1}{2}}$ and the co-ordinate axes.

2. Find the area bounded by the curve $y = \log x$, the line $x = 5$, and the X-axis. *Hint:* Remember that $\log 1 = 0$; why does the lower limit of integration equal 1?

3. Find the area under one arch of the sine curve $y = \sin x$, i.e., from $x = 0$ to $x = \pi$. Find also the area under this curve from $x = \pi$ to $x = 2\pi$. What is the meaning of the minus sign? What is the total area under the curve from $x = 0$ to $x = 2\pi$?

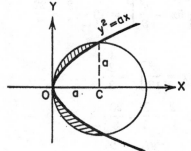

4. Find the area bounded by $y = x^3$, the X-axis, and the lines $x = a$ and $x = b$.

5. Find the area between the curve $y = x^3 - ax^2$ and the X-axis.

6. Find the area between the parabola $y^2 = ax$ and the circle $y^2 = 2ax - x^2$.

15—6. Area under a Curve Given by Parametric Equations. If the equations of a curve are given in parametric form, say $x = f(t)$ and $y = \phi(t)$, it can be proved that, since $dx = f'(t)\, dt$, the area under the curve from $x = a$ to $x = b$ equals

$$\int_{t_1}^{t_2} \phi(t) f'(t)\, dt, \qquad [1]$$

where $t = t_1$ when $x = a$ and $t = t_2$ when $x = b$.

EXAMPLE 1. Find the area of the circle

$$\begin{cases} x = r \cos t, \\ y = r \sin t. \end{cases}$$

Solution. We find the area of the first quadrant from $x = 0$ to $x = r$. Thus, when $x = 0$, $\cos t = 0$, and $t_1 = \dfrac{\pi}{2}$; when $x = r$, $\cos t = 1$, and $t_2 = 0$. Here $\phi(t) = r \sin t$; $f(t) = r \cos t$, and $f'(t) = -r \sin t$. Then, from equation [1]:

$$\text{area of quadrant} = \int_{\pi/2}^{0} r \sin t (-r \sin t)\, dt = -r^2 \int_{\pi/2}^{0} \sin^2 t\, dt$$

$$= -r^2 \int_{\pi/2}^{0} \left(\frac{1}{2} - \cos 2t \right) dt$$

$$= -r^2 \left[\frac{t}{2} - \frac{1}{4} \sin 2t \right]_{\pi/2}^{0}$$

$$= -r^2 \left[-\frac{\pi}{4} + 0 \right] = \frac{\pi r^2}{4}.$$

Therefore the area of entire circle $= 4 \left(\dfrac{\pi r^2}{4} \right) = \pi r^2$.

EXAMPLE 2. Find the area under the curve

$$\begin{cases} x = 4t, \\ y = t^2 + 4, \end{cases}$$

from $x = 0$ to $x = 8$.

Solution. When $x = 0$, $t_1 = 0$; when $x = 8$, $t_2 = 2$. Here

$$\phi(t) = t^2 + 4,$$

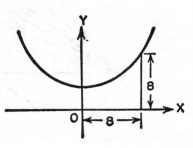

and $f(t) = 4t$. From equation [1]:

$$\text{area} = \int_0^2 (t^2 + 4)(4)\, dt = 4\int_0^2 (t^2 + 4)\, dt$$

$$= 4\left[\frac{t^3}{3} + 4t\right]_0^2 = 4\left(\frac{8}{3} + 8\right) = 42\frac{2}{3}.$$

Alternative method: By eliminating the parameter t, the equation of the curve is found to be $y = \frac{x^2}{16} + 4$; hence, area equals

$$\int_0^8 \left(\frac{x^2}{16} + 4\right) dx = \left[\frac{x^3}{48} + 4x\right]_0^8 = 42\frac{2}{3}.$$

15—7. Area under a Curve with Equation in Polar Coordinates.
Consider the curve whose equation is $\rho = f(\theta)$, where we wish to find the area bounded by the curve and any two radius vectors OR and OS corresponding to $\theta = \alpha$ and $\theta = \beta$. Let the radius vectors be $\rho_1, \rho_2, \cdots$, and let the respective angles of the sectors be $\Delta\theta_1, \Delta\theta_2, \cdots$.

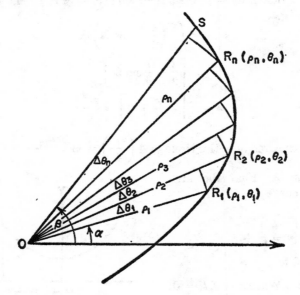

We shall now apply the Fundamental Theorem. With pole O as center, and with $\rho_0, \rho_1, \cdots \rho_{n-1}$ as radii, draw circular arcs; the sum of these sectors is an approximation to the desired area. The greater the value of n, or the smaller the value of $\Delta\theta$, the closer the approximation; as $n \to \infty$, or $\Delta\theta \to 0$, the limit of this sum will be the area sought. Now, the area of a circular sector is equal to $\frac{1}{2}$ radius $\times$ arc; hence the area of the first sector $= \frac{1}{2}\rho_1(\rho_1 \Delta\theta_1) = \frac{1}{2}\rho_1^2 \Delta\theta_1$; the area of the second sector $= \frac{1}{2}\rho_2^2 \Delta\theta_2$; etc. Hence, the entire area OR_1S is given by

$$\frac{1}{2}\rho_1^2 \Delta\theta_1 + \frac{1}{2}\rho_2^2 \Delta\theta_2 + \cdots \frac{1}{2}\rho_n^2 \Delta\theta_n = \sum_{i=1}^{n} \frac{1}{2}\rho_i^2 \Delta\theta_i.$$

Applying the Fundamental Theorem:

$$\lim_{n \to \infty} \sum_{i=1}^{n} \frac{1}{2}\rho_i^2 \Delta\theta_i = \int_{\alpha}^{\beta} \frac{1}{2}\rho^2 \, d\theta. \tag{1}$$

From (1), we see that the area under a polar curve equals

$$\frac{1}{2}\int_{\alpha}^{\beta} \rho^2 \, d\theta, \tag{1}$$

where $\rho = f(\theta)$, and the value of ρ is expressed in terms of θ as given by the equation $\rho = f(\theta)$.

EXAMPLE 1. Find the area of the circle $\rho = a \cos \theta$.

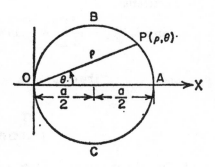

Solution. When the angle θ varies from $\theta = 0$ to $\theta = \pi/2$, the radius vector OP sweeps over the area ABO; from $\theta = \pi/2$ to $\theta = \pi$, the radius vector sweeps over the area OCA. Therefore the limits of integration are 0 and π; hence

$$\frac{1}{2}\int_{\alpha}^{\beta} \rho^2 \, d\theta = \frac{1}{2}\int_{0}^{\pi} \rho^2 \, d\theta = \frac{1}{2}\int_{0}^{\pi} a^2 \cos^2 \theta \, d\theta.$$

But $\int \cos^2 \theta \, d\theta = \dfrac{\theta}{2} + \dfrac{1}{4} \sin 2\theta$; therefore,

$$\text{area} = \frac{1}{2}a^2\left[\frac{\theta}{2} + \frac{1}{4}\sin 2\theta\right]_0^{\pi} = \frac{1}{2}a^2\left(\frac{\pi}{2}\right) = \frac{\pi a^2}{4}.$$

EXAMPLE 2. Find the entire area enclosed by the curve $\rho = a \sin 2\theta$.

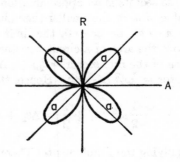

Solution. Since the curve is symmetrical with respect to both *OA* and *OR*, we shall first find the area of one loop. The loop in the first quadrant is formed as the radius vector sweeps out from $\theta = 0$ to $\theta = \dfrac{\pi}{2}$; hence

$$\text{area of loop} = \frac{1}{2} \int_0^{\pi/2} \rho^2 \, d\theta = \frac{1}{2} a^2 \int_0^{\pi/2} (\sin 2\theta)^2 \, d\theta.$$

To find $\int (\sin 2\theta)^2 \, d\theta$, we use formula [89], page 377, making the substitution $z = 2\theta$; thus

$$\int (\sin 2\theta)^2 \, d\theta = \frac{\theta}{2} - \frac{1}{8} \sin 4\theta.$$

Therefore,

$$\text{area of loop} = \frac{a^2}{2} \left[\frac{\theta}{2} - \frac{1}{8} \sin 4\theta \right]_0^{\pi/2} = \frac{a^2}{2} \left(\frac{\pi}{4} - 0 \right) = \frac{\pi a^2}{8} \, ;$$

and $\qquad$ the entire area $= 4 \left(\dfrac{\pi a^2}{8} \right) = \dfrac{\pi a^2}{2} \, .$

EXERCISE 15—2

1. Find the area, from $x = 2$ to $x = 6$, under the curve $x = 2t$, $y = 6/t$.

2. Find the area under the curve $x = t + 4$, $y = \frac{1}{4} t^2$, from $x = 0$ to $x = 6$.

3. Find the area enclosed by the ellipse

$$\begin{cases} x = a \cos \theta, \\ y = b \sin \theta, \end{cases}$$

where the parameter is the eccentric angle θ.

4. Find the area under one arch of the cycloid

$$\begin{cases} x = a(\theta - \sin\theta), \\ y = a(1 - \cos\theta), \end{cases}$$

where θ is the parameter. *Hint:* Remember that, by definition of the cycloid, x varies from 0 to $2\pi a$, where a is the radius of the generating circle; hence angle θ varies from 0 to 2π.

5. Find the area of one loop of the four-leaved rose $\rho = a\cos 2\theta$.

6. Find the area of the cardioid $\rho = a(1 - \cos\theta)$.

7. Find the area under the curve $\rho^2 = a^2\cos 2\theta$.

8. Find the area of the three loops of the curve $\rho = a\sin 3\theta$.

9. Find the area under the curve $\rho = a(\cos 2\theta + \sin 2\theta)$.

10. Find the area swept over as the radius vector of the spiral of Archimedes, $\rho = a\theta$, makes one revolution from $\theta = 0$ to $\theta = 2\pi$.

LENGTH OF A CURVE

15—8. Differential Length of Arc. It will be recalled that in §14—6 we derived a formula for the differential length of arc, namely,

$$\frac{ds}{dx} = \sqrt{1 + \left(\frac{dy}{dx}\right)^2}.$$

We shall now use the Fundamental Theorem to derive a formula for the length of an arc of a curve. By definition, the length of a portion of a

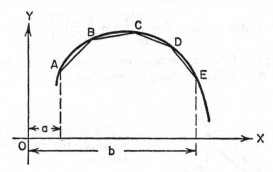

curve means the limit of the sum of the chords as the number of points of division is increased indefinitely in such a manner that the length of each chord, at the same time, separately approaches zero as a limit.

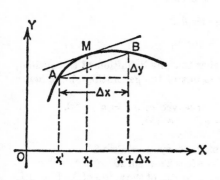

Consider the length of any one of these chords, say AB, where the coordinates of A are (x',y'), and those of B are $(x' + \Delta x, y' + \Delta y)$. It will be seen that

$$AB = \sqrt{(\Delta x)^2 + (\Delta y)^2}$$
$$= \Delta x \left[1 + \left(\frac{\Delta y}{\Delta x}\right)^2 \right]^{\frac{1}{2}}.$$

By the theorem of mean value:

$$\frac{\Delta y}{\Delta x} = f'(x_1)$$

where x_1 is the abscissa of point M on the curve at which the tangent is parallel to the chord.

Hence, $\qquad AB = \Delta x [1 + f'(x_1)^2]^{\frac{1}{2}};$

similarly, $\qquad BC = \Delta x [1 + f'(x_2)^2]^{\frac{1}{2}},$

$$CD = \Delta x [1 + f'(x_3)^2]^{\frac{1}{2}}, \quad \text{etc.}$$

Therefore, the length of the broken line AE is given by

$$\sum_{i=1}^{n} [1 + f'(x_i)^2]^{\frac{1}{2}} \cdot \Delta x^{(i)}.$$

Thus, by the Fundamental Theorem:

$$\text{length of arc} = S = \int_{a}^{b} \left[1 + \left(\frac{dy}{dx}\right)^2 \right]^{\frac{1}{2}} dx.$$

When using this formula, we must remember always to express $\frac{dy}{dx}$ in terms of x, as determined by the given equation.

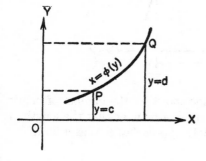

If y is used as the independent variable, the corresponding formula is:

$$S = \int_{c}^{d} \left[1 + \left(\frac{dx}{dy}\right)^2 \right]^{\frac{1}{2}} dy. \quad [2]$$

If the equation of the curve is given in polar coordinates, the analogous formulas for the length of an arc are

$$S = \int_{\theta_1}^{\theta_2} \left[\rho^2 + \left(\frac{d\rho}{d\theta}\right)^2 \right]^{\frac{1}{2}} d\theta, \qquad [3]$$

and
$$S = \int_{\rho_1}^{\rho_2} \left[1 + \rho^2 \left(\frac{d\theta}{d\rho}\right)^2 \right]^{\frac{1}{2}} d\rho. \qquad [4]$$

When using formula [3], remember to express $\sqrt{\rho^2 + \left(\frac{d\rho}{d\theta}\right)^2}$ in terms of θ before integrating; when using [4], the quantity $\sqrt{1 + \rho^2 \left(\frac{d\theta}{d\rho}\right)^2}$ must be expressed in terms of ρ before integrating.

EXAMPLE 1.   Find the length of the arc of the curve $x^2 = 2py$ between the points where $x = 0$ and $x = p$.

Solution. $x^2 = 2py$; hence $\dfrac{dy}{dx} = \dfrac{x}{p}$.

$$S = \int_0^p \left(\sqrt{1 + \left(\frac{dy}{dx}\right)^2} \right) dx = \int_0^p \sqrt{1 + \frac{x^2}{p^2}} \, dx$$

$$= \frac{1}{p} \int_0^p \sqrt{p^2 + x^2} \, dx$$

$$= \frac{1}{p} \left[\frac{x}{2} \sqrt{x^2 + p^2} + \frac{p^2}{2} \log (x + \sqrt{x^2 + p^2}) \right]_0^p$$

$$= \frac{1}{p} \left[\frac{p^2}{2} \sqrt{2} + \frac{p^2}{2} \log (p + p\sqrt{2}) - \frac{p^2}{2} \log p \right]$$

$$= \frac{p}{2} [\sqrt{2} + \log (p)(1 + \sqrt{2}) - \log p]$$

$$= \frac{p}{2} [\sqrt{2} + \log p + \log (1 + \sqrt{2}) - \log p]$$

$$= \frac{p}{2} [\sqrt{2} + \log (1 + \sqrt{2})].$$

EXAMPLE 2.   Find the length of the circle whose equation is $x^2 + y^2 = r^2$.

Solution. Consider the quarter-arc in the first quadrant.

From the equation, $\dfrac{dy}{dx} = -\dfrac{x}{y}$.

$$\text{Length of quadrant} = \int_0^r \sqrt{1 + \left(\frac{dy}{dx}\right)^2}\, dx = \int_0^r \sqrt{1 + \frac{x^2}{y^2}}\, dx$$

$$= \int_0^r \sqrt{\frac{x^2 + y^2}{y^2}}\, dx = r \int_0^r \frac{dx}{\sqrt{r^2 - x^2}}$$

$$= r \left[\arcsin \frac{x}{r}\right]_0^r = r\left(\frac{\pi}{2}\right).$$

Therefore,

$$\text{length of entire circle} = 4\left(\frac{\pi r}{2}\right) = 2\pi r.$$

EXERCISE 15—3

1. Find the length of the arc of $y = \frac{1}{2}x^2$ from the origin to the point whose abscissa is 3.

2. Find the length of the arc on the logarithmic spiral $\rho = e^\theta$ from the point where $\theta = 0$ to the point where $\theta = 1$.

3. Find the length of the arc of $y^2 = 4ax$ between the points whose abscissas are $x = 0$ and $x = 2a$.

4. Find the length of the arc of $y = \log \cos x$ between the points whose abscissas are $x = 0$ and $x = \pi/6$.

5. Find the length of the circle whose equation is $\rho = 2a \cos \theta$.

6. Find the length of the circle whose equation is $y^2 + (x - a)^2 = a^2$. Compare your result with that for Problem 5; explain.

7. Find the length of the arc on the spiral of Archimedes, $\rho = a\theta$, when the radius vector has made one revolution, i.e., from $\theta = 0$ to $\theta = 2\pi$.

8. Find the total length of the curve whose equation is $\rho = a \sec \theta$.

9. Find the length of the arc of $e^y = 1 - x^2$ between the points whose abscissas are $x = 0$ and $x = \frac{1}{2}$.

10. Find the entire length of the curve $\rho = a \sin^3 \dfrac{\theta}{3}$.

SOLIDS OF REVOLUTION

15—9. Volumes of Solids of Revolution. Consider the arc PQ of the curve $y = f(x)$ from $x = a$ to $x = b$ (Fig. I). If we suppose the arc to be revolved about *OX* as an axis, the solid of revolution obtained (Fig. II) may be regarded as composed of a series of right circular cylinders, formed by the revolving rectangles. As $\Delta x \to 0$, that is, as the

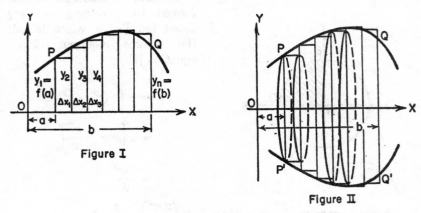

Figure I

Figure II

number of rectangles, and therefore the number of cylinders, increases indefinitely, the sum of the volumes of the cylinders approximates more and more closely to the volume of the solid of revolution. The volume of any one of these cylinders, say the *i*th cylinder, is the area of its base, πy_i^2, multiplied by its altitude, Δx_i; in other words,

$$\text{volume} = \sum_{i=1}^{n} \pi y_i^2 \, \Delta x_i.$$

Applying the Fundamental Theorem:

$$V_x = \pi \int_a^b y^2 \, dx, \tag{1}$$

where V_x denotes the volume of the solid of revolution formed by rotating $y = f(x)$ about the *X*-axis.

If the curve $y = f(x)$ is rotated about the *Y*-axis, then

$$V_y = \pi \int_c^d x^2 \, dy, \tag{2}$$

where the value of x in terms of y, as found from the given equation $y = f(x)$, must be substituted before integrating, and where c and d are the limits of integration in terms of ordinates.

EXAMPLE 1. Find the volume of the sphere generated by revolving the circle $x^2 + y^2 = r^2$ about the X-axis.

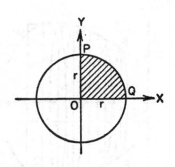

Solution. Consider the volume formed by revolving arc PQ about the X-axis, which is half the volume required. From equation [1]

$$V = \pi \int_0^r (r^2 - x^2)\, dx$$

$$= \pi \left[r^2 x - \frac{x^3}{3} \right]_0^r$$

$$= \frac{2\pi r^3}{3};$$

hence,   volume of entire sphere $= 2 \left(\dfrac{2\pi r^3}{3} \right) = \dfrac{4}{3} \pi r^3$.

EXAMPLE 2. Find the volume generated by the arc of the parabola $x^2 = 4y$ between the points where $y = 0$ and $y = a$ when revolved about the Y-axis.

Solution. Here we use formula [2]:

$$V_y = \pi \int_c^d x^2\, dy = \pi \int_0^a 4y\, dy = \pi [2y^2]_0^a = 2\pi a^2.$$

15—10. Area of a Surface of Revolution. If the arc of a curve $y = f(x)$ between the abscissa $x = a$ and $x = b$ is revolved about the X-axis, the surface S so generated is given by the formula

$$S_x = 2\pi \int_a^b y \left[1 + \left(\frac{dy}{dx} \right)^2 \right]^{\frac{1}{2}} dx. \qquad [3]$$

The corresponding formula for the surface S_y obtained by revolving the arc from $y = c$ to $y = d$ about the Y-axis, is given by

$$S_y = 2\pi \int_c^d x \left[1 + \left(\frac{dx}{dy} \right)^2 \right]^{\frac{1}{2}} dy. \qquad [4]$$

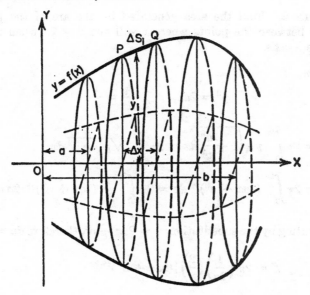

EXAMPLE 1. Find the surface of a sphere generated by revolving a semicircle of radius r about the X-axis, as shown.

Solution.

$$x^2 + y^2 = r^2;$$

$$S = 2\pi \int_{-r}^{+r} y \left[1 + \left(\frac{dy}{dx}\right)^2\right]^{\frac{1}{2}} dx.$$

$$\frac{dy}{dx} = -\frac{x}{y};$$

hence $\sqrt{1 + \left(\frac{dy}{dx}\right)^2} = \sqrt{1 + \frac{x^2}{y^2}} = \sqrt{\frac{y^2 + x^2}{y^2}} = \frac{r}{y};$

$$S = 2\pi \int_{-r}^{+r} y \left[1 + \frac{x^2}{y^2}\right]^{\frac{1}{2}} dx = 2\pi \int_{-r}^{+r} y \left(\frac{r}{y}\right) dx$$

$$= 2\pi r \int_{-r}^{+r} dx = 2\pi r [x]_{-r}^{+r} = 2\pi r(2r) = 4\pi r^2.$$

EXAMPLE 2. Find the area generated by the arc of the parabola $y^2 = 2ax$ between the points where $x = 0$ and $x = 2a$ when revolved about the X-axis.

Solution.

$$y^2 = 2ax; \qquad \frac{dy}{dx} = \frac{a}{y}.$$

$$S = 2\pi \int_0^{2a} y \sqrt{1 + \frac{a^2}{y^2}}\, dx = 2\pi \int_0^{2a} \sqrt{y^2 + a^2}\, dx$$

$$= 2\pi \int_0^{2a} (2ax + a^2)^{\frac{1}{2}}\, dx = 2\pi \int_0^{2a} \frac{1}{2a}(2ax + a^2)^{\frac{1}{2}} \cdot 2a\, dx.$$

(Integrating by the substitution: $u = 2ax + a^2$, therefore $du = 2a\, dx$)

$$S = 2\pi \left(\frac{1}{2a}\right)\left(\frac{2}{3}\right)[(2ax + a^2)^{\frac{3}{2}}]_0^{2a}$$

$$= \frac{2\pi}{3a}[5\sqrt{5}\, a^3 - a^3] = \frac{2\pi a^2}{3}(5\sqrt{5} - 1).$$

EXERCISE 15—4

1. Find the volume generated by revolving the ellipse $b^2x^2 + a^2y^2 = a^2b^2$ about the X-axis.

2. Find the volume generated when the same ellipse (Problem 1) is revolved about the Y-axis. Prove that each result is correct by taking the special case where $a = b$.

3. Find the volume of the cone generated by revolving about the X-axis the segment of the line $y = 3 - x$ cut off by the axes.

4. Find the volume of the solid generated by an arch of the sine curve, $y = \sin x$, from $x = 0$ to $x = \pi$, when revolved about the X-axis.

5. Find the volume of the solid generated by revolving the arc of the parabola $y^2 = 4px$ between the origin and the point for which $x = 2p$ when revolved about the X-axis.

6. Find the volume of the doughnut-like ring, called the *torus*, generated by revolving the circle $x^2 + (y - k)^2 = r^2$ about the X-axis.

7. The line $y = 2x$ is revolved about the X-axis. Find the area of the cone generated by the segment from $x = 0$ to $x = 4$.

8. A right circular cone has an altitude $h = 8$ and a base of radius $r = 4$. Find the lateral surface by integration; check by elementary geometry.

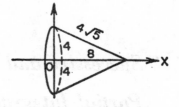

9. Find the total surface generated by revolving the parabola $y^2 = 4x$ about the X-axis from the origin to the point where $x = 8$.

10. The arc of the parabola $y^2 = 4px$ lying between the points where $x = 0$ and $x = 2p$ is revolved about the X-axis; find the surface generated.

11. The segment of the line $y = x + 2$ from $x = 0$ to $x = 3$ is revolved about the X-axis. Find the lateral surface of the frustum of the cone generated, and check your result by elementary geometry.

12. Prove, by integration, that the lateral surface of a right circular cylinder of radius r and altitude h equals $2\pi rh$.

Successive and

Partial Integration;

Approximate Integration

CHAPTER SIXTEEN

MULTIPLE INTEGRALS

16—1. Successive Integration. This is the inverse of the process of successive differentiation. Suppose it is given that

$$\frac{d^3y}{dx^3} = 4x,$$

and we wish to find y. We may then write:

$$\frac{d}{dx}\left(\frac{d^2y}{dx^2}\right) = 4x, \quad \text{or} \quad d\left(\frac{d^2y}{dx^2}\right) = 4x\, dx;$$

integrating:

$$\frac{d^2y}{dx^2} = \int 4x\, dx = 2x^2 + C_1.$$

Again:

$$\frac{d^2y}{dx^2} = \frac{d}{dx}\left(\frac{dy}{dx}\right) = 2x^2 + C_1, \quad \text{or} \quad d\left(\frac{dy}{dx}\right) = (2x^2 + C_1)\, dx;$$

integrating once more:

$$\frac{dy}{dx} = \int (2x^2 + C_1)\, dx = \frac{2x^3}{3} + C_1 x + C_2.$$

Finally:

$$dy = \left(\frac{2x^3}{3} + C_1 x + C_2\right) dx,$$

and integrating,

$$y = \frac{x^4}{6} + \frac{C_1 x^2}{2} + C_2 x + C_3.$$

The above analysis can also be written as follows:

$$\frac{d^2y}{dx^2} = \int 4x\, dx;$$

$$\frac{dy}{dx} = \iint 4x\, dx\, dx \left(\text{or, } \iint 4x\, dx^2\right);$$

$$y = \iiint 4x\, dx\, dx\, dx \left(\text{or, } \iiint 4x\, dx^3\right).$$

These last two are called a *double integral* and a *triple integral*, respectively. It will be seen that there is nothing new about successive integration, except that more than one constant of integration is involved. In general, a *multiple* integral requires two or more successive integrations. The process is also known as repeated integration, or *iterated* integration.

EXAMPLE. Find y, if $y = \iiint 3x^2\, dx\, dx\, dx$.

Solution.

$$y = \iiint 3x^2\, dx\, dx\, dx$$

$$= \iint (x^3 + C_1)\, dx\, dx$$

$$= \int \left(\frac{x^4}{4} + C_1 x + C_2\right) dx$$

$$= \frac{x^5}{20} + \frac{C_1 x^2}{2} + C_2 x + C_3.$$

16—2. Multiple Integrals with Limits of Integration. If successive integrations are performed between limits, the constants of integration disappear.

EXAMPLE.   Evaluate $\int_0^4 \int_1^3 \int_2^4 3x \, dx \, dx \, dx$.

Solution.   Beginning by integrating the "inside" integral first:

$$\int_0^4 \int_1^3 \int_2^4 3x \, dx \, dx \, dx = \int_0^4 \int_1^3 \frac{3x^2}{2} \Big]_2^4 dx \, dx$$

$$= \int_0^4 \int_1^3 18 \, dx \, dx = \int_0^4 18x \Big]_1^3 dx$$

$$= \int_0^4 36 dx = 36x \Big]_0^4 = 144.$$

EXERCISE 16—1

1. Given $\dfrac{d^2y}{dx^2} = 5x^2$; find y.

2. Given $\dfrac{d^3y}{dx^3} = 3x$; find y.

3. Given $\dfrac{d^4y}{dx^4} = 1$; find y.

4. If $y = \displaystyle\iint x^4 \, dx \, dx$, find y.

5. If $y = \displaystyle\iint (x^2 + 2x) \, dx^2$, find y.

6. If $y = \displaystyle\iiint e^x \, dx^3$, find y.

7. Find ρ, if $\rho = \displaystyle\iiint \sin \theta \, d\theta^3$.

8. Find $\displaystyle\int_1^3 \int_0^2 x^3 \, dx \, dx$.

9. Find $\displaystyle\int_1^2 \int_0^2 \int_2^4 3x^2 \, dx \, dx \, dx$.

10. Find $\displaystyle\int_1^4 \int_1^2 \frac{1}{x^2} \, dx \, dx$.

11. Find $\int_0^\pi \int_0^{\pi/2} \cos x \, dx^2$.

12. Find all the curves for which $\dfrac{d^3y}{dx^3} = 0$.

16—3. Successive Partial Integration. Just as we can find partial derivatives of a function of two or more variables, so we can also integrate the function $f(x,y)$ in an analogous inverse process of partial differentiation. In the function $f(x,y)$, where x and y are both independent variables, let us for a moment consider x as a constant, and let y vary; then $f(x,y)$ becomes a function of y only. Now, under these conditions, suppose we integrate between the limits $y = c$ and $y = d$; we then have:

$$\int_c^d f(x,y) \, dy. \tag{1}$$

Now the value of this integral will depend not only upon the value of y, but also upon the value of x; hence the entire expression in (1) may be regarded as a function of x. Under this condition, let us now integrate with respect to x between the limits $x = a$ and $x = b$: the result becomes

$$\int_a^b \left[\int_c^d f(x,y) \, dy \right] dx, \tag{2}$$

which is generally written without the bracket as

$$\int_a^b \int_c^d f(x,y) \, dy \, dx. \tag{3}$$

The expression (3) is read: *"the double integral of $f(x,y)$ from $y = c$ to $y = d$ and from $x = a$ to $x = b$."*

EXAMPLE 1. Find the value of the double integral

$$\int_0^6 \int_2^3 (x^2 - y^2) \, dy \, dx.$$

Solution. We perform the *"y-integration"* first, remembering to *"hold"* x constant:

$$\int_2^3 (x^2 - y^2) \, dy = \left[x^2 y - \frac{1}{3} y^3 \right]_{y=2}^{y=3}$$
$$= 3x^2 - 9 - 2x^2 + \frac{8}{3} = x^2 - \frac{19}{3}. \tag{1}$$

Now we perform the second integration, or the "x-integration," upon the expression in (1), this time "holding" y constant, and integrating with respect to x:

$$\int_0^6 \left(x^2 - \frac{19}{3} \right) dx = \left[\frac{x^3}{3} - \frac{19}{3} x \right]_{x=0}^{x=6} = 72 - 38 = 34.$$

The limits of integration need not necessarily all be constants; very often the limits of y in the first integration are themselves functions of the variable x, as shown in the next two examples.

EXAMPLE 2.   Find the value of

$$\int_0^2 \int_{2x}^{4x} x^3 y \, dy \, dx.$$

Solution.   Integrating first with respect to y, we get:

$$\int_{2x}^{4x} x^3 y \, dx = \left[\frac{x^3 y^2}{2} \right]_{y=2x}^{y=4x} = 6x^5.$$

Now, integrating with respect to x:

$$\int_0^2 6x^5 \, dx = [x^6]_0^2 = 64.$$

EXAMPLE 3.   Find $\displaystyle\int_0^2 \int_0^{2x} (x^2 y + 6y^2 + x) \, dy \, dx.$

Solution.

$$\int_0^{2x} (x^2 y + 6y^2 + x) \, dy = \left[\frac{x^2 y^2}{2} + 2y^3 + xy \right]_{y=0}^{y=2x}$$
$$= 2x^4 + 16x^3 + 2x^2.$$
$$\int_0^2 (2x^4 + 16x^3 + 2x^2) \, dx = \left[\frac{2x^5}{5} + 4x^4 + \frac{2x^3}{3} \right]_{x=0}^{x=2} = 82 \frac{2}{15}.$$

The same ideas may be extended to *triple integrals.*

EXAMPLE 4.   Find the value of

$$\int_1^2 \int_0^x \int_{-2x}^{2x} (x + 2y - z) \, dz \, dy \, dx.$$

Solution. Consider x and y both constant, and integrate with respect to the variable z:

$$\int_{-2x}^{2x} (x + 2y - z)\, dz = \left[xz + 2yz - \frac{z^2}{2} \right]_{z=-2x}^{z=2x}$$
$$= 4x^2 + 8xy.$$

Then perform the y-integration, remembering that x is a constant:

$$\int_0^x (4x^2 + 8xy)\, dy = [4x^2 y + 4xy^2]_{y=0}^{y=x} = 8x^3.$$

Finally, perform the x-integration:

$$\int_1^2 8x^3\, dx = [2x^4]_{x=1}^{x=2} = 30.$$

EXERCISE 16—2

Find the value of each of the following multiple integrals:

1. $\displaystyle\int_1^2 \int_0^1 (2x + y^2)\, dy\, dx$

2. $\displaystyle\int_{-1}^{+1} \int_0^2 xy\, dy\, dx$

3. $\displaystyle\int_1^2 \int_0^x (x + y)^2\, dy\, dx$

4. $\displaystyle\int_0^a \int_0^{a^2-x} y\, dy\, dx$

5. $\displaystyle\int_{-\pi/2}^{\pi/2} \int_0^3 y^2 \cos x\, dy\, dx$

6. $\displaystyle\int_1^k \int_0^y (x + y)\, dx\, dy$

7. $\displaystyle\int_0^1 \int_0^{3x} xy^2\, dy\, dx$

8. $\displaystyle\int_0^\pi \int_0^{\tan \theta} \rho \cos^2 \theta\, d\rho\, d\theta$

9. $\displaystyle\int_0^{\pi/6} \int_0^a \rho^2 \sin \theta\, d\rho\, d\theta$

10. $\displaystyle\int_0^1 \int_0^x e^{x+y}\, dy\, dx$

11. $\displaystyle\int_0^{2\pi} \int_0^a \rho^3\, d\rho\, d\theta$

12. $\displaystyle\int_0^a \int_s^{2s} (st + t^2)\, dt\, ds$

13. $\displaystyle\int_0^2 \int_0^x \int_0^1 (xy + 2yz + 2xz)\, dz\, dy\, dx$

14. $\displaystyle\int_0^a \int_0^a \int_0^a (x^2 + y^2 + z^2)\, dz\, dy\, dx$

15. $\displaystyle\int_0^a \int_0^x \int_0^y xyz\, dz\, dy\, dx$

AREAS AND VOLUMES

16—4.   Plane Areas by Double Integration: Rectangular Coordinates.
It can be shown that the area enclosed by a curve (or several curves)
over a given region is given by the expression

$$A = \underset{\substack{\Delta x \to 0 \\ \Delta y \to 0}}{\text{limit}} \sum\sum \Delta y\, \Delta x.$$

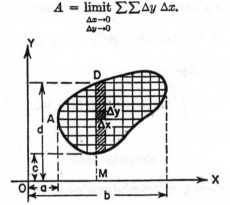

This area can be calculated by evaluating the corresponding integral, or

$$A = \int_a^b \int_c^d dy\, dx.$$

Interpreting this double integral from the figure, we see that when we
keep $x\,(=OM)$ constant and integrate with respect to y, we are summing
up all the elements in the shaded vertical strip, from $y = c$ to $y = d$;
then, by integrating the result of this summation with respect to x, we
are finding the sum of all such vertical strips between the limits $x = a$
and $x = b$, which clearly gives the total area desired. Of course, the
order of integration could be reversed, in which case we should be
summing up a series of horizontal strips, and the limits of both integrations would have to be changed accordingly.

In general, then, a plane area in rectangular coordinates may be found
from the double integrals

$$A = \iint dy\, dx, \quad \text{or} \quad A = \iint dx\, dy,$$

where the limits of integration have been properly selected. Some areas
can be found either by single integration, as we have already seen, or

by double integration; in other cases, the use of double integration affords the only method of finding the area.

EXAMPLE. Find by double integration the area enclosed by the circle $x^2 + y^2 = k^2$.

Solution. The area required is given by finding the area in the first quadrant by a double summation, and then multiplying by 4; hence

$$A = 4 \int_0^k \int_0^y dy \, dx.$$

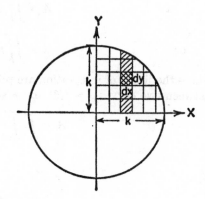

But, from the given equation, $y = \sqrt{k^2 - x^2}$; hence

$$A = 4 \int_0^k \int_0^{\sqrt{k^2 - x^2}} dy \, dx$$

$$= 4 \int_0^k [y]_0^{\sqrt{k^2 - x^2}} \, dx = 4 \int_0^k (k^2 - x^2)^{\frac{1}{2}} \, dx$$

$$= 4 \left[\frac{x}{2} \sqrt{k^2 - x^2} + \frac{k^2}{2} \arcsin \frac{x}{k} \right]_0^k = \pi k^2,$$

by formula [68], page 376.

16—5. Plane Areas by Double Integration: Polar Coordinates. For curves expressed in polar coordinates, the analogous formulas for

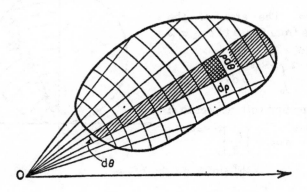

finding the area by double integration are given without proof; their reasonableness may be seen intuitively from the diagram:

$$A = \iint \rho \, d\rho \, d\theta,$$

or

$$A = \iint \rho \, d\theta \, d\rho,$$

where the limits of integration are properly chosen. Similarly, the area between two curves $\rho = f(\theta)$ and $\rho = \phi(\theta)$ is given by the formula

$$A = \int_\alpha^\beta \int_{f(\theta)}^{\phi(\theta)} \rho \, d\rho \, d\theta.$$

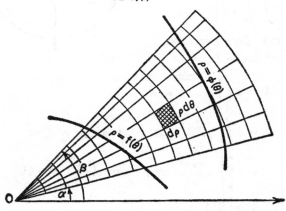

EXAMPLE. Find by double integration the area of the circle $\rho = 2r \sin \theta$.

Solution. The area of the semicircle from $\theta = 0$ to $\theta = \pi/2$ is given by

$$\int_0^{\pi/2} \int_0^{2r \sin \theta} \rho \, d\rho \, d\theta.$$

Integrating first with respect to ρ:

$$\int_0^{2r \sin \theta} \rho \, d\rho = \left[\frac{\rho^2}{2} \right]_0^{2r \sin \theta}$$

$$= 2r^2 \sin^2 \theta.$$

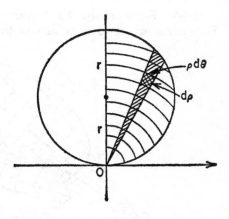

Hence:

$$\int_0^{\pi/2} (2r^2 \sin^2 \theta)\, d\theta = 2r^2 \int_0^{\pi/2} (\sin^2 \theta)\, d\theta$$

$$= 2r^2 \left[\frac{\theta}{2} - \frac{1}{4} \sin 2\theta \right]_0^{\pi/2} = 2r^2 \left(\frac{\pi}{4} \right) = \frac{\pi r^2}{2};$$

therefore the entire circle $= 2 \left(\dfrac{\pi r^2}{2} \right) = \pi r^2$.

16—6. Area between Two Curves. The method of summation is conveniently employed for finding the area included between two plane curves.

EXAMPLE 1. Find the area included between the lines $2y = 3x$, $3y = 2x$, and $x = 6$.

Solution. The area in question is given by

$$\int_0^6 \int_{2x/3}^{3x/2} dy\, dx,$$

since, in summing up a vertical strip like the one shown, y ranges in value from $\dfrac{2x}{3}$ to $\dfrac{3x}{2}$, and in summing up all such vertical strips, x ranges from $x = 0$ to $x = 6$.

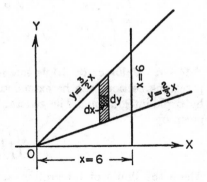

Therefore:

$$A = \int_0^6 \int_{2x/3}^{3x/2} dy\, dx = \int_0^6 [y]_{2x/3}^{3x/2}\, dx = \int_0^6 \left(\frac{5x}{6} \right) dx$$

$$= \frac{5}{6} \left[\frac{x^2}{2} \right]_0^6 = 15.$$

EXAMPLE 2. Find by double integration the area included between $x^2 = 4y$ and $x^2 = 5 - y$.

Solution. The desired area is given by summing up vertical strips such as shown in the figure. To express the area of one such strip, we integrate between the limits given by the equations when solved for y, that is, $y = \dfrac{x^2}{4}$ and $y = 5 - x^2$. To sum up all the vertical strips, we use the limits $x = -2$ to $x = +2$, the abscissas of the points of intersection of the curves.

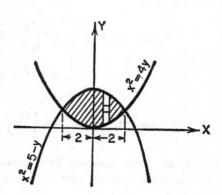

Hence:

$$A = \int_{-2}^{+2} \int_{x^2/4}^{5-x} dy \, dx = \int_{-2}^{+2} [y]_{x^2/4}^{5-x^2} \, dx$$

$$= \frac{5}{4} \int_{-2}^{+2} (4 - x^2) \, dx = \frac{5}{4} \left[4x - \frac{x^3}{3} \right]_{-2}^{+2} = \frac{40}{3} = 13\frac{1}{3}.$$

16—7. Volumes by Triple Integration. The ideas developed in the preceding sections can be extended to the problem of finding volumes by triple integration. The general formula for the volume of a solid is given by

$$V = \iiint dz \, dy \, dx,$$

where the limits of integration are found from the equations of the bounding surfaces. We shall illustrate by the following example.

EXAMPLE. Find the volume of the ellipsoid whose equation is

$$\frac{x^2}{a^2} + \frac{y^2}{b^2} + \frac{z^2}{c^2} = 1.$$

Solution. Consider, for purposes of summation, the portion of the solid lying in the first octant, which is one-eighth of the total volume. To sum up the differential rectangular parallelopipeds with their edges parallel to the respective axes, we take as the limits for z the values $z = 0$ and $z = c\sqrt{1 - \dfrac{x^2}{a^2} - \dfrac{y^2}{b^2}}$; for the limits of y, we take $y = 0$ for

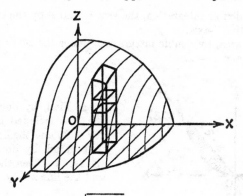

the lower limit, and $y = b\sqrt{1 - \dfrac{x^2}{a^2}}$ for the upper limit; for the limits

of x, we take $x = 0$ and $x = a$. The volume of the entire ellipsoid thus equals

$$V = 8\int_0^a \int_0^{b\sqrt{1 - \frac{x^2}{a^2}}} \int_0^{c\sqrt{1 - \frac{x^2}{a^2} - \frac{y^2}{b^2}}} dz\, dy\, dx$$

$$= 8c\int_0^a \int^{b\sqrt{1 - \frac{x^2}{a^2}}} \left(\sqrt{1 - \frac{x^2}{a^2} - \frac{y^2}{b^2}}\right) dy\, dx.$$

Now to integrate $\sqrt{1 - \dfrac{x^2}{a^2} - \dfrac{y^2}{b^2}}$, we note that formula [68], page 376,

can be applied, where the $\left(1 - \dfrac{x^2}{a^2}\right)$ corresponds to the constant "a^{2}"

of formula [68], and the $\left(\dfrac{y^2}{b^2}\right)$ corresponds to the "x^{2}" of formula [68].

Applying the formula, and substituting the limits, we obtain, finally, the volume $V = \frac{4}{3}\pi abc$. The details are left as an exercise for the reader. It will be noted that in the special case where $a = b = c = r$, the ellipsoid becomes a sphere, and the volume becomes $\frac{4}{3}\pi r^3$.

EXERCISE 16—3

1. Find, by double integration, the area under the parabola $x^2 = 4y$, from $x = 0$ to $x = 4$; verify your result by using single integration.

2. Find, by double integration, the area enclosed by the curve $y^2 = 4 + x$ and the line $x = 4$.

3. Find, by double integration, the area bounded by the curves $y = \sin x$, $y = \cos x$, and the Y-axis.

4. Find the area, by double integration, under the curve $y = e^x$, between the lines $x = 0$ and $x = 2$.

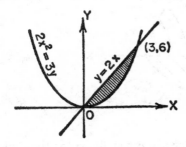

5. Find, by double integration, the area included by the parabola $2x^2 = 3y$ and the line $y = 2x$.

6. Find, by double integration, the area between the curve $8y^2 = x^3$ and the line $x = y$.

7. Find by double integration the area included between the parabolas $y^2 = 4 + x$ and $y^2 = 4 - 2x$.

8. Find by double integration the area of the circle $\rho = 10 \cos \theta$.

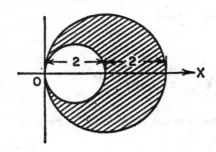

9. Find the area of the upper half of the cardioid

$$\rho = a(1 - \cos \theta).$$

10. Find by double integration the area between the circles

$$\rho = 2 \cos \theta \quad \text{and} \quad \rho = 4 \cos \theta.$$

Hint: Here θ ranges from

$$-\pi/2 \text{ to } \pi/2.$$

APPROXIMATE INTEGRATION

16—8. Need for Approximate Methods. It is frequently impossible or inconvenient to find the value of a definite integral by using the methods of integration discussed in Chapters Twelve and Thirteen. Or, it may be impossible to set up the analytical expression for the desired integration. In such cases we resort to an approximation method, of which there are several. Some approximation methods involve the use of infinite series; others are based on the measurement of an area, which may be found by computation, or by some mechanical device such as a *planimeter* or an *integraph*.

16—9. Integration by the Use of Infinite Series. When the indefinite integral of a given function $f(x) \, dx$ cannot be found, it is frequently possible to expand $f(x)$ into an ascending or descending power series in x. If this power series is convergent for a certain range of values of x, then the series obtained by integrating it term by term is also convergent within the limits for which the original power series is convergent. The more terms we take in the computation, the closer the sum will be to the value of $\int f(x)\, dx$. When integrating a power series term by term within the interval of convergence, it should be remembered that neither limit of integration may be an end point of the convergence interval.

EXAMPLE 1. Find $\int_0^2 e^{-x} \sin x\, dx$.

Solution.

$$e^{-x} = 1 - x + \frac{x^2}{2!} - \frac{x^3}{3!} + \frac{x^4}{4!} - \cdots \tag{1}$$

$$\sin x = x - \frac{x^3}{3!} + \frac{x^5}{5!} - \frac{x^7}{7!} + \cdots \tag{2}$$

Multiplying (1) and (2) term by term, and combining like terms:

$$e^{-x} \sin x = x - x^2 + \frac{x^3}{3} - \frac{x^5}{30} + \cdots \tag{3}$$

Series (3) is convergent for all values of x; hence, integrating (3) term by term:

$$\int e^{-x} \sin x\, dx = \frac{x^2}{2} - \frac{x^3}{3} + \frac{x^4}{12} - \frac{x^6}{180} + \cdots \tag{4}$$

Therefore:

$$\int_0^2 e^{-x} \sin x\, dx = \frac{x^2}{2}\Big]_0^2 - \frac{x^3}{3}\Big]_0^2 + \frac{x^4}{12}\Big]_0^2 - \frac{x^6}{180}\Big]_0^2 + \cdots$$
$$= 2.000 - 2.667 + 1.333 - .356 + \cdots$$
$$= .31, \text{ approx.}$$

EXAMPLE 2. Find $\int_0^{.5} \log \frac{1+x}{1-x}\, dx$.

Solution.

$$\log (1+x) = x - \frac{x^2}{2} + \frac{x^3}{3} - \frac{x^4}{4} + \frac{x^5}{5} - \cdots \tag{1}$$

$$\log (1-x) = -x - \frac{x^2}{2} - \frac{x^3}{3} - \frac{x^4}{4} - \frac{x^5}{5} - \cdots \tag{2}$$

Subtracting (2) from (1):

$$\log \frac{(1+x)}{(1-x)} = 2\left(x + \frac{x^3}{3} + \frac{x^5}{5} + \cdots\right). \tag{3}$$

Upon testing series (3) for convergence, the interval of convergence is found to be $-1 < x < +1$. Integrating (3):

$$\int \log \frac{1+x}{1-x} dx = 2\left[\frac{x^2}{2} + \frac{x^4}{12} + \frac{x^6}{30} + \cdots\right]. \tag{4}$$

Hence

$$\int_0^{0.5} \log \frac{1+x}{1-x} dx = 2\left\{\left[\frac{x^2}{2}\right]_0^{0.5} + \left[\frac{x^4}{12}\right]_0^{0.5} + \left[\frac{x^6}{30}\right]_0^{0.5} + \cdots\right\}$$

$$= 2[.1250 + .0052 + .0005 + \cdots]$$

$$= 2(.1307) = .261, \text{ approx.}$$

16—10. The Trapezoidal Rule. Consider the single-valued function $y = f(x)$, and assume that it is continuous in the interval $a \leqq x \leqq b$. We wish to find the area under the curve from $x = a$ to $x = b$. Suppose

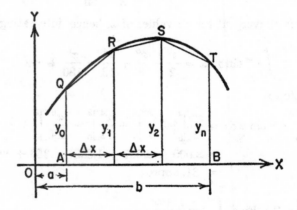

that y is positive throughout the interval from a to b. We may proceed by dividing the segment AB into n equal parts, each equal to Δx; the ordinates drawn through these points of division, $y_0, y_1, y_2, \cdots y_n$, divide the desired area into n strips of equal width. By drawing chords QR, RS, ST, etc., we form n trapezoids, having equal altitudes (Δx), and whose bases are the successive ordinates. Now the area under the curve is approximately equal to the sum of the areas of these inscribed trapezoids. Hence

$$\text{area of 1st trapezoid} = \tfrac{1}{2}\Delta x (y_0 + y_1);$$

$$\text{area of 2nd trapezoid} = \tfrac{1}{2}\Delta x (y_1 + y_2);$$

$$\text{area of 3rd trapezoid} = \tfrac{1}{2}\Delta x (y_2 + y_3);$$

$$\cdots\cdots\cdots\cdots\cdots\cdots\cdots\cdots\cdots\cdots\cdots\cdots\cdots$$

$$\text{area of } n\text{th trapezoid} = \tfrac{1}{2}\Delta x (y_{n-1} + y_n).$$

By adding, we obtain

$$A = \tfrac{1}{2}\Delta x (y_0 + 2y_1 + 2y_2 + 2y_3 + \cdots + 2y_{n-1} + y_n); \quad [1]$$

$$\text{or} \quad A = \Delta x (\tfrac{1}{2}y_0 + y_1 + y_2 + \cdots + y_{n-1} + \tfrac{1}{2}y_n). \quad [2]$$

This is known as the *Trapezoidal Rule*. It may be used not only to determine an area, but also to find the approximate value of any definite integral. The greater the number of intervals taken, the more nearly will the sum of the areas of the trapezoids be equal to the area under the curve.

EXAMPLE 1. Find the approximate area under the curve $y = \sqrt{x^3 - 10}$ from $x = 3$ to $x = 5$.

Solution. Divide the interval in question into any number of equal parts, say five parts. Thus $\Delta x = \tfrac{1}{5}(5 - 3) = 0.4$. Hence:

$$x_0 = 3.0, \qquad y_0 = \sqrt{27 - 10} = 4.12$$

$$x_1 = 3.4, \qquad y_1 = \sqrt{39.304 - 10} = 5.41$$

$$x_2 = 3.8, \qquad y_2 = \sqrt{54.872 - 10} = 6.70$$

$$x_3 = 4.2, \qquad y_3 = \sqrt{73.888 - 10} = 7.99$$

$$x_4 = 4.6, \qquad y_4 = \sqrt{97.336 - 10} = 9.35$$

$$x_5 = 5.0, \qquad y_5 = \sqrt{125 - 10} = 10.72$$

Hence, substituting in formula [1] above:

$$A = \tfrac{1}{2}(0.4)[4.12 + 2(5.41) + 2(6.70) + 2(7.99) + 2(9.35) + 10.72],$$

$$\text{or} \quad A = (0.2)(77.22) = 154.44.$$

To appreciate how close the approximation can come to the exact value, consider the following example.

EXAMPLE 2.   Compute the value of $\int_{1}^{10} x^3\, dx$, using the trapezoidal rule with 9 intervals; compare the result with the value obtained by direct integration.

Solution.

$$\frac{b-a}{n} = \frac{10-1}{9} = 1 = \Delta x. \quad \text{Also:}$$

$$x_0 = 1, \qquad y_0 = (1)^3 = 1$$

$$x_1 = 2, \qquad y_1 = (2)^3 = 8$$

$$x_2 = 3, \qquad y_2 = (3)^3 = 27$$

$$\cdots\cdots\cdots \qquad \cdots\cdots\cdots\cdots\cdots$$

$$x_9 = 10, \qquad y_9 = (10)^3 = 1000$$

Substituting in the formula [2] for the trapezoidal rule:

$$A = (1)[\tfrac{1}{2}+8+27+64+125+216+343+512+729+500],$$

or $A = 2524\tfrac{1}{2}$, approx.

By direct integration: $\int_{1}^{10} x^3\, dx = \dfrac{x^4}{4}\Big]_{1}^{10} = \tfrac{1}{4}(10{,}000 - 1) = 2499\tfrac{3}{4}$; the difference, $2524\tfrac{1}{2} - 2499\tfrac{3}{4}$, represents an error of less than 1%.

The trapezoidal rule may also be stated as follows:

$$\int_{a}^{b} f(x)\, dx = \frac{b-a}{2n}\left\{ f(a) + 2f\left(a + \frac{b-a}{n}\right) \right.$$

$$\left. + 2f\left(a + \frac{2(b-a)}{n}\right) + \cdots + 2f(b-h) + f(b), \quad [3] \right.$$

where the interval $(b-a)$ is divided into n equal parts, and $h = \dfrac{b-a}{n}$.

16—11. Simpson's Rule. Also known as the *parabolic* rule and as the "one-third" rule, this approximation formula is obtained by dividing the interval from a to b into any even number (n) of equal sub-intervals, each equal to Δx; then, instead of drawing chords of the corresponding points on the curve to form trapezoids, we draw *parabolic arcs* through each successive set of three points on the curve, with the axes of these parabolas parallel to the Y-axis. We then find the area under each of these parabolic strips. The area of the first of these parabolic strips, *APQRM*, for example, will be: the area of trapezoid *APRM* + area of

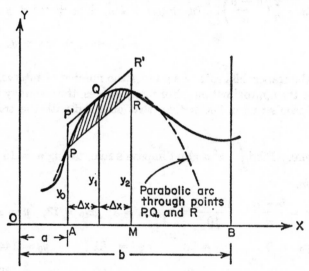

the parabolic segment *PQR*, which equals two-thirds of the area of the circumscribed parallelogram *PP'R'R*; hence the area of the parabolic strip equals

$$A = \frac{1}{2}(2\Delta x)(y_0 + y_2) + \frac{2}{3}(2\Delta x)\left[y_1 - \frac{1}{2}(y_0 + y_2)\right]$$

$$= \Delta x(y_0 + y_2) + \frac{2}{3}\Delta x(2y_1 - y_0 - y_2)$$

$$= \frac{\Delta x}{3}(y_0 + 4y_1 + y_2).$$

Similarly, the areas of the succeeding parabolic strips are given by

$$A = \frac{\Delta x}{3}(y_2 + 4y_3 + y_4),$$

$$A = \frac{\Delta x}{3}(y_4 + 4y_5 + y_6), \quad \text{etc.}$$

Upon adding, we arrive at Simpson's rule:

$$A = \frac{\Delta x}{3}(y_0 + 4y_1 + 2y_2 + 4y_3 + 2y_4 + \cdots + y_n). \qquad [4]$$

This may also be written, for convenience in computation, as:

$$\int_a^b f(x)\, dx = \frac{2}{3}\left(\frac{b-a}{n}\right)\left[\frac{1}{2}(y_0 + y_n) + 2(y_1 + y_3 + \cdots + y_{n-1})\right.$$
$$\left. + (y_2 + y_4 + \cdots + y_{n-2})\right] \quad [5]$$

As with the trapezoidal rule, the greater the number of intervals taken, the better the approximation. For many curves, the accuracy obtained with Simpson's rule is better than that obtained with the trapezoidal rule.

EXAMPLE. Find $\int_3^{13} x^2\, dx$ by Simpson's rule, taking $n = 10$.

Solution.

$$\Delta x = \frac{b-a}{n} = \frac{13-3}{10} = 1; \quad x_0 = 3; \quad x_{10} = 13; \quad y = x^2.$$

$y_0 = 9$	$y_3 = 36$	$y_6 = 81$	$y_9 = 144$
$y_1 = 16$	$y_4 = 49$	$y_7 = 100$	$y_{10} = 169$
$y_2 = 25$	$y_5 = 64$	$y_8 = 121$	

Hence, by Simpson's rule, we have:

$$\text{Area} = \tfrac{1}{3}(1)[9 + 4(16) + 2(25) + 4(36) + 2(49)$$
$$+ 4(64) + 2(81) + 4(100)$$
$$+ 2(121) + 4(144) + 169]$$

or $\qquad\qquad A = \tfrac{1}{3}(2170) = 723.3.$

NOTE. By the trapezoidal rule, $A = 725.0$; by direct integration, $\int_3^{13} x^2\, dx = \dfrac{x^3}{3}\bigg]_3^{13} = 723.3$, so that in this particular instance the use of Simpson's rule, by coincidence, gives the exact value.

EXERCISE 16—4

1. Find $\int_1^6 \dfrac{dx}{x}$ by the trapezoidal rule; take $n = 10$. Check for accuracy by direct integration; remember that $\log N = (2.3026)(\log_{10} N)$.

2. Find $\int_2^{10} x^2\, dx$ by the trapezoidal rule; take $n = 8$; check by direct integration.

3. Find $\int_0^8 x^4\, dx$ by Simpson's rule for $n = 8$; check by direct integration.

4. Find $\int_0^3 \dfrac{dx}{x^2 + 1}$ by Simpson's rule for $n = 6$; check by direct integration.

5. Calculate $\int_{-20°}^{+20°} \cos\theta\, d\theta$ by the trapezoidal rule for 5-degree intervals. Check by direct integration; remember that $\Delta x = 5° = \dfrac{\pi}{36}$.

6. Calculate $\int_4^{14} \log_{10} x\, dx$, taking unit intervals, by Simpson's rule. Check by direct integration; note that $\log_{10} x = \dfrac{1}{2.3026} \int \log_e x$.

16—12. The Integraph.

To find the area bounded by a curve, whether we know its equation or not, we may use a mechanical device known as the integraph. Its use depends upon the properties of *integral curves*, which will be briefly explained again (see §14—13).

Let $\phi(x)$ and $f(x)$ be two functions related as follows:

$$\text{If} \qquad \frac{d}{dx}\phi(x) = f(x), \tag{1}$$

then the curve
$$y = \phi(x) \tag{2}$$

is called an *integral curve* of the curve

$$y = f(x). \tag{3}$$

The curve $y = f(x)$ is called the original curve, or the *fundamental* curve; the curve $y = \phi(x)$ is more precisely known as the *first* integral curve, since there are also other integral curves associated with a given original curve.

This relation may also be expressed:

$$\int_0^x f(x)\, dx = \phi(x); \qquad \phi(0) = 0. \tag{4}$$

From the figure, it will be noted that:

(1) zero values of ordinates of the fundamental curve correspond to maximum or minimum values of ordinates of the first integral curve;

(2) maximum or minimum values of ordinates of the fundamental curve correspond to points of inflection on the first integral curve.

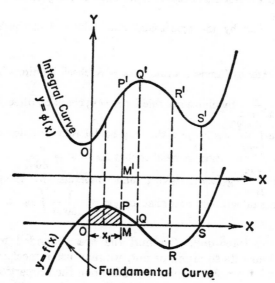

Fundamental Curve

Since, in general, an area is given by $\int_a^b y \, dx$, we see that the shaded area equals $\int_0^{x_1} f(x) \, dx$; but, from (4) above, we may say:

$$\text{shaded area} = \int_0^{x_1} f(x) \, dx = \phi(x) \Big]_0^{x_1} ;$$

since $\phi(0) = 0$, shaded area $= \int_0^{x_1} f(x) \, dx = \phi(x_1)$. But $\phi(x_1) = M'P'$. We may therefore conclude that, for a given abscissa x_1, the number which expresses the length of the ordinate of the integral curve $y = \phi(x)$ is the same as the number which expresses the area between the original curve $y = f(x)$, the axes, and the ordinate corresponding to this abscissa.

The integraph is a mechanical contrivance for drawing the first integral curve from its fundamental curve; as one movable part of a sliding framework follows the original curve, another part traces out the integral curve. Thus with the aid of this device we can find the area under, or enclosed by, a given empirical curve, simply by setting the instrument appropriately and reading the scales accurately.

16—13. The Planimeter. There are, generally speaking, three types of planimeters: the *polar* planimeter, the *disc* type, and the *rolling* type. The polar planimeter, for example, consists of (1) a horizontal bar OT known as the *tracer arm*; (2) a wheel or *roller* (R) of radius r turning on

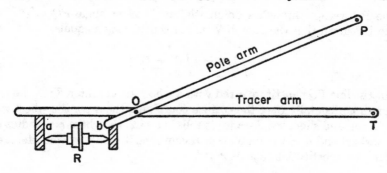

its axle *ab*; and (3) another horizontal bar *OP* called the *pole arm*. One end of the pole arm is hinged or pivoted at *O*, and acts as the center of rotation for the tracer arm. The other end of the pole arm (*P*) is known as the *pole*, and remains fixed throughout the operation of the planimeter; it serves as a center about which the entire instrument rotates. Any area circumscribed by the tracer point equals the product of the length of the tracer arm and the distance actually rolled over by the rim of the measuring roller.

The principle upon which the polar planimeter operates involves the calculation of the area swept over by a moving line of constant length (*l*). Let the area *MNPN'M'QM* be the area swept over by the line *MN*, of

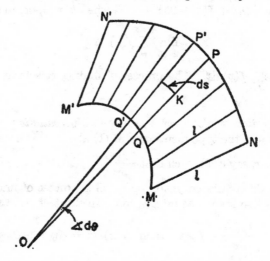

fixed length *l*; *PQ* and *P'Q'* are two successive positions of *MN*, and *ds* the circular arc described about *O* by the mid-point *K* of *PQ*, as the line PQ sweeps through a differential angle $d\theta$ = angle $PQOQ'P'$. It can be shown that the area MN' under consideration equals

$$\int l\,ds = l\int ds = ls.$$

For the line PQ substitute a rod with a wheel at its center K; as the rod moves horizontally over the surface containing the area, the wheel will both roll and slide; the distance it rolls is $s = 2\pi rn$, where r = radius of the wheel and n = the number of revolutions it makes. Thus the area MN', by substitution, equals $2\pi rnl$.

PRACTICAL APPLICATIONS

16—14. Statement of Principles. We shall now see how the Integral Calculus may be applied to practical problems in mechanics, physics, engineering, etc. Before studying specific illustrations, the following restatement of principles may prove helpful.

I. *Integration is a process of summation.* If $f(x)$ is any function of x, the values which $f(x)$ takes as x increases from a to b by equal increments h are: $f(a)$, $f(a + h)$, $f(a + 2h)$, $\cdots$. The limit of the sum $h[f(a) + f(a + h) + f(a + 2h) + \cdots]$ when $h = 0$, or, in symbols,

$$\lim_{\Delta x \to 0} \sum_{x=a}^{x=b} f(x) \cdot \Delta x,$$

is written as $\int_a^b f(x)\,dx$. The process of finding this limit is known as integration.

II. *Integration is also a process of anti-differentiation.* To find the limit of a sum of the type mentioned in (I) above, it is necessary to find a function of x, say $\phi(x)$, such that $\dfrac{d\phi(x)}{dx} = f(x)$, that is, a function $\phi(x)$ which when differentiated yields $f(x)$. This process of finding an anti-derivative is also known as integration. In short, if

$$\frac{d\phi(x)}{dx} = f(x), \quad \text{then} \quad f(x)\,dx = \phi(x) + C.$$

III. *Any definite integral may be interpreted as an area, even when the elements to be summed up represent quantities other than areas.* Since

$\int_{a}^{b} f(x)\, dx$ is the area bounded by the curve $y = f(x)$, the X-axis and the ordinate $x = a$ and $x = b$, it follows that, even when the value of the definite integral cannot be found by the ordinary methods of integration, its *approximate value* can nevertheless be determined by drawing the curve $y = f(x)$ between $x = a$ and $x = b$ and finding the area by approximation methods mentioned in the preceding section.

16—15. Speed-Time-Distance Relationship. Any curve drawn to represent the speed of a moving point (or object) is known as a *speed curve*. Suppose that the segment AB represents the total time interval t

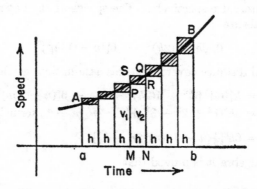

during which the motion is considered. Let AB be divided into a *large* number of small, equal parts; the ordinate aA represents the speed at the beginning of the interval from a to b, and bB the speed at the end of the interval. Ordinates such as MP and NQ represent the instantaneous values of the speed at the beginnings (or ends) of the various sub-intervals.

Now consider one of these sub-intervals, say MN. If the speed throughout this interval had *remained the same as at the beginning of the interval*, say v_1, the distance covered in that interval would equal $v_1 \times MN =$ the area of the rectangle $MPRN$, since distance = *rate* $\times$ *time* where the rate is constant. If the speed throughout the interval had been the same as that at the end of the interval, say v_2, the distance covered would have been $v_2 \times MN =$ the area of rectangle $MSQN$. The actual distance covered in the interval MN, however, lies *between* these two values. Similarly, if the number of intervals be increased indefinitely, the small shaded areas will decrease without limit, and the area under the curve AB bounded by aA, bB and ab will represent the *actual distance* covered in the interval ab. Thus the determination of distance covered when the relation between speed and time is known represents a problem in finding an area, and can be looked upon either as a problem of summation or of finding an anti-derivative.

EXAMPLE 1. A body moves according to the law $v = 6t^2$; find the distance covered in 8 seconds from rest.

Solution. First consider the problem as a summation. Let us suppose the entire interval of 8 seconds divided into n equal sub-intervals of h seconds each, so that $nh = 8$, and suppose the speed to remain constant throughout each sub-interval, being equal to the speed at the *beginning* of each sub-interval respectively. The speeds at the beginning of successive intervals are

$$0, \ 6h^2, \ 6(2h)^2, \ \cdots \ 6[(n-1)h]^2;$$

hence the total distance covered, on this assumption, would be

$$S_1 = h\{0 + 6h^2 + 6(2h)^2 + \cdots + 6[(n-1)h]^2\}$$
$$= 6h^3[1^2 + 2^2 + 3^2 + \cdots + (n-1) \text{ terms}],$$

or $\quad S_1 = 6h^3[\tfrac{1}{6}(n-1)(n)(2n-1)].$ \hfill (1)

NOTE. In algebra it is proved that

$$\sum_{n=1}^{n=n} (1^2 + 2^2 + 3^2 + \cdots + n^2) = \tfrac{1}{6}(n)(n+1)(2n+1).$$

Since $nh = 8$, or $n = \dfrac{8}{h}$, equation (1) becomes

$$S_1 = h^3[2n^3 - 3n^2 + n] = h^3\left[\frac{1024}{h^3} - \frac{192}{h^2} + \frac{8}{h}\right],$$

or $\quad S_1 = 1024 - 192h + 8h^2.$ \hfill (2)

Now let us assume that the speed during each sub-interval remained constant but equal to the speed at the *end* of each sub-interval. The total distance would then be

$$S_2 = h[6h^2 + 6(2h)^2 + \cdots + 6(nh)^2]$$
$$= 6h^3[1^2 + 2^2 + 3^2 + \cdots \text{ to } n \text{ terms}]$$
$$= 6h^3[\tfrac{1}{6}(n)(n+1)(2n+1)]$$
$$= h^3[2n^3 + 3n^2 + n],$$

or $\quad S_2 = 1024 + 192h + 8h^2.$ \hfill (3)

The actual distance $S > S_1$, but $< S_2$; as $h \to 0$, $S_1 \to S$, and $S_2 \to S$, and the limiting value $S = 1024$.

The same result is obtained if the problem is treated as one of finding the anti-derivative. For,

$$\text{since} \quad v = 6t^2, \quad \text{then} \quad \frac{dS}{dt} = 6t^2;$$

$$\text{hence} \quad S = \int_0^8 \frac{dS}{dt}\, dt = \int_0^8 6t^2\, dt = 2t^3 \Big]_0^8 = 1024.$$

The latter method, obviously, is the more convenient when the given function can be integrated. If it cannot be integrated, we may use an approximation method for finding the area instead of the algebraic method of summation used above.

EXAMPLE 2. If a body moves so that $v = 4t + 6t^2$, find the distance covered between the beginning of the third second and the end of the sixth second.

Solution. Here $v = 4t + 6t^2$; hence

$$S = \int_0^3 (4t + 6t^2)\, dt = [2t^2 + 2t^3]_3^6 = 432.$$

16—16. Force and Work. In physics we learn that when a force acts on a body, the product of the force by the distance through which it acts in the direction of the force is called the *work* done by the force.

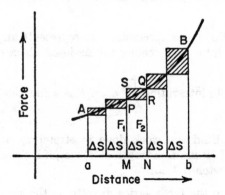

If the acting force is constant, the work done simply equals the force $\times$ displacement. But if the force is variable, we must use a method similar to the speed-time curve solution of §16—15. Now we are dealing with a *force-distance* curve, such as that shown here. The ordinates represent the varying values of the force, and the intervals ab, MN, etc. represent displacements, or distances through which the body is moved by the force. Precisely the same reasoning applies here as in §16—15; thus the work done while the body suffers a displacement MN is $F_1 \times MN =$ area of rectangle $MPRN$, or $F_2 \times MN =$ rectangle $MSQN$, depending upon whether we assume a force of F_1 or of F_2 is constant. The actual total work done is $\sum F \cdot \Delta S$, where $MN = \Delta S$. In general, therefore, the work done by a variable force when moving a body any distance, say x, is given by the area under a force-distance curve; or

$$W = \int_a^b F \, dx.$$

The same result is obtained by using the anti-derivative method of reasoning. Thus, to move the body an additional distance Δx requires additional work ΔW; this equals the average force $\bar{F}$ acting during Δx, multiplied by the distance Δx:

$$\Delta W = \bar{F} \cdot \Delta x,$$

or

$$\frac{\Delta W}{\Delta x} = \bar{F}.$$

Hence the instantaneous rate at which W is increasing is

$$\frac{dW}{dx} = F, \quad \text{or} \quad W = \int F \, dx.$$

NOTE 1. In the above discussion, Δs represents any one of a large number of *equal* intervals, whereas the Δx-intervals need not be equal.

NOTE 2. Before integrating $\int F \, ds$, if F is a function of s, it must be so expressed.

EXAMPLE 1. Find the work done in stretching a spring from its original length of 12 cm. to a length of 20 cm., if it is known that a force of 1.5 kg. will stretch it 1 cm.

Solution. Here the force varies directly as the elongation (Hooke's Law), or $F = ks$, where s is the elongation; hence, since $s = 1$ when $F = 1.5$, we have $1.5 = k(1)$, or $k = 1.5$. Therefore, $F = \frac{3}{2}s$.

Now, $$W = \int_0^8 F\, ds = \int_0^8 \frac{3}{2} s\, ds = \frac{3}{4} s^2 \Big]_0^8$$
$$= 48.$$

NOTE. Since the units used were kg. and cm., the work done equals 48 cm.-kg., 48,000 cm.-g.

EXAMPLE 2. The force (F lb.) driving a piston varies with the piston displacement (x in.) (distance moved) according to the law $F = 3000/x$. Find the work done from $x = 8$ to $x = 12$.

Solution.
$$W = \int_a^b F\, ds = \int_8^{12} \left(\frac{3000}{x}\right) dx = 3000 \int_8^{12} \frac{1}{x}\, dx$$
$$= 3000[\log x]_8^{12}$$
$$= 3000(\log 12 - \log 8) = 3000(\log 1.5)$$
$$= 3000(.4055) = 1217 \text{ in.-lb., approx.}$$

NOTE 1. When finding logarithms here, the table of *natural* logarithms must be used, since the relations $\dfrac{d}{dx}(\log x) = \dfrac{1}{x}$ and $\displaystyle\int \dfrac{dx}{x} = \log x$ are based upon logarithms to the base e, not the base 10.

NOTE 2. If a table of natural logarithms is not available, the value of $\log_e N$ can be found from a table of *common* logarithms by means of the relation $\log_e N = (2.3026) \log_{10} N$.

16—17. Liquid Pressure. From physics we know that the pressure exerted by a liquid on the walls of an open vessel is due to the *head* of liquid, that is, the height of liquid above that point. Hence, the pressure on any *horizontal* surface simply equals the weight of the column of liquid standing on that surface as a base and having a height equal to the distance that this surface is below the free surface of the liquid. For a horizontal surface of area A at a distance h below the surface, the total pressure P is given by

$$P = whA,$$

where w = weight of liquid in pounds per cubic unit.

To find the pressure on a surface that is not horizontal, we must remember that the pressure at different points varies with the distance

below the free surface, and so integration must be used. For a vertical
surface, the pressure increases with the depth; the differential pressure
equals the differential of the area multiplied by wh, or

$$dP = w \cdot h \cdot dA;$$

hence
$$P = w \int_{h_1}^{h_2} h \, dA,$$

where dA is expressed as a function of h to make integration possible,
and where the limits of integration h_1 and h_2 are respectively the smallest
and greatest heads on the surface in question.

EXAMPLE 1. Find the total pressure exerted on a vertical wall of an
open tank two-thirds filled with water ($w = 62.5$ lb./cu. ft.), if the
dimensions are those given in the diagram.

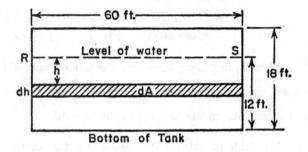

Solution. From the diagram,

$$dA = (RS) \, dh = 60dh.$$

$$P = w \int_0^{12} h \, dA$$

$$= 62.5 \int_0^{12} 60h \, dh$$

$$= (62.5)(60) \left[\frac{h^2}{2} \right]_0^{12}$$

$$= 270{,}000 \text{ lb., or } 135 \text{ tons, total pressure.}$$

EXAMPLE 2. Find the total pressure on the vertical wall of an open
tank filled with water if the wall is a trapezoid standing on the smaller
base with dimensions as given.

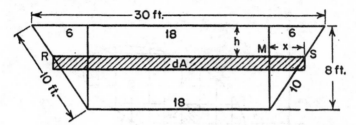

Solution. Here, to find RS, we note that if $MS = x$, then $\dfrac{x}{6} = \dfrac{8-h}{8}$, or $x = 6 - \dfrac{3h}{4}$.

Hence, $\qquad RS = 18 + 2x,\quad$ or $\quad RS = 30 - \dfrac{3h}{2}$.

Now, $\qquad dA = (RS)\, dh = \left(30 - \dfrac{3h}{2}\right) dh;$

and $\qquad P = w \displaystyle\int_0^8 \left(30 - \dfrac{3h}{2}\right) h\, dA.$

$$P = w \int_0^8 \left(30h - \dfrac{3h^2}{2}\right) dA,$$

$$P = 62.5 \left[15h^2 - \dfrac{h^3}{2}\right]_0^8$$

$$= (62.5)(704) = 44{,}000 \text{ lb.}$$

16—18. Center of Gravity. If we think of a plane area as divided into many small rectangles such as PQ, where the coordinates of P are (x,y), and the dimensions of the rectangle are Δx and Δy, then the area of PQ is $(\Delta x)(\Delta y)$, and the product of the area and its distance x from the Y-axis is called the *moment* of PQ with respect to the Y-axis. If we sum up all such moments throughout the entire area S, we have the *moment of the area* with respect to the Y-axis; thus

$$M_y = \lim_{\Delta x \to 0} \Sigma \left(\lim_{\Delta y \to 0} \Sigma x\, \Delta y\, \Delta x \right),$$

or $\qquad M_y = \displaystyle\iint x\, dy\, dx,$ \hfill (1)

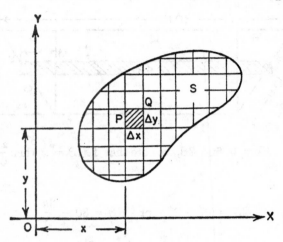

where the limits of integration are found from the equation of the boundary curve. Similarly,

$$M_x = \iint y \, dy \, dx. \tag{2}$$

Now if the moment of an area with respect to an axis is divided by the entire area, the quotient represents the average distance at which the entire area could be concentrated *and still give the same moment.* Let us denote these average distances by $\bar{x}$ and $\bar{y}$; then

$$\bar{x} = \frac{\iint x \, dy \, dx}{\iint dy \, dx}, \qquad \bar{y} = \frac{\iint y \, dy \, dx}{\iint dy \, dx}; \tag{1}$$

the point whose coordinates are $\bar{x}$ and $\bar{y}$ is known as the *center* of gravity of the area S. An alternative form of [1] is to write

$$A\bar{x} = \int x \, dA; \qquad A\bar{y} = \int y \, dA, \tag{2}$$

a form which is convenient for figures whose areas can be expressed or determined without first integrating.

EXERCISE 16—5

Review

1. Evaluate the following:

(a) $\displaystyle\int_{-2}^{0} (1 - 3x + 6x^2)\, dx$

(b) $\displaystyle\int_{0}^{a} x(b^2 - x^2)\, dx$

(c) $\displaystyle\int_{0}^{\pi/2} \sin^2 x\, dx$

(d) $\displaystyle\int_{0}^{\pi/4} \frac{\sin x\, dx}{\cos x}$

(e) $\displaystyle\int_{0}^{\infty} e^{-ax}\, dx$

(f) $\displaystyle\int_{1}^{e} x \log x\, dx$

2. Find the length of that part of each curve indicated:
   (a) $\rho = 2a \sin \theta$, from $\theta = 0$ to $\theta = \pi$.
   (b) $\rho = e^{a\theta}$, from $\rho = 0$ to $\rho = 2a$.

3. Find the area enclosed by the cardioid $\rho = a(1 - \cos \theta)$.

4. Find the value of:

(a) $\displaystyle\int_{0}^{2} \int_{-3x}^{3x} dy\, dx$

(b) $\displaystyle\int_{0}^{\pi/4} \int_{\sin x}^{\cos x} dy\, dx$

5. Find the total differential:
   (a) $z = ax^2 y^3$

   (b) $z = x^y$

6. If $z = \dfrac{x}{y} + \dfrac{y}{x}$, find:

(a) $\dfrac{\partial z}{\partial x}$;

(b) $\dfrac{\partial z}{\partial y}$

7. Determine the interval of convergence:
   (a) $1 + 2x^2 + 3x^3 + 4x^4 + \cdots$.

   (b) $2\left(x + \dfrac{x^3}{3} + \dfrac{x^5}{5} + \dfrac{x^7}{7} + \cdots\right)$.

8. (a) Using Taylor's theorem, prove that

$$(x + k)^n = x^n + nx^{n-1}k + \frac{n(n-1)}{2!} x^{n-2}k^2 + \cdots.$$

(b) Using Maclaurin's series, prove that

$$a^x = 1 + x \log a + \frac{(\log a)^2}{2!} x^2 + \frac{(\log a)^3}{3!} x^3 + \cdots.$$

9. Find the following:

    (a) $\lim_{x \to 0} (\log x)^x$ (b) $\lim_{x \to 0} (x + e^x)^{1/x}$

10. Find the following:

    (a) $\int x^2 \log x \, dx$ (b) $\int e^x \cos x \, dx$

Tables

INTEGRALS

A. Standard Elementary Forms

[1] $\displaystyle\int dx = x + C$

[2] $\displaystyle\int a\,dv = a\int dv$

[3] $\displaystyle\int (du + dv - dw) = \int du + \int dv - \int dw$

[4] $\displaystyle\int v^n\,dv = \frac{v^{n+1}}{n+1} + C \qquad (n \neq -1)$

[5] $\displaystyle\int \frac{dv}{v} = \log v + C = \log v + \log c = \log cv \qquad (C = \log c)$

[6] $\displaystyle\int a^v\,dv = \frac{a^v}{\log a} + C$

[7a] $\displaystyle\int e^v\,dv = e^v + C$ 　　　　　　　　　[7b] $\displaystyle\int e^{av}\,dv = \frac{e^{av}}{a} + C$

[8] $\displaystyle\int \sin v\,dv = -\cos v + C$

[9] $\displaystyle\int \cos v \, dv = \sin v + C$

[10] $\displaystyle\int \sec^2 v \, dv = \tan v + C$

[11] $\displaystyle\int \csc^2 v \, dv = -\cot v + C$

[12] $\displaystyle\int \sec v \tan v \, dv = \sec v + C$

[13] $\displaystyle\int \csc v \cot v \, dv = -\csc v + C$

[14] $\displaystyle\int \tan v \, dv = \log \sec v + C = -\log \cos v + C$

[15] $\displaystyle\int \cot v \, dv = \log \sin v + C$

[16] $\displaystyle\int \sec v \, dv = \log (\sec v + \tan v) + C$

[17] $\displaystyle\int \csc v \, dv = \log (\csc v - \cot v) + C$

[18] $\displaystyle\int \frac{dv}{v^2 + a^2} = \frac{1}{a} \arctan \frac{v}{a} + C$

[19] $\displaystyle\int \frac{dv}{v^2 - a^2} = \frac{1}{2a} \log \frac{v - a}{v + a} + C$

[20] $\displaystyle\int \frac{dv}{a^2 - v^2} = \frac{1}{2a} \log \frac{a + v}{a - v} + C$

[21] $\displaystyle\int \frac{dv}{\sqrt{v^2 \pm a^2}} = \log (v + \sqrt{v^2 \pm a^2}) + C$

[22] $\displaystyle\int \frac{dv}{\sqrt{a^2 - v^2}} = \arcsin \frac{v}{a} + C$

[23] $\displaystyle\int \frac{dv}{v\sqrt{v^2 - a^2}} = \frac{1}{a} \text{arc sec} \frac{v}{a} + C$

[24] $\displaystyle\int \frac{dv}{\sqrt{2av - v^2}} = \text{arc vers} \frac{v}{a} + C, \quad$ where $\text{vers} \dfrac{v}{a} = 1 - \cos \dfrac{v}{a}$

B. Forms Containing $(a + bx)$

[25] $\displaystyle\int \frac{dx}{a + bx} = \frac{1}{b}\log (a + bx) + C$

[26] $\displaystyle\int (a + bx)^n \, dx = \frac{(a + bx)^{n+1}}{b(n + 1)} + C,$ where $n \neq -1$

[27] $\displaystyle\int \frac{x \, dx}{a + bx} = \frac{1}{b^2}[a + bx - a \log (a + bx)] + C$

[28] $\displaystyle\int \frac{x^2 \, dx}{a + bx} = \frac{1}{b^3}\left[\frac{1}{2}(a + bx)^2 - 2a(a + bx) + a^2 \log (a + bx)\right] + C$

[29] $\displaystyle\int \frac{dx}{x(a + bx)} = -\frac{1}{a}\log \frac{a + bx}{x} + C$

[30] $\displaystyle\int \frac{dx}{x^2(a + bx)} = -\frac{1}{ax} + \frac{b}{a^2}\log \frac{a + bx}{x} + C$

[31] $\displaystyle\int \frac{x \, dx}{(a + bx)^2} = \frac{1}{b^2}\left[\log (a + bx) + \frac{a}{a + bx}\right] + C$

[32] $\displaystyle\int \frac{x^2 \, dx}{(a + bx)^2} = \frac{1}{b^3}\left[a + bx - 2a \log (a + bx) - \frac{a^2}{a + bx}\right] + C$

[33] $\displaystyle\int \frac{dx}{x(a + bx)^2} = \frac{1}{a(a + bx)} - \frac{1}{a^2}\log \frac{a + bx}{x} + C$

[34] $\displaystyle\int \frac{x \, dx}{(a + bx)^3} = \frac{1}{b^2}\left[-\frac{1}{a + bx} + \frac{a}{2(a + bx)^2}\right] + C$

C. Forms Containing $(a^2 + x^2)$, $(a^2 - x^2)$, $(a + bx^2)$

[35a] $\displaystyle\int \frac{dx}{a^2 + x^2} = \frac{1}{a}\arctan \frac{x}{a} + C$ [35b] $\displaystyle\int \frac{dx}{1 + x^2} = \arctan x + C$

[36a] $\displaystyle\int \frac{dx}{a^2 - x^2} = \frac{1}{2a}\log \frac{a + x}{a - x} + C$ [36b] $\displaystyle\int \frac{dx}{x^2 - a^2} = \frac{1}{2a}\log \frac{x - a}{x + a} + C$

[37] $\displaystyle\int \frac{dx}{a + bx^2} = \frac{1}{\sqrt{ab}}\arctan \sqrt{\frac{b}{a}}\,x + C,$ where a and b are positive

[38] $\displaystyle\int \frac{dx}{a^2 - b^2x^2} = \frac{1}{2ab}\log \frac{a + bx}{a - bx} + C$

[39] $\displaystyle\int \frac{x\,dx}{a+bx^2} = \frac{1}{2b}\log\left(x^2+\frac{a}{b}\right)+C$

[40] $\displaystyle\int \frac{x^2\,dx}{a+bx^2} = \frac{x}{b} - \frac{a}{b}\int\frac{dx}{a+bx^2}$

[41] $\displaystyle\int \frac{dx}{x(a+bx^2)} = \frac{1}{2a}\log\frac{x^2}{a+bx^2}+C$

[42] $\displaystyle\int \frac{dx}{x^2(a+bx^2)} = -\frac{1}{ax} - \frac{b}{a}\int\frac{dx}{a+bx^2}$

[43] $\displaystyle\int \frac{dx}{(a+bx^2)^2} = \frac{x}{2a(a+bx^2)} + \frac{1}{2a}\int\frac{dx}{a+bx^2}$

D. Forms Containing $\sqrt{x^2+a^2}$

[44] $\displaystyle\int (x^2+a^2)^{\frac{1}{2}}\,dx = \frac{x}{2}\sqrt{x^2+a^2} + \frac{a^2}{2}\log(x+\sqrt{x^2+a^2})+C$

[45] $\displaystyle\int (x^2+a^2)^{\frac{3}{2}}\,dx = \frac{x}{8}(2x^2+5a^2)\sqrt{x^2+a^2} + \frac{3a^4}{8}\log(x+\sqrt{x^2+a^2})+C$

[46] $\displaystyle\int x^2(x^2+a^2)^{\frac{1}{2}}\,dx = \frac{x}{8}(2x^2+a^2)\sqrt{x^2+a^2} - \frac{a^4}{8}\log(x+\sqrt{x^2+a^2})+C$

[47] $\displaystyle\int \frac{dx}{(x^2+a^2)^{\frac{1}{2}}} = \log(x+\sqrt{x^2+a^2})+C$

[48] $\displaystyle\int \frac{dx}{(x^2+a^2)^{\frac{3}{2}}} = \frac{x}{a^2\sqrt{x^2+a^2}}+C$

[49] $\displaystyle\int \frac{x\,dx}{(x^2+a^2)^{\frac{1}{2}}} = \sqrt{x^2+a^2}+C$

[50] $\displaystyle\int \frac{x^2\,dx}{(x^2+a^2)^{\frac{1}{2}}} = \frac{x}{2}\sqrt{x^2+a^2} - \frac{a^2}{2}\log(x+\sqrt{x^2+a^2})+C$

[51] $\displaystyle\int \frac{x^2\,dx}{(x^2+a^2)^{\frac{3}{2}}} = -\frac{x}{\sqrt{x^2+a^2}} + \log(x+\sqrt{x^2+a^2})+C$

[52] $\displaystyle\int \frac{dx}{x(x^2+a^2)^{\frac{1}{2}}} = \frac{1}{a}\log\frac{x}{a+\sqrt{x^2+a^2}}+C$

[53] $\displaystyle\int \frac{dx}{x^2(x^2+a^2)^{\frac{1}{2}}} = -\frac{\sqrt{x^2+a^2}}{a^2x}+C$

[54] $\displaystyle\int \frac{(x^2 + a^2)^{\frac{1}{2}}\, dx}{x} = \sqrt{x^2 + a^2} - a \log \frac{a + \sqrt{x^2 + a^2}}{x} + C$

[55] $\displaystyle\int \frac{(x^2 + a^2)^{\frac{1}{2}}\, dx}{x^2} = -\frac{\sqrt{x^2 + a^2}}{x} + \log\,(x + \sqrt{x^2 + a^2}) + C$

E. Forms Containing $\sqrt{x^2 - a^2}$

[56] $\displaystyle\int (x^2 - a^2)^{\frac{1}{2}}\, dx = \frac{x}{2}\,(x^2 - a^2)^{\frac{1}{2}} - \frac{a^2}{2} \log\,(x + \sqrt{x^2 - a^2}) + C$

[57] $\displaystyle\int (x^2 - a^2)^{\frac{3}{2}}\, dx = \frac{x}{8}\,(2x^2 - 5a^2)\sqrt{x^2 - a^2} + \frac{3a^4}{8} \log\,(x + \sqrt{x^2 - a^2}) + C$

[58] $\displaystyle\int x^2(x^2 - a^2)^{\frac{1}{2}}\, dx = \frac{x}{8}\,(2x^2 - a^2)\sqrt{x^2 - a^2} - \frac{a^4}{8} \log\,(x + \sqrt{x^2 - a^2}) + C$

[59] $\displaystyle\int \frac{dx}{(x^2 - a^2)^{\frac{1}{2}}} = \log\,(x + \sqrt{x^2 - a^2}) + C$

[60] $\displaystyle\int \frac{dx}{(x^2 - a^2)^{\frac{3}{2}}} = -\frac{x}{a^2\sqrt{x^2 - a^2}} + C$

[61] $\displaystyle\int \frac{x\, dx}{(x^2 - a^2)^{\frac{1}{2}}} = \sqrt{x^2 - a^2} + C$

[62] $\displaystyle\int \frac{x^2\, dx}{(x^2 - a^2)^{\frac{1}{2}}} = \frac{x}{2}\sqrt{x^2 - a^2} + \frac{a^2}{2} \log\,(x + \sqrt{x^2 - a^2}) + C$

[63] $\displaystyle\int \frac{x^2\, dx}{(x^2 - a^2)^{\frac{3}{2}}} = -\frac{x}{\sqrt{x^2 - a^2}} + \log\,(x + \sqrt{x^2 - a^2}) + C$

[64a] $\displaystyle\int \frac{dx}{x(x^2 - a^2)^{\frac{1}{2}}} = \frac{1}{a} \operatorname{arc\,sec} \frac{x}{a} + C$ [64b] $\displaystyle\int \frac{dx}{x(x^2 - 1)^{\frac{1}{2}}} = \operatorname{arc\,sec} x + C$

[65] $\displaystyle\int \frac{dx}{x^2(x^2 - a^2)^{\frac{1}{2}}} = \frac{\sqrt{x^2 - a^2}}{a^2 x} + C$

[66] $\displaystyle\int \frac{(x^2 - a^2)^{\frac{1}{2}}\, dx}{x} = \sqrt{x^2 - a^2} - a \operatorname{arc\,cos} \frac{a}{x} + C$

[67] $\displaystyle\int \frac{(x^2 - a^2)^{\frac{1}{2}}\, dx}{x^2} = -\frac{\sqrt{x^2 - a^2}}{x} + \log\,(x + \sqrt{x^2 - a^2}) + C$

F. Forms Containing $\sqrt{a^2 - x^2}$

[68] $\int (a^2 - x^2)^{1/2}\, dx = \dfrac{x}{2}\sqrt{a^2 - x^2} + \dfrac{a^2}{2}\arcsin\dfrac{x}{a} + C$

[69] $\int (a^2 - x^2)^{3/2}\, dx = \dfrac{x}{8}(5a^2 - 2x^2)\sqrt{a^2 - x^2} + \dfrac{3}{8}a^4\arcsin\dfrac{x}{a} + C$

[70] $\int x^2(a^2 - x^2)^{1/2}\, dx = \dfrac{x}{8}(2x^2 - a^2)\sqrt{a^2 - x^2} + \dfrac{a^4}{8}\arcsin\dfrac{x}{a} + C$

[71] $\int \dfrac{dx}{(a^2 - x^2)^{1/2}} = \arcsin\dfrac{x}{a} + C$

[72] $\int \dfrac{dx}{\sqrt{1 - x^2}} = \arcsin x + C$

[73] $\int \dfrac{dx}{(a^2 - x^2)^{3/2}} = \dfrac{x}{a^2\sqrt{a^2 - x^2}} + C$

[74] $\int \dfrac{x\, dx}{(a^2 - x^2)^{1/2}} = -\sqrt{a^2 - x^2} + C$

[75] $\int \dfrac{x^2\, dx}{(a^2 - x^2)^{1/2}} = -\dfrac{x}{2}\sqrt{a^2 - x^2} + \dfrac{a^2}{2}\arcsin\dfrac{x}{a} + C$

[76] $\int \dfrac{x^2\, dx}{(a^2 - x^2)^{3/2}} = \dfrac{x}{\sqrt{a^2 - x^2}} - \arcsin\dfrac{x}{a} + C$

[77] $\int \dfrac{dx}{x(a^2 - x^2)^{1/2}} = \dfrac{1}{a}\log\dfrac{x}{a + \sqrt{a^2 - x^2}} + C$

[78] $\int \dfrac{dx}{x^2(a^2 - x^2)^{1/2}} = -\dfrac{\sqrt{a^2 - x^2}}{a^2 x} + C$

[79] $\int \dfrac{(a^2 - x^2)^{1/2}}{x}\, dx = \sqrt{a^2 - x^2} - a\log\dfrac{a + \sqrt{a^2 - x^2}}{x} + C$

[80] $\int \dfrac{(a^2 - x^2)^{1/2}}{x^2}\, dx = -\dfrac{\sqrt{a^2 - x^2}}{x} - \arcsin\dfrac{x}{a} + C$

G. Forms Containing $\sqrt{a + bx}$

[81] $\int x\sqrt{a + bx}\, dx = -\dfrac{2(2a - 3bx)\sqrt{(a + bx)^3}}{15b^2} + C$

[82] $\displaystyle\int x^2\sqrt{a+bx}\,dx = \frac{2(8a^2 - 12abx + 15b^2x^2)\sqrt{(a+bx)^3}}{105b^3} + C$

[83] $\displaystyle\int \frac{x\,dx}{\sqrt{a+bx}} = -\frac{2(2a-bx)}{3b^2}\sqrt{a+bx} + C$

[84] $\displaystyle\int \frac{x^2\,dx}{\sqrt{a+bx}} = \frac{2(8a^2 - 4abx + 3b^2x^2)}{15b^3}\sqrt{a+bx} + C$

[85] $\displaystyle\int \frac{dx}{x\sqrt{a+bx}} = \frac{1}{\sqrt{a}}\log\frac{\sqrt{a+bx}-\sqrt{a}}{\sqrt{a+bx}+\sqrt{a}} + C, \quad \text{if } a > 0$

[86] $\displaystyle\int \frac{dx}{x\sqrt{a+bx}} = \frac{2}{\sqrt{-a}}\,\text{arc tan}\sqrt{\frac{a+bx}{-a}} + C, \quad \text{if } a < 0$

[87] $\displaystyle\int \frac{dx}{x^2\sqrt{a+bx}} = \frac{-\sqrt{a+bx}}{ax} - \frac{b}{2a}\int \frac{dx}{x\sqrt{a+bx}}$

[88] $\displaystyle\int \frac{\sqrt{a+bx}\,dx}{x} = 2\sqrt{a+bx} + a\int \frac{dx}{x\sqrt{a+bx}}$

H. Forms Containing Trigonometric Functions

(See also formulas [8] *to* [17])

[89] $\displaystyle\int \sin^2 x\,dx = \frac{x}{2} - \frac{1}{4}\sin 2x + C$

[90] $\displaystyle\int \cos^2 x\,dx = \frac{x}{2} + \frac{1}{4}\sin 2x + C$

[91] $\displaystyle\int \sin mx \sin nx\,dx = -\frac{\sin(m+n)x}{2(m+n)} + \frac{\sin(m-n)x}{2(m-n)} + C$

[92] $\displaystyle\int \cos mx \cos nx\,dx = \frac{\sin(m+n)x}{2(m+n)} + \frac{\sin(m-n)x}{2(m-n)} + C$

[93] $\displaystyle\int \sin mx \cos nx\,dx = -\frac{\cos(m+n)x}{2(m+n)} - \frac{\cos(m-n)x}{2(m-n)} + C$

[94] $\displaystyle\int e^{ax} \sin nx\,dx = \frac{e^{ax}(a\sin nx - n\cos nx)}{a^2 + n^2} + C$

[95] $\displaystyle\int e^{ax} \cos nx\,dx = \frac{e^{ax}(n\sin nx + a\cos nx)}{a^2 + n^2} + C$

I. Forms Containing Exponential and Logarithmic Functions

[96a] $\displaystyle\int e^x\,dx = e^x + C$ [96b] $\displaystyle\int e^{-x}\,dx = -e^{-x} + C$

[97] $\displaystyle\int e^{ax}\,dx = \frac{e^{ax}}{a} + C$

[98] $\displaystyle\int a^x\,dx = \frac{a^x}{\log a} + C$

[99] $\displaystyle\int a^{bx}\,dx = \frac{a^{bx}}{b\log a} + C$

[100] $\displaystyle\int xe^{ax}\,dx = \frac{e^{ax}}{a^2}\,(ax - 1) + C$

[101] $\displaystyle\int xa^x\,dx = \frac{a^x x}{\log a} - \frac{a^x}{(\log a)^2} + C$

[102] $\displaystyle\int \log x\,dx = x\log x - x + C$

[103] $\displaystyle\int \frac{dx}{x\log x} = \log\,(\log x) + C$

[104] $\displaystyle\int x\log x\,dx = \frac{1}{2}x^2\left(\log x - \frac{1}{2}\right) + C$

[105] $\displaystyle\int x^2\log x\,dx = \frac{1}{3}x^3\left(\log x - \frac{1}{3}\right) + C$

VALUES OF e^x AND e^{-x}

VALUES OF e^x FROM $x = 0$ TO $x = 5.9$

x	0.0	0.1	0.2	0.3	0.4	0.5	0.6	0.7	0.8	0.9
0	1.00	1.11	1.22	1.35	1.49	1.65	1.82	2.01	2.23	2.46
1	2.72	3.00	3.32	3.67	4.06	4.48	4.95	5.47	6.05	6.69
2	7.39	8.17	9.03	9.97	11.0	12.2	13.5	14.9	16.4	18.2
3	20.1	22.2	24.5	27.1	30.0	33.1	36.6	40.4	44.7	49.4
4	54.6	60.3	66.7	73.7	81.5	90.0	99.5	109.9	121.5	134.3
5	148.4	164.0	181.3	200.3	224.4	244.7	270.4	298.9	330.3	365.0

VALUES OF e^{-x} FROM $x = 0$ TO $x = 5.9$

x	0.0	0.1	0.2	0.3	0.4	0.5	0.6	0.7	0.8	0.9
0	1.00	0.90	0.82	0.74	0.67	0.61	0.55	0.50	0.45	0.41
1	0.37	0.33	0.30	0.27	0.25	0.22	0.20	0.18	0.17	0.15
2	0.14	0.12	0.11	0.10	0.09	0.08	0.07	0.07	0.06	0.06
3	0.05	0.05	0.04	0.04	0.03	0.03	0.03	0.02	0.02	0.02
4	0.02	0.02	0.01	0.01	0.01	0.01	0.01	0.01	0.01	0.01
5	0.007	0.006	0.005	0.005	0.005	0.004	0.004	0.003	0.003	0.003

COMMON LOGARITHMS

N	0	1	2	3	4	5	6	7	8	9
10	0000	0043	0086	0128	0170	0212	0253	0294	0334	0374
11	0414	0453	0492	0531	0569	0607	0645	0682	0719	0755
12	0792	0828	0864	0899	0934	0969	1004	1038	1072	1106
13	1139	1173	1206	1239	1271	1303	1335	1367	1399	1430
14	1461	1492	1523	1553	1584	1614	1644	1673	1703	1732
15	1761	1790	1818	1847	1875	1903	1931	1959	1987	2014
16	2041	2068	2095	2122	2148	2175	2201	2227	2253	2279
17	2304	2330	2355	2380	2405	2430	2455	2480	2504	2529
18	2553	2577	2601	2625	2648	2672	2695	2718	2742	2765
19	2788	2801	2833	2856	2878	2900	2923	2945	2967	2989
20	3010	3032	3054	3075	3096	3118	3139	3160	3181	3201
21	3222	3243	3263	3284	3304	3324	3345	3365	3385	3404
22	3424	3444	3464	3483	3502	3522	3541	3560	3579	3598
23	3617	3636	3655	3674	3692	3711	3729	3747	3766	3784
24	3802	3820	3838	3856	3874	3892	3909	3927	3945	3962
25	3979	3997	4014	4031	4048	4065	4082	4099	4116	4133
26	4150	4166	4183	4200	4216	4232	4249	4265	4281	4298
27	4314	4330	4346	4362	4378	4393	4409	4425	4440	4456
28	4472	4487	4502	4518	4533	4548	4564	4579	4594	4609
29	4624	4639	4654	4669	4683	4698	4713	4728	4742	4757
30	4771	4786	4800	4814	4829	4843	4857	4871	4886	4900
31	4914	4928	4942	4955	4969	4983	4997	5011	5024	5038
32	5051	5065	5079	5092	5105	5119	5132	5145	5159	5172
33	5185	5198	5211	5224	5237	5250	5263	5276	5289	5302
34	5315	5328	5340	5353	5366	5378	5391	5403	5416	5428
35	5441	5453	5465	5478	5490	5502	5514	5527	5539	5551
36	5563	5575	5587	5599	5611	5623	5635	5647	5658	5670
37	5682	5694	5705	5717	5729	5740	5752	5763	5775	5786
38	5798	5809	5821	5832	5843	5855	5866	5877	5888	5899
39	5911	5922	5933	5944	5955	5966	5977	5988	5999	6010
40	6021	6031	6042	6053	6064	6075	6085	6096	6107	6117
41	6128	6138	6149	6160	6170	6180	6191	6201	6212	6222
42	6232	6243	6253	6263	6274	6284	6294	6304	6314	6325
43	6335	6345	6355	6365	6375	6385	6395	6405	6415	6425
44	6435	6444	6454	6464	6474	6484	6493	6503	6513	6522
45	6532	6542	6551	6561	6571	6580	6590	6599	6609	6618
46	6628	6637	6646	6656	6665	6675	6684	6693	6702	6712
47	6721	6730	6739	6749	6758	6767	6776	6785	6794	6803
48	6812	6821	6850	6839	6848	6857	6866	6875	6884	6893
49	6902	6911	6920	6928	6937	6946	6955	6964	6972	6981
50	6990	6998	7007	7016	7024	7033	7042	7050	7059	7067
51	7076	7084	7093	7101	7110	7118	7126	7135	7143	7152
52	7160	7168	7177	7185	7193	7202	7210	7218	7226	7235
53	7243	7251	7259	7267	7275	7284	7292	7300	7308	7316
54	7324	7332	7340	7348	7356	7364	7372	7380	7388	7396

N	0	1	2	3	4	5	6	7	8	9
55	7404	7412	7419	7427	7435	7443	7451	7459	7466	7474
56	7482	7490	7497	7505	7513	7520	7528	7536	7543	7551
57	7559	7566	7574	7582	7589	7597	7604	7612	7619	7627
58	7634	7642	7649	7657	7664	7672	7679	7686	7694	7701
59	7709	7716	7723	7731	7738	7745	7752	7760	7767	7774
60	7782	7789	7796	7803	7810	7818	7825	7832	7839	7846
61	7853	7860	7868	7875	7882	7889	7896	7903	7910	7917
62	7924	7931	7938	7945	7952	7959	7966	7973	7980	7987
63	7993	8000	8007	8014	8021	8028	8035	8041	8048	8055
64	8062	8069	8075	8082	8089	8096	8102	8109	8116	8122
65	8129	8136	8142	8149	8156	8162	8169	8176	8182	8189
66	8195	8202	8209	8215	8222	8228	8235	8241	8248	8254
67	8261	8267	8274	8280	8287	8293	8299	8306	8312	8319
68	8325	8331	8338	8344	8351	8357	8363	8370	8376	8382
69	8388	8395	8401	8407	8414	8420	8426	8432	8439	8445
70	8451	8457	8463	8470	8476	8482	8488	8494	8500	8506
71	8513	8519	8525	8531	8537	8543	8549	8555	8561	8567
72	8573	8579	8585	8591	8597	8603	8609	8615	8621	8627
73	8633	8639	8645	8651	8657	8663	8669	8675	8681	8686
74	8692	8698	8704	8710	8716	8722	8727	8733	8739	8745
75	8751	8756	8762	8768	8774	8779	8785	8791	8797	8802
76	8808	8814	8820	8825	8831	8837	8842	8848	8854	8859
77	8865	8871	8876	8882	8887	8893	8899	8904	8910	8915
78	8921	8927	8932	8938	8943	8949	8954	8960	8965	8971
79	8976	8982	8987	8993	8998	9004	9009	9015	9020	9025
80	9031	9036	9042	9047	9053	9058	9063	9069	9074	9079
81	9085	9090	9096	9101	9106	9112	9117	9122	9128	9133
82	9138	9143	9149	9154	9159	9165	9170	9175	9180	9186
83	9191	9196	9201	9206	9212	9217	9222	9227	9232	9238
84	9243	9248	9253	9258	9263	9269	9274	9279	9284	9289
85	9294	9299	9304	9309	9315	9320	9325	9330	9335	9340
86	9345	9350	9355	9360	9365	9370	9375	9380	9385	9390
87	9395	9400	9405	9410	9415	9420	9425	9430	9435	9440
88	9445	9450	9455	9460	9465	9469	9474	9479	9484	9489
89	9494	9499	9504	9509	9513	9518	9523	9528	9533	9538
90	9542	9547	9552	9557	9562	9566	9571	9576	9581	9586
91	9590	9595	9600	9605	9609	9614	9619	9624	9628	9633
92	9638	9643	9647	9652	9657	9661	9666	9671	9675	9680
93	9685	9689	9694	9699	9703	9708	9713	9717	9722	9727
94	9731	9736	9741	9745	9750	9754	9759	9763	9768	9773
95	9777	9782	9786	9791	9795	9800	9805	9809	9814	9818
69	9823	9827	9832	9836	9841	9845	9850	9854	9859	9863
97	9868	9872	9877	9881	9886	9890	9894	9899	9903	9908
98	9912	9917	9921	9926	9930	9934	9939	9943	9948	9952
99	9956	9961	9965	9969	9974	9978	9983	9987	9991	9996

Reference Material from Other Branches of Mathematics

ALGEBRA

FACTORS:

$$a^2 \pm 2ab + b^2 = (a \pm b)^2$$
$$a^2 - b^2 = (a + b)(a - b)$$
$$a^3 \pm b^3 = (a \pm b)(a^2 \mp ab + b^2)$$
$$a^4 - b^4 = (a^2 + b^2)(a + b)(a - b)$$
$$a^4 + b^4 = (a^2 + b^2 + \sqrt{2}\,ab)(a^2 + b^2 - \sqrt{2}\,ab)$$
$$a^{2n} - b^{2n} = (a^n + b^n)(a^n - b^n)$$

EXPONENTS:

$$a^n = a \cdot a \cdot a \cdots \text{ to } n \text{ factors} \qquad a^{-n} = \frac{1}{a^n}$$

$$a^m \cdot a^n = a^{m+n} \qquad \left(\frac{a}{b}\right)^n = \frac{a^n}{b^n}$$

$$(a^m)^n = a^{mn}$$

$$\frac{a^{-m}}{b^{-n}} = \frac{b^n}{a^m}$$

$$\frac{a^m}{a^n} = a^{m-n}$$

$$\frac{a^n}{a^n} = a^{n-n} = a^0 = 1$$

$$(ab)^n = a^n b^n$$

$$a^n \cdot a^{-n} = a^{n-n} = a^0 = 1$$

ROOTS:

$$a^{1/n} = \sqrt[n]{a}$$

$$\sqrt[n]{ab} = \sqrt[n]{a} \cdot \sqrt[n]{b}$$

$$a^{m/n} = \sqrt[n]{a^m}$$

$$\sqrt[n]{\frac{1}{a^m}} = \sqrt[n]{a^{-m}} = a^{-m/n}$$

$$\sqrt[n]{a^m} = (\sqrt[n]{a})^m$$

$$\sqrt[n]{\frac{a}{b}} = \frac{\sqrt[n]{a}}{\sqrt[n]{b}}$$

$$\sqrt[n]{a^n} = (\sqrt[n]{a})^n = a$$

$$\sqrt[n]{\sqrt[m]{a}} = \sqrt[mn]{a} = a^{1/mn}$$

LOGARITHMS:

$$\text{If } b^x = N, \text{ then } x = \log_b N$$

$$\log \frac{1}{a} = -\log a$$

$$b^{\log_b N} = N$$

$$\log (MN) = \log M + \log N$$

$$\log_b a \cdot \log_a b = 1$$

$$\log \left(\frac{M}{N}\right) = \log M - \log N$$

$$\log_b b = 1$$

$$\log M^n = n \log M$$

$$\log_b 1 = 0$$

$$\log \sqrt[n]{M} = \frac{1}{n} \log M$$

$$\log_a N = \frac{\log_b N}{\log_b a} = \log_b N \cdot \log_a b$$

QUADRATIC EQUATIONS:

If r_1 and r_2 are the roots of the equation

$$ax^2 + bx + c = 0,$$

then $\quad r_1 = \dfrac{-b + \sqrt{b^2 - 4ac}}{2a}, \qquad r_2 = \dfrac{-b - \sqrt{b^2 - 4ac}}{2a}.$

The discriminant $\Delta = b^2 - 4ac$;

if $\Delta > 0$, the roots are real and distinct;

if $\Delta = 0$, the roots are real and equal;

if $\Delta < 0$, the roots are complex.

$$(x - r_1)(x - r_2) = ax^2 + bx + c = 0;$$

$$r_1 + r_2 = -\frac{b}{a}; \qquad r_1 \cdot r_2 = \frac{c}{a}.$$

PROGRESSIONS:

Arithmetic. $\quad l = a + (n - 1)\, d;$

$$S_n = \frac{n}{2}\,(a + l) = \frac{n}{2}[2a + (n - 1)\, d];$$

$$A.M. = \frac{a + b}{2}.$$

Geometric. $\quad l = ar^{n-1};$

$$S_n = \frac{rl - a}{r - 1} = \frac{a(r^n - 1)}{r - 1};$$

$$S_\infty = \frac{a}{1 - r}, \text{ where } |r| < 1;$$

$$G.M. = \sqrt{ab}.$$

COMPLEX NUMBERS:

$$i = \sqrt{-1}; \qquad i^2 = -1; \qquad i^3 = -i; \qquad i^4 = +1.$$

If $\qquad a + bi = x + yi, \quad$ then $a = x$ and $b = y.$

If $\qquad a + bi = r(\cos\theta + i\sin\theta),$

then $\qquad a = r\cos\theta, \qquad\qquad b = r\sin\theta,$

$$r = \sqrt{a^2 + b^2}, \qquad \tan\theta = \frac{b}{a}.$$

$$[r(\cos\theta + i\sin\theta)]^n = r^n(\cos n\theta + i\sin n\theta).$$

FACTORIALS:

$$n! = 1 \cdot 2 \cdot 3 \cdot 4 \cdots \text{ to } n \text{ factors.}$$

$$_nP_n = n! \qquad 0! = 1$$

$$_nC_r = \frac{n(n-1)(n-2)\cdots(n-r+1)}{r!} = \frac{n!}{r!(n-r)!}.$$

$$_nC_r = {_nC_{n-r}}$$

$$_nC_n = {_nC_0} = 1$$

$$_nC_1 + {_nC_2} + {_nC_3} + \cdots + {_nC_n} = 2^n - 1;$$

$$_nC_0 + {_nC_1} + {_nC_2} + \cdots + {_nC_n} = 2^n.$$

BINOMIAL EXPANSION:

$$(a + b)^n = (a + b)(a + b)(a + b) \cdots \text{ to } n \text{ factors.}$$

$$(a + b)^n = {_nC_0}a^n + {_nC_1}a^{n-1}b + {_nC_2}a^{n-2}b^2 + \cdots$$
$$+ {_nC_{n-1}}ab^{n-1} + {_nC_n}b^n.$$

$$(a + b)^n = a^n + na^{n-1}b + \frac{n(n-1)}{2!}a^{n-2}b^2$$

$$+ \frac{n(n-1)(n-2)}{3!}a^{n-3}b^3 + \cdots;$$

$$+ \frac{n(n-1)(n-2)\cdots(n-r+2)}{(r-1)!}a^{n-r+1}b^{r-1}$$

$$+ \cdots + b^n.$$

The rth term of $(a + b)^n$ is:

$$\frac{n(n-1)(n-2)\cdots(n-r+2)}{(r-1)!}a^{n-r+1}b^{r-1}.$$

DETERMINANTS:

$$\begin{vmatrix} a_1b_1 \\ a_2b_2 \end{vmatrix} = a_1b_2 - a_2b_1$$

$$\Delta = \begin{vmatrix} a_1 b_1 c_1 \\ a_2 b_2 c_2 \\ a_3 b_3 c_3 \end{vmatrix} = \begin{aligned} &(a_1 b_2 c_3 + b_1 c_2 a_3 + c_1 b_3 a_2) \\ &\quad - (a_3 b_2 c_1 + a_2 b_1 c_3 + a_1 b_3 c_2) \end{aligned}$$

$$\Delta = (a_1) \begin{vmatrix} b_2 c_2 \\ b_3 c_3 \end{vmatrix} - (b_1) \begin{vmatrix} a_2 c_2 \\ a_3 c_3 \end{vmatrix} + (c_1) \begin{vmatrix} a_2 b_2 \\ a_3 b_3 \end{vmatrix}$$

Zero and Infinity:

The symbols 0 and ∞ are not to be regarded as "numbers" in these formulas:

$$a \cdot 0 = 0 \qquad\qquad 0^a = 0$$

$$a \cdot \infty = \infty \qquad\qquad \infty^a = \infty$$

$$\frac{0}{a} = 0 \qquad\qquad a^\infty = \infty \text{ if } |a| > 1$$

$$\frac{a}{0} = \infty \qquad\qquad a^\infty = 0 \text{ if } |a| < 1$$

$$\frac{a}{\infty} = 0 \qquad\qquad a^{-\infty} = 0 \text{ if } |a| > 1$$

$$\frac{\infty}{a} = \infty \qquad\qquad a^{-\infty} = \infty \text{ if } |a| < 1$$

$$\infty \pm a = \infty$$

GEOMETRY

In the following formulas,

b = length of base	P = perimeter
h = altitude	C = circumference
l = slant height	S = arc length
r = radius	s = semiperimeter
m = median	K = area
d = diameter	B = area of base
θ = angle (in radians)	V = volume

$$\pi \text{ radians} = 180°, \qquad \frac{\pi}{3} \text{ radians} = 60°$$

$$2\pi \text{ radians} = 360°, \qquad \frac{\pi}{4} \text{ radians} = 45°$$

$$\frac{\pi}{2} \text{ radians} = 90°, \qquad \frac{\pi}{6} \text{ radians} = 30°$$

Triangle: $\quad P = a + b + c$
$$K = \tfrac{1}{2}bh = s\sqrt{(s - a)(s - b)(s - c)}$$

Trapezoid: $\quad m = \tfrac{1}{2}(b_1 + b_2)$
$$K = \tfrac{1}{2}h(b_1 + b_2) = \tfrac{1}{2}mh$$

Circle: $\quad C = \pi d = 2\pi r$
$$K = \pi r^2 = \frac{\pi d^2}{4}$$

Sector: $\quad S = r\theta$
$$K = \tfrac{1}{2}r^2\theta$$

Prism: $\quad V = Bh$

Cylinder: $\quad$ (*right circular*) $\quad K = 2\pi rh + 2\pi r^2$
$$V = Bh = \pi r^2 h$$

Pyramid: $\quad V = \tfrac{1}{3}Bh$

Cone: $\quad$ (*right circular*) $\quad K = \pi rl + \pi r^2; \; l = \sqrt{r^2 + h^2}$
$$V = \tfrac{1}{3}Bh = \tfrac{1}{3}\pi r^2 h$$

Sphere: $\quad K = 4\pi r^2$
$$V = \tfrac{4}{3}\pi r^3$$

Spherical triangle: $\quad K = \dfrac{\pi r^2 E}{180}$, where $E = (a° + b° + c°) - 180°$.

TRIGONOMETRY

FUNDAMENTAL IDENTITIES:

$$\sin x = \frac{1}{\csc x}; \qquad \cos x = \frac{1}{\sec x}; \qquad \tan x = \frac{1}{\cot x}; \qquad \tan x = \frac{\sin x}{\cos x}.$$

$$\sin^2 x + \cos^2 x = 1; \qquad \sec^2 x = 1 + \tan^2 x; \qquad \csc^2 x = 1 + \cot^2 x.$$

Transformations:

Functions of the Sum and Difference of Two Angles:

$$\sin (x \pm y) = \sin x \cos y \pm \cos x \sin y,$$
$$\cos (x \pm y) = \cos x \cos y \mp \sin x \sin y,$$
$$\tan (x \pm y) = \frac{\tan x \pm \tan y}{1 \mp \tan x \tan y}.$$

Multiple Angle Formulas:

$$\sin 2x = 2 \sin x \cos x,$$
$$\cos 2x = \cos^2 x - \sin^2 x = 2 \cos^2 x - 1 = 1 - 2 \sin^2 x,$$
$$\tan 2x = \frac{2 \tan x}{1 - \tan^2 x},$$
$$\sin \frac{x}{2} = \sqrt{\frac{1 - \cos x}{2}},$$
$$\cos \frac{x}{2} = \sqrt{\frac{1 + \cos x}{2}},$$
$$\tan \frac{x}{2} = \sqrt{\frac{1 - \cos x}{1 + \cos x}} = \frac{1 - \cos x}{\sin x} = \frac{\sin x}{1 + \cos x}.$$

Sum and Product Formulas:

$$\sin x + \sin y = 2 \sin \tfrac{1}{2}(x + y) \cos \tfrac{1}{2}(x - y),$$
$$\sin x - \sin y = 2 \cos \tfrac{1}{2}(x + y) \sin \tfrac{1}{2}(x - y),$$
$$\cos x + \cos y = 2 \cos \tfrac{1}{2}(x + y) \cos \tfrac{1}{2}(x - y),$$
$$\cos x - \cos y = -2 \sin \tfrac{1}{2}(x + y) \sin \tfrac{1}{2}(x - y),$$
$$\sin x \sin y = \tfrac{1}{2} \cos (x - y) - \tfrac{1}{2} \cos (x + y),$$
$$\sin x \cos y = \tfrac{1}{2} \sin (x - y) + \tfrac{1}{2} \sin (x + y),$$
$$\cos x \cos y = \tfrac{1}{2} \cos (x - y) + \tfrac{1}{2} \cos (x + y).$$

$$\sin^2 x - \sin^2 y = \sin (x + y) \sin (x - y),$$
$$\cos^2 x - \cos^2 y = -\sin (x + y) \sin (x - y),$$
$$\cos^2 x - \sin^2 y = \cos (x + y) \cos (x - y),$$

$$\sin^2 x = \tfrac{1}{2} - \tfrac{1}{2} \cos 2x,$$
$$\cos^2 x = \tfrac{1}{2} + \tfrac{1}{2} \cos 2x.$$

Plane Triangles:

a, b, c = sides of the triangle

A, B, C = opposite angles of the triangle

$s = \dfrac{a + b + c}{2}$ = semiperimeter

$r = \sqrt{\dfrac{(s - a)(s - b)(s - c)}{s}}$ = radius of the inscribed circle

R = radius of the circumscribed circle

K = area of the triangle

$$K = \frac{1}{2}\, ab \sin C = \sqrt{s(s - a)(s - b)(s - c)} = rs = \frac{abc}{4R}\cdot$$

$$K = \frac{\frac{1}{2}a^2 \sin B \sin C}{\sin (B + C)} = \frac{a^2 \sin B \sin C}{2 \sin A}$$

Law of sines: $\dfrac{a}{\sin A} = \dfrac{b}{\sin B} = \dfrac{c}{\sin C} = 2R.$

Law of cosines: $a^2 = b^2 + c^2 - 2bc \cos A.$

Law of Tangents: $\dfrac{a + b}{a - b} = \dfrac{\tan \frac{1}{2}(A + B)}{\tan \frac{1}{2}(A - B)}\cdot$

Tangent of Half-angle: $\tan \dfrac{1}{2} A = \dfrac{r}{s - a}\cdot$

FUNCTIONS OF SPECIAL ANGLES:

$\theta°$	θ radians	$\sin\theta$	$\cos\theta$	$\tan\theta$	$\cot\theta$
0°	0	0	1	0	∞
30°	$\dfrac{\pi}{6}$	$\dfrac{1}{2}$	$\dfrac{\sqrt{3}}{2}$	$\dfrac{\sqrt{3}}{3}$	$\sqrt{3}$
45°	$\dfrac{\pi}{4}$	$\dfrac{\sqrt{2}}{2}$	$\dfrac{\sqrt{2}}{2}$	1	1
60°	$\dfrac{\pi}{3}$	$\dfrac{\sqrt{3}}{2}$	$\dfrac{1}{2}$	$\sqrt{3}$	$\dfrac{\sqrt{3}}{3}$
90°	$\dfrac{\pi}{2}$	1	0	∞	0
120°	$\dfrac{2\pi}{3}$	$\dfrac{\sqrt{3}}{2}$	$-\dfrac{1}{2}$	$-\sqrt{3}$	$-\dfrac{\sqrt{3}}{3}$
135°	$\dfrac{3\pi}{4}$	$\dfrac{\sqrt{2}}{2}$	$-\dfrac{\sqrt{2}}{2}$	-1	-1
150°	$\dfrac{5\pi}{6}$	$\dfrac{1}{2}$	$-\dfrac{\sqrt{3}}{2}$	$-\dfrac{\sqrt{3}}{3}$	$-\sqrt{3}$
180°	π	0	-1	0	∞
210°	$\dfrac{7\pi}{6}$	$-\dfrac{1}{2}$	$-\dfrac{\sqrt{3}}{2}$	$\dfrac{\sqrt{3}}{3}$	$\sqrt{3}$
225°	$\dfrac{5\pi}{4}$	$-\dfrac{\sqrt{2}}{2}$	$-\dfrac{\sqrt{2}}{2}$	1	1
240°	$\dfrac{4\pi}{3}$	$-\dfrac{\sqrt{3}}{2}$	$-\dfrac{1}{2}$	$\sqrt{3}$	$\dfrac{\sqrt{3}}{3}$
270°	$\dfrac{3\pi}{2}$	-1	0	∞	0
300°	$\dfrac{5\pi}{3}$	$-\dfrac{\sqrt{3}}{2}$	$\dfrac{1}{2}$	$-\sqrt{3}$	$-\dfrac{\sqrt{3}}{3}$
315°	$\dfrac{7\pi}{4}$	$-\dfrac{\sqrt{2}}{2}$	$\dfrac{\sqrt{2}}{2}$	-1	-1
330°	$\dfrac{11\pi}{6}$	$-\dfrac{1}{2}$	$\dfrac{\sqrt{3}}{2}$	$-\dfrac{\sqrt{3}}{3}$	$-\sqrt{3}$
360°	2π	0	1	0	∞

PLANE ANALYTIC GEOMETRY

FUNDAMENTAL RELATIONS:

Distance: $P_1(x_1,y_1)$ to $P_2(x_2,y_2) = \sqrt{(x_1 - x_2)^2 + (y_1 - y_2)^2}$.

Slope of Line: $P_1P_2 = m = \dfrac{y_1 - y_2}{x_1 - x_2}$.

Divided Segment: $\dfrac{P_1P}{PP_2} = \dfrac{r_1}{r_2}$; $x = \dfrac{r_1x_2 + r_2x_1}{r_1 + r_2}$, $y = \dfrac{r_1y_2 + r_2y_1}{r_1 + r_2}$ for any

point $P(x,y)$ on P_1P_2.

Related Lines: If parallel, $m_1 = m_2$; if perpendicular, $m_1 = -\dfrac{1}{m_2}$.

Angle Between Two Lines whose slopes are m_1 and m_2:

$$\tan \theta = \frac{m_1 - m_2}{1 + m_1m_2}.$$

EQUATIONS OF A STRAIGHT LINE:

General form: $ax + by + c = 0$

Point-slope form: $y - y_1 = m(x - x_1)$

Two-point form: $\dfrac{y - y_1}{x - x_1} = \dfrac{y_2 - y_1}{x_2 - x_1}$

Intercept form: $\dfrac{x}{a} + \dfrac{y}{b} = 1$

Slope-intercept form: $y = mx + b$

Normal form: $x \cos \omega + y \sin \omega - p = 0$

Distance d of $P_1(x_1,y_1)$ *from* $Ax + By + C = 0$ is given by

$$d = \frac{Ax_1 + By_1 + C}{\pm \sqrt{A^2 + B^2}}$$

Lines parallel to X-axis and to Y-axis,   respectively:

$$y = k; \qquad x = k.$$

POLAR COORDINATES:

If $P(x,y) \equiv P(r,\theta)$, then

$$x = r \cos \theta; \qquad y = r \sin \theta; \qquad x^2 + y^2 = r^2; \qquad \theta = \arc \tan \frac{y}{x}.$$

CONIC SECTIONS:

Circle, center at (h,k): $(x - h)^2 + (y - k)^2 = r^2$.

Parabolas, vertex at (h,k): $(y - k)^2 = \pm 2p(x - h)$; $(x - h)^2 = \pm 2p(y - k)$.

Ellipse, center at (h,k): $\dfrac{(x - h)^2}{a^2} + \dfrac{(y - k)^2}{b^2} = 1$.

Hyperbolas, center at (h,k): $\dfrac{(x - h)^2}{a^2} - \dfrac{(y - k)^2}{b^2} = \pm 1$.

Equilateral Hyperbola: $xy = c$, where asymptotes are $xy = 0$.

CURVES FOR REFERENCE

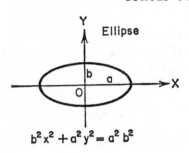

Ellipse

$$b^2 x^2 + a^2 y^2 = a^2 b^2$$

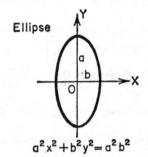

Ellipse

$$a^2 x^2 + b^2 y^2 = a^2 b^2$$

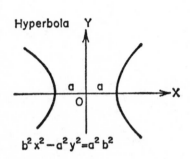

Hyperbola

$$b^2 x^2 - a^2 y^2 = a^2 b^2$$

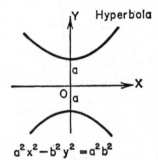

Hyperbola

$$a^2 x^2 - b^2 y^2 = a^2 b^2$$

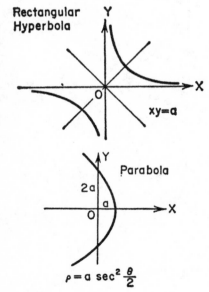

Rectangular Hyperbola

$$xy = a$$

Parabola

$$x^2 = 2py$$

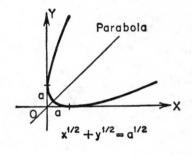

Parabola

$$\rho = a \sec^2 \frac{\theta}{2}$$

Parabola

$$x^{1/2} + y^{1/2} = a^{1/2}$$

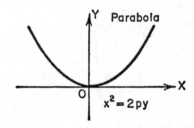

The Cissoid of Diocles

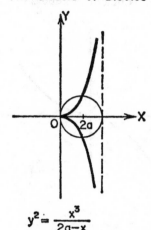

$$y^2 = \frac{x^3}{2a-x}$$

The Witch of Agnesi

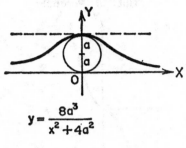

$$y = \frac{8a^3}{x^2 + 4a^2}$$

The Folium of Descartes

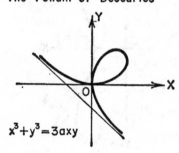

$$x^3 + y^3 = 3axy$$

The Conchoid of Nicomedes

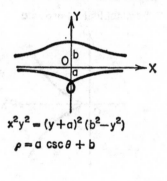

$$x^2 y^2 = (y+a)^2 (b^2 - y^2)$$

$$\rho = a \csc\theta + b$$

The Cycloid

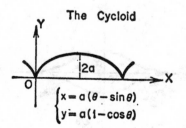

$$\begin{cases} x = a(\theta - \sin\theta) \\ y = a(1 - \cos\theta) \end{cases}$$

The Cycloid

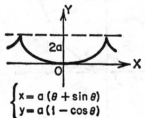

$$\begin{cases} x = a(\theta + \sin\theta) \\ y = a(1 - \cos\theta) \end{cases}$$

Cubical Parabola

$y = ax^3$

Cubical Parabola

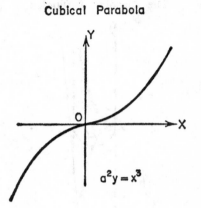

$a^2 y = x^3$

Semicubical Parabola

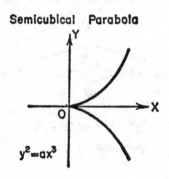

$y^2 = ax^3$

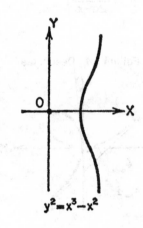

$y^2 = x^3 - x^2$

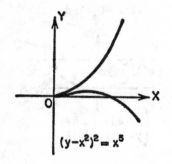

$(y - x^2)^2 = x^5$

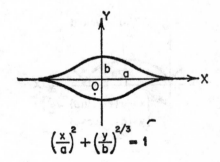

$$\left(\frac{x}{a}\right)^2 + \left(\frac{y}{b}\right)^{2/3} = 1$$

The Sine Curve

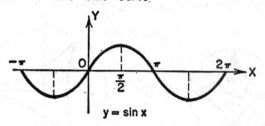

$y = \sin x$

The Cosine Curve

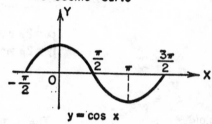

$y = \cos x$

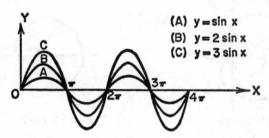

(A) $y = \sin x$
(B) $y = 2 \sin x$
(C) $y = 3 \sin x$

The Secant Curve   The Tangent Curve

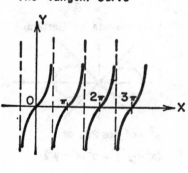

$y = \sec x$ $y = \tan x$

Hypocycloid of Four Cusps

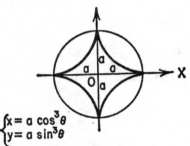

$$\begin{cases} x = a \cos^3\theta \\ y = a \sin^3\theta \end{cases}$$

$$x^{2/3} + y^{2/3} = a^{2/3}$$

Hypocycloid of Three Cusps

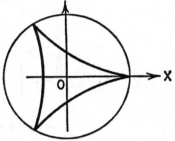

$$\begin{cases} x = 2\rho \cos\theta + \rho \cos 2\theta \\ y = 2\rho \sin\theta - \rho \sin 2\theta \end{cases}$$

The Cardioid

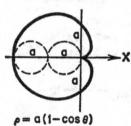

$$\rho = a(1 - \cos\theta)$$

The Limaçon

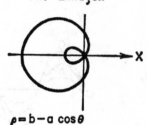

$$\rho = b - a \cos\theta$$

The Lemniscate of Bernoulli

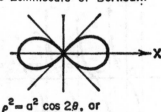

$$\rho^2 = a^2 \cos 2\theta, \text{ or}$$
$$(x^2 + y^2)^2 = a^2(x^2 - y^2)$$

The Strophoid

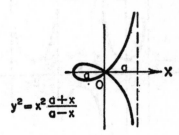

$$y^2 = x^2 \frac{a+x}{a-x}$$

Three-Leaved Rose

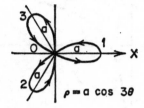

$\rho = a \cos 3\theta$

Three-Leaved Rose

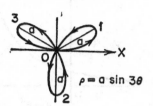

$\rho = a \sin 3\theta$

Four-Leaved Rose

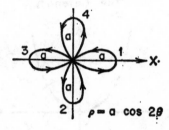

$\rho = a \cos 2\theta$

Four-Leaved Rose

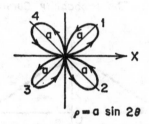

$\rho = a \sin 2\theta$

Two-Leaved Rose Lemniscate

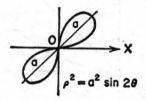

$\rho^2 = a^2 \sin 2\theta$

Eight-Leaved Rose

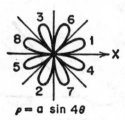

$\rho = a \sin 4\theta$

The Exponential Function

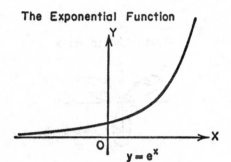

$y = e^x$

The Logarithmic Function

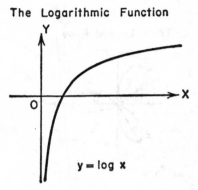

$y = \log x$

The Probability Curve

The Catenary

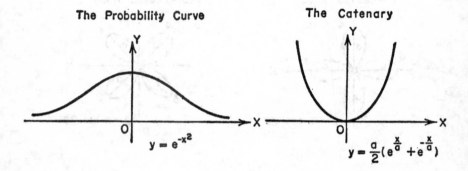

$y = e^{-x^2}$

$y = \dfrac{a}{2}(e^{\frac{x}{a}} + e^{-\frac{x}{a}})$

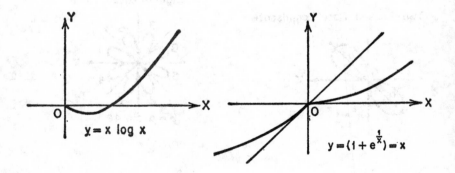

$y = x \log x$

$y = (1 + e^{\frac{1}{x}}) - x$

**The Equiangular
or Logarithmic Spiral**

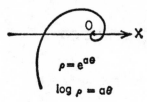

$\rho = e^{a\theta}$

$\log \rho = a\theta$

Spiral of Archimedes

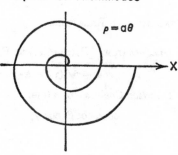

$\rho = a\theta$

The Parabolic Spiral

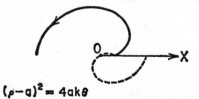

$(\rho - a)^2 = 4ak\theta$

Involute of a circle

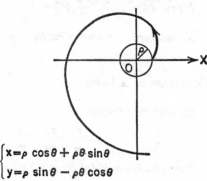

$$\begin{cases} x = \rho \cos\theta + \rho\theta\sin\theta \\ y = \rho \sin\theta - \rho\theta\cos\theta \end{cases}$$

**The Reciprocal or
Hyperbolic Spiral**

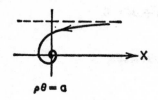

$\rho\theta = a$

The Lituus

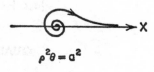

$\rho^2\theta = a^2$

SOLID ANALYTIC GEOMETRY

FUNDAMENTAL RELATIONS:

Distance from $P_1(x_1,y_1,z_1)$ to $P_2(x_2,y_2,z_2)$:

$$d = \sqrt{(x_1 - x_2)^2 + (y_1 - y_2)^2 + (z_1 - z_2)^2}.$$

Direction Cosines of a line:

$$\cos \alpha, \cos \beta, \cos \gamma; \qquad \cos^2 \alpha + \cos^2 \beta + \cos^2 \gamma = 1.$$

EQUATIONS OF A PLANE:

General form: $ax + by + cz + d = 0.$

Intercept form: $\dfrac{x}{a} + \dfrac{y}{b} + \dfrac{z}{c} = 1.$

Normal form: $\cos \alpha x + \cos \beta y + \cos \gamma z - p = 0.$

EQUATIONS OF A LINE:

Through Two Points:

$$\frac{x - x_1}{x_2 - x_1} = \frac{y - y_1}{y_2 - y_1} = \frac{z - z_1}{z_2 - z_1}.$$

Through a Given Point (x_1,y_1,z_1) with direction numbers a, b, c:

$$\frac{x - x_1}{a} = \frac{y - y_1}{b} = \frac{z - z_1}{c}.$$

Intersection of Two Planes:

$$\begin{cases} a_1x + b_1y + c_1z + d_1 = 0, \\ a_2x + b_2y + c_2z + d_2 = 0. \end{cases}$$

QUADRIC SURFACES

(1) Sphere with center at $(0,0,0)$:

$$x^2 + y^2 + z^2 = r^2.$$

(2)   Ellipsoid with center at (0,0,0):

$$\frac{x^2}{a^2} + \frac{y^2}{b^2} + \frac{z^2}{c^2} = 1.$$

(3)   Hyperboloid of one sheet:

$$\frac{x^2}{a^2} + \frac{y^2}{b^2} - \frac{z^2}{c^2} = 1.$$

(4)   Hyperboloid of two sheets:

$$\frac{x^2}{a^2} - \frac{y^2}{b^2} - \frac{z^2}{c^2} = 1.$$

(5)   Elliptic Paraboloid:

$$\frac{x^2}{a^2} + \frac{y^2}{b^2} = cz.$$

(6)   Hyperbolic Paraboloid (saddle surface):

$$\frac{x^2}{a^2} - \frac{y^2}{b^2} = cz.$$

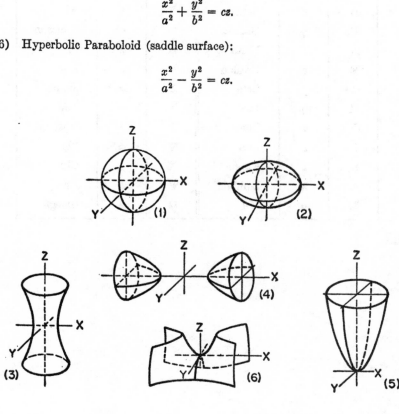

GREEK ALPHABET

LETTERS		NAMES	LETTERS		NAMES
Capitals	Lower Case		Capitals	Lower Case	
A	α	Alpha	N	ν	Nu
B	β	Beta	Ξ	ξ	Xi
Γ	γ	Gamma	O	o	Omicron
Δ	δ	Delta	Π	π	Pi
E	ϵ	Epsilon	P	ρ	Rho
Z	ζ	Zeta	Σ	σ	Sigma
H	η	Eta	T	τ	Tau
Θ	θ	Theta	Υ	υ	Upsilon
I	ι	Iota	Φ	ϕ	Phi
K	κ	Kappa	X	χ	Chi
Λ	λ	Lambda	Ψ	ψ	Psi
M	μ	Mu	Ω	ω	Omega

Answers to Problems

1. $-1, 8, 23, 44$.
2. The positive integers and zero.
3. $7, 11, 15, 19, \cdots$; $1, 2, 3, \cdots$; $7, 11, 15, 19, \cdots$.
4. $u = 0, 1, 2, \cdots 9$; $t = 0, 1, 2, \cdots 9$; $h = 1, 2, 3, \cdots 9$.
5. (a) $\frac{1}{4}\%, \frac{1}{2}\%, \frac{3}{4}\%, 1\%, \cdots 3\frac{1}{2}\%, 3\frac{3}{4}\%, 4\%$;
   (b) $1, 2, 3, \cdots$; (c) $\frac{1}{12}, \frac{2}{12}, \frac{3}{12}, \cdots \frac{12}{12}, \frac{13}{12}, \frac{14}{12} \cdots$;
   $\frac{1}{360}, \frac{2}{360}, \frac{3}{360}, \cdots$, or $\frac{1}{365}, \frac{2}{365}, \frac{3}{365}, \cdots$.
7. Functions: $a, c, e, g, i, k, o, p, q, r$;
   non-functions: $b, d, f, h, j, l, m, n, s, t, u$.

1. $a_0x^5 + a_1x^4 + a_2x^3 + a_3x^2 + a_4x^5 + a_5$
2. $a_0x^9 + a_2x^7 + a_4x^5 + a_6x^3 + a_8x$
3. $f(m) = am^2 - bm + c$; $f(5) = 25a - 5b + c$
4. $F(-1) = -7$; $F(0) = -6$
5. $f(x + h) = 2x^2 + (4h + 5)x + (2h^2 + 5h - 3)$;
   absolute term $= 2h^2 + 5h - 3$
6. $\frac{1}{9}$; $\frac{1}{3}$; 1; $\frac{1}{27}$
7. $x^3 - x^2 + 3x - 3$;
   $x^3 - 3x^2h + 3xh^2 - h^3 - 4x^2 + 8xh - 4h^2 + 8x - 8h - 8$

8. $\dfrac{-x^2 - x + 1}{x(x + 1)}$

9. $ax^2 + 2ahx + bx$

10. $\log \dfrac{u}{v} + \log \dfrac{v - m}{u - m}$

11. Range of y: all y's $\geqq -6$.

12. (a) $\{2, 5, 8, 11, 14, 17, 20\}$

(b) $\{0, 2, 8, 18, 32, 50, 72\}$

(c) $\{0, 3, 6, 9, 12, 15, 18\}$

Ex. 1—3:

1. 9	**2.** ∞	**3.** 0
4. ∞	**5.** ∞	**6.** 2
7. -6	**8.** ∞	**9.** $\frac{3}{2}$
10. -2	**11.** $\frac{1}{5}$	**12.** 2

Ex. 2—1:

1. 2	**2.** $-2x$	**3.** $4x + 3$	**4.** $20x$
5. $2x - 1$	**6.** $6x - 2$	**7.** $6x^2$	**8.** $3x^2 - 2$
9. $2x - 1$	**10.** $12x + 1$	**11.** 2	**12.** $-\dfrac{1}{x^2}$
13. $\dfrac{2}{x^2}$	**14.** $\dfrac{1}{(x + 1)^2}$	**15.** $\dfrac{-2}{(x - 1)^2}$	**16.** $-\dfrac{x + 2}{x^3}$

Ex. 2—2:

1. 12	**2.** 8	**3.** 5	**4.** -8
5. 6	**6.** 33	**7.** 12	**8.** $\frac{1}{2}$

Ex. 3—1:

1. $6x^2 - 5$

2. $5x^4 - 12x^3 + 6x^2 + 1$

3. $\dfrac{1}{2} + \dfrac{2x}{3} - \dfrac{x^2}{2}$

4. $\dfrac{5}{2\sqrt{x}}$

5. $v_1 + gt$

6. $-\dfrac{3}{x^4} + \dfrac{6}{x^3} - \dfrac{4}{x^2}$

7. $\dfrac{3}{\sqrt{6x + 3}}$

8. $4t^2 + 8t + \dfrac{2}{t^2}$

9. $2(x^2 - 5x)(2x - 5)$

10. $6k(a + 2t)^2$

11. $mx^{m-1} - 3mx^2$

12. $\dfrac{2x - 3}{2\sqrt{x^2 - 3x}}$

13. $4(x^3 - 2x)^3(3x^2 - 2)$

14. $\dfrac{3}{2\sqrt{3x}} - \dfrac{5}{x^2}$

15. $\dfrac{-x}{\sqrt{m^2 - x^2}}$

16. $\dfrac{-x}{(m^2 + x^2)^{3\!/\!4}}$

Ex. 3—2:

1. $2x - 1$
2. $9x^2 + 8x + 3$
3. $8x^3 - 9x^2 - 2x + 3$
4. $6x^2(x^3 - 1)$
5. $6x(x^2 + 1)^2$

6. $4t^3 - 3t^2 + 4t - 2$
7. $48x^3 + 15x^2 - 4x$
8. $3x^2 - 7$
9. $3(2t - 3)(t^2 - 3t + 4)^2$
10. $(x + 2)(x + 3)^2(5x + 12)$

Ex. 3—3:

1. $\dfrac{1}{x^2}$

2. $\dfrac{1}{(1 + x)^2}$

3. $\dfrac{2k}{(x + k)^2}$

4. $\dfrac{b - a}{(b + x)^2}$

5. $\dfrac{-2}{(x + k)^3}$

6. $\dfrac{2x^2(3 - x)}{(2 - x)^2}$

7. $\dfrac{-36x}{(x^2 - 9)^2}$

8. $\dfrac{-9x^2}{(x^3 + 1)^2}$

9. $\dfrac{1 - 2x - x^2}{(1 + x^2)^2}$

10. $\dfrac{8m^2x^3 - 4x^5}{(m^2 - x^2)^2}$

Ex. 3—4:

1. $3x^2 + 6x - 2$

2. $b - a - 2x$

3. $(k + x)(3x + k)$

4. $\dfrac{2x^4 - 3x^2(x^2 + 1)}{x^6}$

5. $\dfrac{7}{(x + 2)^2}$

6. $\dfrac{1}{x^2}$

7. $\dfrac{-1}{(x - 1)^2}$

8. $\dfrac{3(3x + 2)}{2\sqrt{3x - 2}} + 3\sqrt{3x - 2}$

9. $\dfrac{\sqrt{x^2 - 1}}{(x + 1)(x^2 - 1)}$

10. $\dfrac{ax^2 - c}{x^2}$

11. $-50x^4 - 6x^2 + 10x$

12. $\dfrac{b - x}{2\sqrt{a + x}} - \sqrt{a + x}$

13. $\dfrac{-x(m^2 + x^2)}{\sqrt{m^2 - x^2}} + 2x\sqrt{m^2 - x^2}$

14. $\dfrac{2a}{(x+a)^2}$

15. $\dfrac{-2x^4 - 9x^2 - 8x}{(x^3 - 2)^2}$

16. $\dfrac{(x+3)(x-1)}{(x+1)^2}$

17. $\dfrac{6m^2t^2 - 2t^4}{(m^2 - t^2)^2}$

18. $3t^2 - 2t - 2$

19. $\dfrac{3t^2}{(1-t)^4}$

20. $\dfrac{\sqrt{x+1}}{2x^3(x+1)} - \dfrac{3\sqrt{x+1}}{x^4}$

Ex. 3—5:

1. $-\dfrac{x}{y}$

2. $\dfrac{x}{2p}$

3. $-\dfrac{y}{x}$

4. $\dfrac{2x+1}{2y+1}$

5. $\dfrac{-2x-y}{x+2y}$

6. $\dfrac{4\sqrt{x+y} - 1}{1 - 2\sqrt{x+y}}$

7. $\dfrac{x^3 + 3x^2(x+a)}{2y}$

8. $\dfrac{2y-1}{1-2x}$

9. $-\dfrac{b^2x}{a^2y}$

10. $\dfrac{b^2x}{a^2y}$

11. $\dfrac{3-y^2}{2xy}$

12. $\dfrac{-2x-y}{x-2y+2}$

13. $\dfrac{y+2\sqrt{xy}}{-x+2\sqrt{xy}}$

14. $\dfrac{\sqrt{y}}{\sqrt{x}} - \sqrt{y}$

15. $\dfrac{ay - x^2}{y^2 - ax}$

16. $\dfrac{-3x^2 - 4xy}{2x^2 + 9y^2}$

Ex. 3—6:

1. $\dfrac{\sqrt{px}}{x}$

2. $\dfrac{-3x}{\sqrt{k^2 - 3x^2}}$

3. $\dfrac{ax}{b\sqrt{x^2 + a^2}}$

4. $\dfrac{1}{\sqrt[3]{(3x+1)^2}}$

5. $\dfrac{-k}{t^2} + 3at^2$

6. $3bt(a + bt^2)^{\frac{1}{2}}$

7. $3(x^3 + 3x^2 + 2)^2(3x^2 + 6x)$

8. $15x^2(x^3 - 2)^4$

9. $-\dfrac{k}{x^{k+1}}$

10. $\dfrac{x^2 - 3}{2x^2}$

11. $3x^2 - 2x - 2$

12. $36x^5 + 16x^3 - 6x$

13. $\dfrac{x+1}{2-y}$

14. $\dfrac{y - 3x^2}{2y - x}$

15. $\dfrac{2x(y-1)}{1-x^2}$

16. $\dfrac{-4x^3 - 2xy^2}{2x^2y + 4y^3}$

17. $\dfrac{a^2x}{b^2y}$

18. $\dfrac{x(2 - y^2)}{y(x^2 + 2)}$

19. $\dfrac{-2Ax - D}{2Cy + E}$

20. $-\dfrac{y^{\frac{1}{3}}}{x^{\frac{2}{3}}}$

Ex. 4—1:

1. 20.6π sq. in./in.
2. 576π cu. in./in.
3. 9 units
4. 48 cal./amp.
5. $-1\frac{1}{2}$ units

6. 2997 cu. ft. per sec./ft.
7. -494 B.T.U. per unit concentration
8. 16,000 H.P.; 2400 H.P./knot
9. 48 B.T.U./degree
10. $V = \dfrac{7.5}{\sqrt{d}}$; -3750

Ex. 4—2:

1. (a) $20x^3 + 18x$ (b) $100 - 40t_3$ (c) $40x^{-6}$ (d) $8\pi r$
2. (a) $\dfrac{6}{x^4}$ (b) -6
3. (a) $360x^2 + 1200x^{-6}$ (b) $120t$
4. (a) $s = 14, v = 15, a = 12$
   (b) $s = 5, v = 0, a = -10$
   (c) $s = -4, v = -6, a = 0$
5. (a) 322 ft./sec.;   (b) 35.9 ft./sec.   (c) 32.2 ft./sec./sec.
   (d) 257.6 ft.
7. (a) $\frac{1}{4}$ mi./min./min.;   (b) 2 miles
8. (a) 1760 ft./min./min.;   (b) 3520 ft.
9. (a) $v = v_1 - 32.2t; a = -32.2$
   (b) $v = 103.4$ ft./sec. upward; $a = 32.2$ ft./sec./sec. downward
   (c) $v = 186.4$ ft./sec. downward; $a = 32.2$ ft./sec./sec. downward
10. (a) 717 ft./sec.;   (b) $37.2 +$ sec. (approx.)

Ex. 4—3:

1. $x = 2$ gives min. $y = -4$
2. $x = -3$ gives min. $y = -1$
3. $x = +2$ gives min. $y = -16$; $x = -2$ gives max. $y = +16$
4. $x = 0$ gives max. $y = 15$; $x = +3$ gives min. $y = -66$;
   $x = -3$ gives min. $y = -66$
5. Neither max. nor min.
6. 20; 20
7. $\frac{1}{2}$
8. 1

Ex. 4—4:

1. $\dfrac{a}{4}$

2. $\dfrac{a}{6}$

3. $h = \dfrac{4a}{3}$; $V = \dfrac{32a^3\pi}{81}$

4. $h = \dfrac{1}{3}\sqrt{6}\,d; w = \dfrac{1}{3}\sqrt{3}\,d$

5. $h = a\sqrt{2}$

6. $r = \dfrac{a\sqrt{6}}{3}$; $h = \dfrac{a\sqrt{3}}{3}$; $V = \dfrac{2\pi a^3 \sqrt{3}}{27}$ **7.** $l = r$

Ex. 4—5:

1. $v = \dfrac{ds}{dt} = \dfrac{1}{2} t^{-\frac{1}{2}};\ \dfrac{d^2 s}{dt^2} = -\dfrac{1}{4} t^{-\frac{3}{2}} = -\dfrac{1}{2}\left[\dfrac{1}{2}\,(t^{-\frac{1}{2}})^3\right] = -\dfrac{1}{2}\,v^3$

2. $a = (4t + 8)$ ft./sec./sec.; 28 ft./sec./sec.

3. Max., $x = -2$; min., $x = +4$

4. $r = 3\sqrt{6}$

5. $v = 0$; $a = -20$ ft./sec./sec.

6. Max., $x = -1$; min., $x = +5/3$

7. Max., $x = 0$; min., $x = +1$ and $x = -1$

8. 7 mi./hr./hr.

9. Length $= 30$ in.; circumference $= 60$ in.

10. Square; area $= 2k^2$

Ex. 5—1:

1. $\dfrac{1}{k + x}$

2. $\dfrac{2}{x}$

3. $2 \log x \left(\dfrac{1}{x}\right)$

4. $\dfrac{2}{x}$

5. $\dfrac{a}{ax + b}$

6. $\dfrac{3x + 2k}{x^2 + kx}$

7. $\dfrac{2x + 3}{x^2 + 3x - 2}$

8. $\dfrac{1}{x} + \dfrac{3}{x}$, or $\dfrac{4}{x}$

9. $\dfrac{x^2}{x^3 + 1}$

10. $\dfrac{2}{1 - x^2}$

11. $1 + \log x$

12. $\dfrac{2x + a}{x^2 + ax}$

13. $\dfrac{3}{x}$

14. $3 (\log x)^2 \left(\dfrac{1}{x}\right)$

15. $-\dfrac{1}{2(1 - x)}$

16. $\dfrac{x}{1 + x^2}$

17. $\log_a e \left(\dfrac{6x - 2}{3x^2 - 2x}\right)$

18. $\log_a e \left(\dfrac{3x^2}{x^3 + k}\right)$

19. $x + 2x \log x$

20. $\dfrac{x^2 + 1}{x(1 - x^2)}$

Ex. 5—2:

1. $4 \log k \,(k^{4x})$

2. $2x \,(\log a)\,(a^{x^2 - 1})$

3. ke^{kx}

4. ke^x

5. $-3e^{5 - 3t}$

6. $2x \,(\log k)\,(k^{m^2 + x^2})$

7. $(3x^2 + 4)\,(\log 5)\,(5^{x^3 + 4x})$

8. $2t \cdot e^{a^2 + t^2}$

9. $\dfrac{ke^{\sqrt{x}}}{2\sqrt{x}}$

10. $-2xe^{-x^2}$

11. $\log a \cdot a^{\theta}$

12. $\dfrac{\log a \cdot a^{\log \theta}}{\theta}$

13. $ae^{a\theta}$

14. $-\dfrac{e^{1/x}}{x^2}$

15. $e^x(x+1)$

16. $2x \cdot e^{2x} \cdot (x+1)$

17. $nx^{n-1} + n^x \log n$

18. $a^x x^{a-1}(a + x \log a)$

19. $e^x x^2$

20. $e^x \cdot x$

Ex. 5—3:

1. $(x+1)(x^x) + \log x \, (x^{x+1})$

2. $x(x^2+1)^{x-1} + [\log (x^2+1)](x^2+1)^x$

3. $-x^{-3x}(3 + \log x^3)$

4. $2x^{2x}(1 + \log x)$

5. $2x^3(2x)^{x^3-1} + 3x^2(\log 2x)(2x)^{x^3}$

6. $x^{x^2+1}(1 + 2\log x)$

7. $\dfrac{x^{1/x}(1 - \log x)}{x^2}$

8. $\log x^2 \cdot x^{\log x - 1}$

Ex. 5—4:

1. $\left(\dfrac{k}{x}\right)^x (\log k - \log x - 1)$

2. $x^{x^n + n - 1}(n \log x + 1)$

3. $e^{x^x}(\log x + 1)x^x$

4. $\dfrac{9x \cdot e^{3x}}{(3x+1)^2}$

5. $\dfrac{3 - 2x}{x(x-1)}$

6. $\dfrac{2x - 3}{2(x-1)(x-2)}$

7. $\dfrac{-2}{x^2 - 4}\sqrt{\dfrac{x+2}{x-2}}$;

    or $\dfrac{-2}{(x+2)^{1/2}(x-2)^{3/2}}$

8. $\dfrac{-x^2 - 6x - 7}{(x+1)^2(x+2)^2}$

Ex. 5—5:

1. $3 \cos 3x$

2. $-5a \sin 5ax$

3. $\sin x(2 \cos^2 x - \sin^2 x)$

4. $-2a \sin 2\theta$

5. $b \cos (a + bx)$

6. $e^{\tan x} \cdot \sec^2 x$

7. $-2 \tan x$

8. $(\sin x)^x \cdot [(\log \sin x + x \cot x)]$

9. $x^{\cos x - 1} \cdot [\cos x - x \log x \sin x]$

10. $e^x (\cot x + \log \sin x)$

11. $\dfrac{2}{\sin 2\theta}$

12. $(\cos x)^{x-1} [\cos x (\log \cos x) - x \sin x]$

13. $\sin \dfrac{x}{2} \cos \dfrac{x}{2}$

14. $\dfrac{2 \cos x}{(1 - \sin x)^2}$

16. $\dfrac{x \cos x - \sin x}{x^2}$

18. $2 \cos 2x$

15. $x^{m-1}(1 + m \log x)$

17. $2 \sin^2 \theta$

19. $\dfrac{1}{x} + \cos x$

20. $2 \cot 2\theta$

Ex. 5—6:

1. $\dfrac{2x}{\sqrt{1 - x^4}}$

5. $\dfrac{3}{10x^2 - 2x + 1}$

2. $-\dfrac{1}{\sqrt{4 - x^2}}$

6. $-\dfrac{2}{\sqrt{-4x^2 - 4x + 3}}$

3. $\dfrac{1}{2x^2 - 2x + 1}$

7. $\dfrac{-2ax}{a^2 + x^4}$

4. $\dfrac{x}{\sqrt{1 - x^2}} + \text{arc} \sin x$

8. $\dfrac{-ax^2}{x^2 + a^2} + 2x \ \text{arc} \tan \dfrac{a}{x}$

Ex. 5—7:

1. $2e^x \cos x$

5. $-\dfrac{4p^2}{y^3}$

2. $2 \cot x \csc^2 x$

6. $\dfrac{2}{(1 - 2y)^3}$

3. (a) $\dfrac{6}{x}$ (b) $a^4 \sin ax = a^4 y$

7. $e^x (x^2 + 6x + 6)$

4. $-\dfrac{k^2}{y^3}$

Ex. 5—8:

1. $\dfrac{1 - x \log x}{x e^x}$

2. $e^t (1 + t)^2$

3. $\dfrac{6x}{3x^2 - 2}$

4. $\log c \cdot c^{e^x} \cdot e^x$

5. $\dfrac{1}{x \log x}$

6. $2x$

7. $y = \dfrac{2}{2 + e^x}$

8. $e^{x \log x}(1 + \log x)$

9. $-2e^{\cos 2x}(\sin 2x)$

10. $\cot x$

11. $\sec^2 x (1 + 2 \tan x)$

12. $\dfrac{3}{\sqrt{3x}\,(x + 3)}$

13. $e^{x \sin x}[x \cos x + \sin x]$

14. $\dfrac{2}{x^2 + 1}$

15. $\dfrac{1}{a^2 + x^2}$

16. $-2 \tan x$

17. $-\dfrac{2}{x^2 - 1}$

18. $\dfrac{1 - x \log x}{x \cdot e^x}$

19. $\dfrac{2e^x}{(e^x + 1)^2}$

20. $\left(\dfrac{a}{x}\right)^x \left(\log \dfrac{a}{x} - 1\right)$

Ex. 6—1:

1. $\dfrac{dy}{dx} = \dfrac{y}{x}$

2. Slope of tangent $= -\frac{1}{2}$

3. 5

4. $x = \frac{3}{4}, \ y = \frac{9}{8}$

5. $\phi = \arc\tan\left(\mp\dfrac{\sqrt{3}}{5}\right)$

6. $x = \frac{5}{13}k, \ y = \frac{12}{13}k$

7. At origin, $90°$; at $(4p, 4p)$, $\arc\tan \frac{3}{4} = 36°50'$, approx.

8. $\dfrac{dy}{dx} = \dfrac{1}{x}$

Ex. 6—2:

1. 3

2. Tangent, $4x + y = 0$; normal, $4y - x = 34$

3. Subtangent, $2x$; subnormal, $2p$

4. $3x + 4y = 25$; $-\frac{16}{3}$

5. $y = x$

6. nk

7. $\dfrac{1}{\log k}$

8. Intercept on x-axis $= 2x$; intercept on y-axis $= 2y$;
area $= \frac{1}{2}(2x_1)(2y_1) = 2x_1 y_1 = a^2$.

Ex. 6—3:

1. $(2, -114)$; $(-3, -294)$ **2.** None **3.** $(\sqrt{2}, -20)$; $(-\sqrt{2}, -20)$
4. $(0,0)$ **5.** None **6.** $(0,0)$

7. None **8.** $-2, -\dfrac{2}{e^2}$ **9.** (a,b)

10. $(-1,0)$

Ex. 6—5:

1. (a) $\left(y - 40 + \dfrac{g}{2}\right) = \left(1 - \dfrac{g}{40}\right)(x - 40)$; (b) $\dfrac{-g}{1600}$

2. (a) $4x - y = 17$; (b) $-\frac{1}{8}$

3. $\dfrac{dy}{dx} = -\cot \phi$; (a) when $\phi = \pi/2$, $x_1 = 0$ and $y_1 = r$; hence $y - r = (-\cot \pi/2)(x - 0)$, or $y = r$; (b) when $\phi = \pi/4$, $x_1 = y_1 = r\dfrac{\sqrt{2}}{2}$; hence subnormal $= \left(r\dfrac{\sqrt{2}}{2}\right)\left(-\cot \dfrac{\pi}{4}\right) = -r\dfrac{\sqrt{2}}{2}$

4. $-\dfrac{b}{a^2} \csc^3 \theta$ **6.** $-\dfrac{g}{v_0^2 \cos^2 \phi}$

5. $\dfrac{\sec^4 \theta}{3a \sin \theta}$

Ex. 6—6:

1. $2\frac{10}{13}$ mi./hr. **6.** $8\sqrt{5}$ ft./min.

2. $\dfrac{16}{\pi}$ ft./min.; $9\sqrt{5}\pi$ sq. ft./min. **7.** $2\sqrt{2}$, or 2.83 ft./min.

3. $2\frac{2}{3}$ in./min.; $(4,8)$ **8.** $\dfrac{10}{\pi}$ cm./min.

4. 100 mi./hr. **9.** $60°$

5. $\dfrac{r}{2}$ **10.** $4\sqrt{5}$; $2\sqrt{73}$; $10\sqrt{10}$

Ex. 6—7:

1. $3\sqrt{3}$ cm./sec. **6.** $h = 2r$

2. $\dfrac{5\sqrt{3}}{3}$ sq. ft./sec. **7.** Altitude $= \dfrac{S}{\sqrt{3}}$

3. (a) $(6,12)$; (b) $(24,24)$ **8.** $2\pi R^2$
4. $(0,0)$ **9.** Altitude of cone $= \frac{4}{3}R$
5. Base $= a\sqrt{2}$, altitude $= b\sqrt{2}$ **10.** 10 sq. in./sec.

Ex. 7—1:

1. $dy = (3x^2 + 10x)\, dx$
2. $dy = 15(5x + 2)^2\, dx$
3. $dy = x(x^2 - 1)^{-\frac{1}{2}}\, dx$
4. $dy = (kte^{tx})\, dx$
5. $dy = e^x \cdot \left(\dfrac{1}{x} + \log x\right) dx$
6. $d\rho = 2(\cos 2\theta - \sin \theta)\, d\theta$
7. $f'(x)\, dx = \dfrac{4(\log x)^3\, dx}{x}$
8. $d\rho = (3a^2 \sec 3\theta \tan 3\theta)\, d\theta$

9. $dy = -\dfrac{x}{y}\, dx$
10. $dy = \dfrac{b^2 x}{a^2 y}\, dx$
11. $dy = -\dfrac{y}{x}\, dx$
12. $d^3y = (24x - 30)(dx)^3$
13. $d^3y = 2e^x(\cos x - \sin x)(dx)^3$
14. $d^3y = (-\frac{3}{8}x^{-\frac{5}{2}})(dx)^3$
15. $d^3y = -6x^{-4}(dx)^3$

Ex. 7—2:

1. .72 sq. cm.; 3.6 cu. cm.
2. $.04\pi$ in.; $.2\pi$ sq. in.
3. 1.60π sq. in.; 8π cu. in.
4. $.4\pi$ cu. cm.
5. 17.059
6. 3.917
7. .025 cm.

8. (a) $-\dfrac{p}{v}(\Delta v)$; (b) $-\dfrac{v}{p}(\Delta p)$
9. $\frac{27}{2}\pi$ sq. in.
10. 960 calories
11. 16,875 H.P.
12. $\Delta C = 2\pi(\Delta r); \dfrac{1}{2\pi} = .16$ in.;
same amount, .16 in.

Ex. 7—3:

1. (a) $-\sqrt{\dfrac{y}{x}}$; (b) $-\sqrt{\dfrac{x}{y}}$
2. (a) $-\dfrac{2(x+1)}{2y-5}$; (b) $\dfrac{x(1 - 2x^2 + 2y^2)}{y(-1 - 2x^2 + 2y^2)}$
3. (a) -1; (b) $\dfrac{5}{4}$
4. 576 feet; 6 seconds
5. (a) $\dfrac{1}{1+e^x}$; (b) $\dfrac{2a}{a^2-x^2}$; (c) $\dfrac{2px+q}{px^2+qx+r}$; (d) $\dfrac{x}{x^2-1}$
6. (a) $-\dfrac{b^4}{a^2y^3}$; (b) $-\dfrac{1}{y^3}$; (c) $\dfrac{3y}{(x^2-1)^2}$

Ex. 8—1:

1. $\dfrac{-\sqrt{5}}{25p}$

2. $\dfrac{-a}{b^2}$

3. 2

4. $\dfrac{-e}{(e^2+1)^{3/2}}$

5. 6

6. 18

7. 2

8. 2

9. -4

10. $\dfrac{y^2-x^2}{(y^2+x^2)^{3/2}}$

11. $\dfrac{1}{\rho(1+a^2)^{1/2}}$

12. $\dfrac{3x^2}{(1+x^6)^{3/2}}$

13. $\dfrac{\rho^2+2a^2}{(\rho^2+a^2)^{3/2}}$

Ex. 8—2:

1. $-\frac{5}{4}\sqrt{10}$

2. $\dfrac{(41)^{3/2}}{40}$

3. $-\dfrac{b^2}{a}$

4. p

5. -1

6. $\dfrac{5\sqrt{5}}{4}$

7. $\dfrac{(1+9x^4)^{3/2}}{6x}$

8. $\dfrac{a}{2}$

9. $\dfrac{(4y^2+9x^4)^{3/2}}{24xy^2-18x^4}$

10. $\dfrac{(2-2\cos\theta)^{3/2}}{3(1-\cos\theta)}$

11. $-\frac{4}{3}$

12. ∞

Ex. 8—3:

1. $\alpha=2,\ \beta=3$
2. $\alpha=\frac{9}{2},\ \beta=\log 2-5$

3. $\alpha=\beta=0$
4. $\alpha=5,\ \beta=-2$
5. $\alpha=\dfrac{\pi}{2},\ \beta=0$

Ex. 8—4:

1. (a) $\dfrac{1}{3(1-y^2)}$; (b) $-\dfrac{3x^2+y^2}{2xy+3y^2}$

2. (a) $\dfrac{1}{y^3}(y^2-x^2)$; (b) $\dfrac{y(y-2x)}{(y-x)^3}$

3. (a) $\dfrac{1}{2\sqrt{x(1-x)}}$; (b) $\dfrac{1}{1-e^x}$

4. (a) Max. $=e$; (b) max. $=\dfrac{\pi}{3}$; (c) no max. or min.

5. (a) $R=2$; (b) $R=1$

6. (a) $2y-3x+a=0$; (b) $x+4y=8$; (c) $y-x=1$

7. A square, each side of which is $\dfrac{a+b}{\sqrt{2}}$

8. $2x-y=2$

Ex. 9—1:

1. -1 2. n 3. $-\frac{3}{2}$ 4. 0

5. $\frac{1}{6}$ 6. 1 7. $\frac{1}{2}$ 8. $\frac{1}{2}$

9. 2 10. 2

Ex. 9—2:

1. 0 2. 0 3. 0 4. 1

5. 0 6. 0 7. 0 8. ∞

9. $\dfrac{1}{2\pi}$ 10. 0 11. 0 12. 0

13. -3 14. ∞ 15. $-\frac{1}{2}$

Ex. 9—3:

1. 1 2. 1 3. 1 4. e

5. e^3 6. 1 7. e^{mn} 8. $\dfrac{1}{e}$

9. e 10. e

Ex. 9—4:

1. $3x + 2y - 10 = 0$ 6. $\dfrac{9}{2\pi}$ ft. per min.

2. $2y = 2x + 3$ 7. $(-1,14)$

3. (a) $\dfrac{1}{e}$;   (b) e^a;   (c) 1 8. $\dfrac{4e^{2x}}{(1 + 4e^{4x})^{3/2}}$

4. $-\dfrac{2x + y}{x + 2y}$; 0 9. (a) y^2;

                                 (b) 5.27 or $\frac{5}{3}\sqrt{10}$

5. $t = 1$

Ex. 10—1:

1. $\dfrac{\partial z}{\partial x} = 3x^2 + 10xy; \dfrac{\partial z}{\partial y} = 5x^2 + 3y^2$

2. $\dfrac{\partial z}{\partial x} = e^x \cdot \log y; \dfrac{\partial z}{\partial y} = \dfrac{e^x}{y}$

3. $\dfrac{\partial u}{\partial x} = 2Ax + By; \dfrac{\partial u}{\partial y} = Bx + 2Cy$

4. $\dfrac{\partial z}{\partial x} = ae^{ax+y}; \dfrac{\partial z}{\partial y} = e^{ax+y}$

5. $\dfrac{\partial u}{\partial x} = \dfrac{1}{y}\cos\dfrac{x}{y}; \dfrac{\partial u}{\partial y} = -\dfrac{x}{y^2}\cos\dfrac{x}{y}$

6. $\dfrac{\partial u}{\partial x} = yx^{y-1}; \dfrac{\partial u}{\partial y} = x^y \log x$

7. $\dfrac{\partial z}{\partial x} = \dfrac{x}{b^2c}; \dfrac{\partial z}{\partial y} = \dfrac{y}{a^2c}$

8. $\dfrac{\partial z}{\partial x} = \dfrac{y}{x}; \dfrac{\partial z}{\partial y} = \log x$

9. $\dfrac{\partial z}{\partial x} = \dfrac{x}{z}; \dfrac{\partial z}{\partial y} = \dfrac{y}{z}$

10. $\dfrac{\partial z}{\partial x} = -\dfrac{x(y^2 + z^2)}{z(x^2 + y^2)}; \dfrac{\partial z}{\partial y} = -\dfrac{y(x^2 + z^2)}{z(x^2 + y^2)}$

Ex. 10—2:

1. $\dfrac{\partial^2 u}{\partial x^2} = 2x; \dfrac{\partial^2 u}{\partial y^2} = 2y; \dfrac{\partial^2 u}{\partial x\,\partial y} = \dfrac{\partial^2 u}{\partial y\,\partial x} = 1$

2. $\dfrac{\partial^2 z}{\partial x^2} = -y^2 \sin xy; \dfrac{\partial^2 z}{\partial y^2} = -x^2 \sin xy; \dfrac{\partial^2 z}{\partial x\,\partial y} = -xy \sin xy$

3. $\dfrac{\partial^2 u}{\partial x^2} = e^x; \dfrac{\partial^2 u}{\partial y^2} = e^y; \dfrac{\partial^2 z}{\partial x\,\partial y} = \dfrac{\partial^2 z}{\partial y\,\partial x} = 0$

4. $\dfrac{\partial^2 z}{\partial x^2} = y^2 e^{xy}; \dfrac{\partial^2 z}{\partial y^2} = x^2 e^{xy}; \dfrac{\partial^2 z}{\partial x\,\partial y} = \dfrac{\partial^2 z}{\partial y\,\partial x} = xye^{xy} + e^{xy}$

5. $\dfrac{\partial^2 z}{\partial x^2} = 6xy^2; \dfrac{\partial^2 z}{\partial y^2} = 2x^3 + 18xy; \dfrac{\partial^2 z}{\partial x\,\partial y} = 6x^2 y + 9y^2$

6. $\dfrac{\partial^2 z}{\partial x^2} = 6x + 8y; \dfrac{\partial^2 z}{\partial y^2} = -6y; \dfrac{\partial^2 z}{\partial x\,\partial y} = 8x$

7. $\dfrac{\partial^2 u}{\partial x\,\partial y} = -\sin(x + y)$

8. $\dfrac{\partial^2 u}{\partial x\,\partial y} = \dfrac{1}{x}$

9. $\dfrac{\partial^2 u}{\partial x\,\partial y} = -\left(\dfrac{1}{x^2} + \dfrac{1}{y^2}\right)$

10. $\dfrac{\partial^2 u}{\partial x\,\partial y} = -\sin y$

Ex. 10—4:

1. (a) 1;   (b) n;   (c) 0

2. (a) $dz = 2ax\,dx + 3by^2\,dy$

(b) $dz = 2ay^3x\,dx + 3ax^2y^2\,dy$

(c) $dz = yx^{y-1}\,dx + x^y \log x\,dy$

3. Length, $r\sqrt{2}$; width, $\dfrac{r}{2}\sqrt{2}$

4. $x \cdot \dfrac{d}{dx}[f(x)] + f(x) = 0$

5. (a) $\left(\dfrac{a}{3}, \dfrac{2a^3}{27}\right)$;   (b) $(-1, 0)$;   (c) $(a, -b)$

6. (a) $\dfrac{-2(px + p^2)^{3/2}}{p^2}$;   (b) $\dfrac{(\cos^2 y + e^{2x})^{1/2}}{e^x}$

7. $x = \dfrac{\pi}{4}$

Ex. 11—1:

1. $\dfrac{1}{2} + \dfrac{2}{2^2} + \dfrac{3}{2^3} + \dfrac{4}{2^4} + \dfrac{5}{2^5} + \cdots$

2. $1 - 2^2 + 3^2 - 4^2 + 5^2 - \cdots$

3. $1 + \dfrac{1}{3} + \dfrac{1}{3^2} + \dfrac{1}{3^3} + \dfrac{1}{3^4} + \cdots$

4. $1 + 2k + 4k^2 + 8k^3 + 16k^4 + \cdots$

5. $1 + 6 + 15 + 28 + 45 + \cdots$

6. $-\dfrac{1}{2} + \dfrac{1}{8} - \dfrac{1}{26} + \dfrac{1}{80} - \dfrac{1}{242} + \cdots$

7. $1 + \dfrac{1}{2!} + \dfrac{1}{3!} + \dfrac{1}{4!} + \dfrac{1}{5!} + \cdots$

8. $1 + \dfrac{3}{2} + \dfrac{5}{2^2} + \dfrac{7}{2^3} + \cdots$

9. $1 - \dfrac{x}{2} + \dfrac{x^2}{2^2} - \dfrac{x^3}{2^3} + \cdots$

10. $-1 + \dfrac{k}{2} - \dfrac{k^2}{2^2} + \dfrac{k^3}{2^3} - \cdots$

11. $\dfrac{1}{3\cdot4} + \dfrac{1}{4\cdot5} + \dfrac{1}{5\cdot6} + \dfrac{1}{6\cdot7} + \cdots$

12. $1 + \dfrac{\sqrt{2}}{2!} + \dfrac{\sqrt{3}}{3!} + \dfrac{\sqrt{4}}{4!} + \cdots$

13. $\dfrac{2^2}{3^3} + \dfrac{2^3}{3^5} + \dfrac{2^4}{3^7} + \dfrac{2^5}{3^9} + \cdots$

14. $1 + \dfrac{1}{3} + \dfrac{1}{5^2} + \dfrac{1}{7^3} + \cdots$

15. $\dfrac{1}{2} - \dfrac{2}{3^2} + \dfrac{3}{4^3} - \dfrac{4}{5^4} + \cdots$

16. $\dfrac{1}{2n - 1}$

17. $(-1)^{n+1} \dfrac{1}{n(n + 1)}$

18. $\dfrac{2n + 1}{n(n + 1)}$

19. $\dfrac{1}{(2n)!}$

20. $\dfrac{1 + n}{1 + n^2}$

21. $(-1)^{n+1}\left(\dfrac{1}{2n}\right)$

22. $\dfrac{1}{(2n + 1)^n}$

23. $\dfrac{2^n + 1}{2^n}$

24. $\dfrac{n}{3^n}$

25. $\dfrac{n}{(n + 1)^n}$

Ex. 11—2:

1. Divergent   **3.** Divergent   **5.** Convergent   **7.** Divergent
2. Convergent   **4.** Convergent   **6.** Divergent   **8.** Convergent

                9. Convergent
              10. Convergent

Ex. 11—3:

1. Convergent   **3.** Convergent   **5.** Convergent   **7.** Divergent
2. No test   **4.** Convergent   **6.** No test   **8.** Convergent

                9. Convergent
              10. Divergent

Ex. 11—4:

1. $-1 < x < 1$ **5.** $-2 < x < 2$
2. $-1 < x < 1$ **6.** $1 > x \geqq -1$
3. $-1 \leqq x \leqq 1$ **7.** $-1 \leqq x < 1$
4. $-1 \leqq x < 1$ **8.** All values of x

Ex. 11—5:

1. Convergent for all values of x **4.** Convergent for $-1 \leqq x < 1$
2. Convergent for all values of x **5.** Convergent for all values of x
3. Convergent for all values of x **6.** Convergent for all values of x

Ex. 11—6:

1. $e^x = e^{-3}\left[1 + (x+3) + \dfrac{1}{2!}(x+3)^2 + \dfrac{1}{3!}(x+3)^3 + \cdots \right]$

2. $e^{-x} = \dfrac{1}{e}\left[1 - (x-1) + \dfrac{1}{2!}(x-1)^2 - \dfrac{1}{3!}(x-1)^3 + \cdots \right]$

3. $\sin x = \sin a + \cos a \,(x-a) - \dfrac{\sin a \,(x-a)^2}{2!} - \dfrac{\cos a \,(x-a)^3}{3!} + \cdots$

4. $\cos x = \cos a - \sin a \,(x-a) - \dfrac{\cos a \,(x-a)^2}{2!} + \dfrac{\sin a \,(x-a)^3}{3!} + \cdots$

5. $\log x = \log a + \dfrac{(x-a)}{a} - \dfrac{(x-a)^2}{2a^2} + \dfrac{(x-a)^3}{3a^3} - \dfrac{(x-a)^4}{4a^4} + \cdots$

6. $\log x = \log 3 + \dfrac{1}{3}(x-3) - \dfrac{1}{18}(x-3)^2 + \dfrac{1}{81}(x-3)^3 - \cdots$

7. $\log(x+h) = \log h + \dfrac{x}{h} - \dfrac{x^2}{2h^2} + \dfrac{x^3}{3h^3} - \dfrac{x^4}{4h^4} + \cdots$

8. $\sin(x+h) = \sin h + \dfrac{\cos h}{1!}x - \dfrac{\sin h}{2!}x^2 - \dfrac{\cos h}{3!}x^3 + \cdots$

9. $\cos(x+h) = \cos x - h\sin x - \dfrac{h^2}{2!}\cos x\, \dfrac{h^3}{3!}\sin x + \cdots$

10. $e^{x+h} = e^x\left[1 + h + \dfrac{h^2}{2!} + \dfrac{h^3}{3!} + \cdots\right]$

Ex. 11—8:

1. (a) $\dfrac{1}{2},\ 1,\ 2,\ 4,\ 8,\ \cdots$

(b) $\dfrac{1}{\sqrt{2}},\ \dfrac{4}{\sqrt{3}},\ \dfrac{9}{\sqrt{4}},\ \dfrac{16}{\sqrt{5}},\ \dfrac{25}{\sqrt{6}},\ \cdots$

(c) $\dfrac{x}{a},\ \dfrac{x^2}{2a^2},\ \dfrac{x^3}{3a^3},\ \dfrac{x^4}{4a^4},\ \dfrac{x^5}{5a^5},\ \cdots$

(d) $\dfrac{e^x}{\sin x},\ -\dfrac{e^{2x}}{\sin 2x},\ +\dfrac{e^{3x}}{\sin 3x},\ -\dfrac{e^{4x}}{\sin 4x},\ +\cdots$

2. (a) Convergent; (b) convergent

3. (a) For all values of x
(b) For all values of x
(c) For $x \leqq -1$ and $x > 1$

4. (a) 0; (b) 1

5. $x = \dfrac{1}{2}h$

6. $\dfrac{dy}{dx} = \dfrac{x}{y} - \dfrac{2x^3}{y} = \pm\dfrac{1}{\sqrt{1-x^2}} \mp \dfrac{2x^2}{\sqrt{1-x^2}}$; when $x = 0$, $\dfrac{dy}{dx} = \pm 1$;

hence, a multiple point at the origin.

7. (a) $\dfrac{du}{dy} = 2ay + \dfrac{1}{x} + \dfrac{2ax}{\cos y}$

(b) $\dfrac{du}{dx} = (2y + z)\cos x - (4z^3 + y)\sin x$

8. (a) $x = 4$, minimum; $x = -2$, maximum
(b) $x = e$, maximum

(c) $x = \dfrac{\pi}{4}$, maximum

9. (a) Double point at the origin
(b) Double point at the origin

10. (a) Cusp at $(0,0)$; $\dfrac{dy}{dx} = \pm 1$

    (b) Cusp at $(1,1)$

Ex. 12—1:

1. $2x^4 + C$

2. $2ax^5 + C$

3. $\dfrac{y^5}{5} + C$

4. $\dfrac{4x^{7/4}}{7} + C$

5. $\dfrac{4}{3} x^{3/2} + C$

6. $\dfrac{2}{5} x^{5/4} + C$

7. $\dfrac{2}{5} z^{5/2} + C$

8. $\dfrac{25}{7} y^{7/5} + C$

9. $\dfrac{nx^{m/n+1}}{m + n} + C$

10. $-\dfrac{2}{x^3} + C$

11. $\dfrac{2}{3} x\sqrt{2px} + C$

12. $\dfrac{y^{p+q+1}}{p + q + 1} + C$

13. $z^{5/6} + C$

14. $2z^{5/2} + C$

15. $\dfrac{3kz^{5/3}}{4} + C$

16. $-\dfrac{2}{x} + C$

17. $\dfrac{2ax^{5/2}}{5} + C$

18. $\dfrac{-6}{\sqrt[3]{y^2}} + C$

19. $\dfrac{x^3}{12} - \dfrac{x^2}{3} + C$

20. $\dfrac{x^4}{2} - \dfrac{8}{3} x^3 + \dfrac{x^2}{8} + C$

21. $\dfrac{10}{3\sqrt[5]{x}} + C$

22. $-\dfrac{12}{x} + \dfrac{1}{x^3} + C$

23. $\dfrac{x^3}{3} + 3x^2 + 9x + C$

24. $9a^2x - 2ax^3 + \dfrac{x^5}{5} + C$

25. $\dfrac{5}{4} (x + 2)^4 + C$

Ex. 12—2:

1. $-2\sqrt{1 - x} + C$

2. $-\dfrac{3}{x + a} + C$

3. $-\dfrac{1}{2(x + a)^2} + C$

4. $-\dfrac{1}{4a(ax - b)^2} + C$

5. $\dfrac{2}{3} (a + x)^{3/2} + C$

6. $-\dfrac{2}{9} (1 - 3x)^{3/2} + C$

7. $-\dfrac{k}{a(ax + b)} + C$

8. $-\dfrac{1}{3} (m^2 - x^2)^{3/2} + C$

9. $\dfrac{a}{(b - x)^4} + C$

10. $+\dfrac{1}{3} (x^2 + a^2)^{3/2} + C$

11. $-\dfrac{3}{4} (1 - x^2)^{2/3} + C$

12. $\dfrac{2}{3} (z - 2)^{3/2} + C$

13. $\dfrac{1}{3} (z^2 - 2)^{3/2} + C$

14. $amx + \dfrac{1}{2} (a + m)x^2 + \dfrac{x^3}{3} + C$

15. (a) $\displaystyle\int (x^4 - 1)^2 x^3\, dx = \int (x^8 - 2x^4 + 1)x^3\, dx$

$$= \int (x^{11} - 2x^7 + x^3)\, dx$$

$$= \frac{x^{12}}{12} - \frac{x^8}{4} + \frac{x^4}{4} + C$$

(b) $\displaystyle\int (x^4 - 1)^2 x^3\, dx = \frac{1}{4}\int (x^4 - 1)\, d(x^4 - 1)$

$$= \frac{1}{4}\cdot\frac{(x^4 - 1)^3}{3} = \frac{1}{12}(x^4 - 1)^3 + C$$

$$= \frac{x^{12}}{12} - \frac{x^8}{4} + \frac{x^4}{4} - \frac{1}{12} + C$$

Check by differentiation.

Ex. 12—3:

1. $3 \log x + C$

2. $\frac{4}{3} \log x + C$

3. $5 \log x + C$

4. $2 \log x + C$

5. $\frac{1}{2} \log (2x + 3) + C$

6. $-\frac{1}{3} \log (2 - 3z) + C$

7. $\log (x^2 + 2) + C$

8. $\frac{1}{2} \log (x^2 - 4) + C$

9. $\frac{1}{3} \log (z^3 - 1) + C$

10. $\log z + \dfrac{1}{2z^2} + C$

11. $y^2 - \log y^2 + C$

12. $\frac{1}{2}[x^2 + \log (x^2 - 1)] + C$

13. $\frac{1}{3} \log (x^3 - 1) + C$

14. $x^3 - \log (x^3 + 2)^2 + C$

15. $\log (x^3 - 2x + 1) + C$

Ex. 12—4:

1. $\dfrac{1}{5} e^{4x} + C$

2. $\dfrac{ka^{mx}}{m \log a} + C$

3. $e^{\sin x} + C$

4. $\dfrac{1}{2} e^{2\sin x} + C$

5. $-\dfrac{1}{3} e^{-3x} + C$

6. $\dfrac{na^{x/n}}{\log a} + C$

7. $\dfrac{a^{2x-1}}{2 \log a} + C$

8. $\dfrac{1}{6} e^{2x^3} + C$

9. $\dfrac{1}{2} e^{2x} + e^x + C$

10. $-\dfrac{4}{a^{2x} \log a} + C$

11. $\dfrac{k^{x^2}}{2 \log k} + C$

12. $2e^{x/2} - 2e^{-x/2} + C$

Ex. 12—5:

1. $\dfrac{1}{4} \sin 4x + C$

2. $-\dfrac{5}{3} \cos \dfrac{3}{5} x + C$

3. $2 \log \sec \dfrac{\theta}{2} + C$

4. $\dfrac{1}{a} \tan a\theta + C$

5. $3 \log \sin \dfrac{\theta}{3} + C$

6. $\dfrac{1}{a} \log \sec a\theta + C$

7. $\dfrac{m}{2} \log \sin \dfrac{x^2}{m} + C$

8. $-\dfrac{1}{m} \cot m\theta + C$

9. $\log \sin e^\theta + C$

10. $\sin (\log x) + C$

11. $\dfrac{1}{4} \tan x^4 + C$

12. $m \log \left(\sec \dfrac{\theta}{m} + \tan \dfrac{\theta}{m} \right) + C$

13. $\tan x - \cot x + C$

14. $\tan x + C$

Ex. 12—6:

1. $\dfrac{1}{40} \log \dfrac{5x - 4}{5x + 4} + C$

2. $\dfrac{3}{2b} \arctan \dfrac{x}{2b} + C$

3. $\log (x + \sqrt{x^2 + a^2 b^2}) + C$

4. $\dfrac{1}{m} \log (mx + \sqrt{m^2 x^2 - 25}) + C$

5. $\dfrac{1}{6} \arctan \dfrac{2x}{3} + C$

6. $\dfrac{1}{3} \arcsin \dfrac{3x}{4} + C$

7. $\dfrac{1}{3} \log (3x + \sqrt{9x^2 + 16}) + C$

8. $\dfrac{1}{2ab} \log \dfrac{ax - b}{ax + b} + C$

9. $\dfrac{1}{\sqrt{5}} \operatorname{arc\,sec} \dfrac{3x}{\sqrt{5}} + C$

10. $\dfrac{1}{\sqrt{21}} \arctan \dfrac{\sqrt{21}}{7} x + C$

11. $\dfrac{1}{2m^2} \arctan \dfrac{x^2}{m^2} + C$

12. $\dfrac{1}{2} \operatorname{arc\,sec} \dfrac{\sqrt{3}\,x}{2} + C$

Ex. 12—7:

1. $\dfrac{2}{3} t^{3/2}$

2. $\dfrac{a^{3x}}{3 \log a}$

3. $\dfrac{1}{2} (2x + 3)^{3/2}$

4. $-\dfrac{1}{2 e^{x^2}}$

5. $\dfrac{1}{5} (\log x)^5$

6. $\log (x + 5)$

7. $\frac{1}{2} \log (z^2 - 1)$

8. $\frac{1}{3} x^3 - 8 \log x$

9. $\frac{1}{6} x^2 - \frac{2}{9} x + \frac{4}{27} \log (3x + 2)$

10. $\frac{1}{3m} (mx + b)^3$

11. $\frac{1}{12} \tan^{-1} \left(\frac{3x}{4} \right)$

12. $\frac{1}{24} \log \frac{3x - 4}{3x + 4}$

13. $-e^{\cos x}$

14. $y = 2x^3 - 5x^2 + 8x - 2$

15. $y = 2x^3 + 3x^2 - 8x + 2$

Ex. 13—1:

1. $x \sin x + \cos x + C$
2. $x \log x - x + C$
3. $\frac{1}{2} e^{2x} (x - \frac{1}{2}) + C$
4. $e^x (x^2 - 2x + 2) + C$
5. $\frac{1}{2} e^x (\sin x + \cos x) + C$
6. $x (\log^2 x - 2 \log x + 2) + C$
7. $x^2 \sin x + 2x \cos x - 2 \sin x + C$
8. $\frac{1}{4} x^4 (\log x - \frac{1}{4}) + C$
9. $\theta \tan \theta - \log \sec \theta + C$
10. $x \arctan x - \frac{1}{2} \log (1 + x^2) + C$
11. $-e^{-x} (x^2 + 2x + 2) + C$
12. $\frac{1}{3} x^3 (\log x - \frac{1}{3}) + C$
13. $\theta \tan \theta + \log \cos \theta - \frac{1}{2} \theta^2 + C$
14. $z \tan z + \log \cos z + C$
15. $\frac{1}{5} e^x (2 \sin 2x + \cos 2x) + C$
16. $\sin x (\log \sin x - 1) + C$
17. $\frac{1}{2} x \sin 2x - \frac{1}{2} x^2 \cos 2x + \frac{1}{4} \cos 2x + C$
18. $\frac{1}{2} x \sin 2x + \frac{1}{4} \cos 2x + C$

Ex. 13—2:

1. $\frac{1}{2} \theta + \frac{1}{4} \sin 2\theta + C$
2. $\sin \theta - \frac{\sin^3 \theta}{3} + C$
3. $- \cos \theta + \frac{2}{3} \cos^3 \theta - \frac{1}{5} \cos^5 \theta + C$
4. $\frac{\theta}{8} - \frac{\sin 4\theta}{32} + C$
5. $\frac{3}{8} \theta - \frac{\sin 2\theta}{4} + \frac{\sin 4\theta}{32} + C$
6. $- \cot \theta - \theta + C$
7. $\frac{\sin^4 x}{4} + C$
8. $\frac{\sin^3 \theta}{3} + C$

9. $-\dfrac{\cos^4 x}{4} + C$

10. $\dfrac{\tan^2 \theta}{2} + \log \cos \theta + C$

11. $\frac{1}{3} \cos^3 \theta - \frac{1}{5} \cos^5 \theta + C$

12. $\dfrac{\tan 2\theta}{2} - \theta + C$

Ex. 13—3:

1. $\frac{2}{75}(3 + 5x)^{3/2} - \frac{6}{25}\sqrt{3 + 5x} + C$

2. $\log \dfrac{\sqrt{x+1} - 1}{\sqrt{x+1} + 1} + C$

3. $4x^{1/4} - \arctan x^{1/4} + C$

4. $\frac{3}{5}(x-1)^{5/3} + \frac{3}{2}(x-1)^{2/3} + C$

5. $2[\sqrt{x} - \log(\sqrt{x} + 1)] + C$

6. $\frac{2}{3}(x-1)^{3/2} + 2\sqrt{x-1} + C$

7. $\frac{2}{15}(3t-2)(t+1)^{3/2} + C$

8. $-\dfrac{\sqrt{x^2 + 1}}{x} + C$

Ex. 13—5:

1. $-\dfrac{1}{3x} + \dfrac{5}{9}\log\left(\dfrac{3 + 5x}{x}\right) + C$

2. $\frac{1}{2}\sqrt{2x^2 - 9} + C$

3. $\dfrac{e^{3x}(3 \sin 2x - 2 \cos 2x)}{13} + C$

4. $\dfrac{-\sqrt{9 - 4x^2}}{4} + C$

5. $\dfrac{1}{2\sqrt{6}} \log \dfrac{\sqrt{3} + \sqrt{2}\,x}{\sqrt{3} - \sqrt{2}\,x} + C$

6. $\dfrac{x}{\sqrt{2x^2 + 1}} + C$

7. $\dfrac{x}{2}\sqrt{x^2 - 5} + \dfrac{5}{2}\log(x + \sqrt{x^2 - 5}) + C$

8. $\dfrac{1}{4} \operatorname{arc\,sec} \dfrac{3x}{4} + C$

9. $\dfrac{1}{\sqrt{3}} \log(x + \sqrt{x^2 + \frac{5}{3}}) + C$

10. $\sqrt{2} \arctan \sqrt{\dfrac{3x - 2}{2}} + C$

11. $\frac{1}{2} \log \left(\frac{x^2}{1 + 7x^2} \right) + C$

12. $\sqrt{2x^2 - 9} - 3 \arccos \frac{3\sqrt{2}}{2x} + C$

13. $\frac{1}{2} \log \frac{\sqrt{4 + 3x} - 2}{\sqrt{4 + 3x} + 2} + C$

14. $-\frac{\sqrt{3 - x^2}}{3x} + C$

15. $-\frac{\cos 9x}{18} - \frac{\cos 3x}{6} + C$

Ex. 13—6:

1. $8 \log x + 4 \log (x + 2) + C$
2. $2 \log x + \log (x - 3) + C$
3. $2 \log (x - 4) + 5 \log (x + 3) + C$
4. $4 \log (x - 5) - 3 \log (x + 5) + C$
5. $2 \log (x - 1) - 3 \log (x - 2) + C$
6. $\log x + 2 \log (x + 4) + 3 \log (x - 1) + C$
7. $\frac{9}{2} \log (2x + 1) - 4 \log (x - 3) + C$
8. $3 \log x + 7 \log (x + 4) - \frac{5}{3} \log (3x - 1) + C$
9. $\log \frac{(p + x)^p}{(q + x)^q} + C$
10. $\log \frac{(x - a)(x + b)}{x} + C$

Ex. 13—7:

1. $2 \log x - \frac{1}{x} + \frac{3}{5} \log (5x + 3) + C$
2. $2 \log x + 4 \log (x + 5) - \frac{1}{x + 5} + C$
3. $5 \log x - \frac{5}{x} + 2 \log (2x - 3) + C$
4. $3 \log (x + 2) + \frac{1}{x + 2} + \frac{1}{(x + 2)^2} + C$
5. $3 \log (x - 2) - \frac{2}{x - 2} + \log (x + 3) + C$

6. $8 \log x - \dfrac{6}{x} - 2 \log (x + 2) + C$

7. $3 \log (x - 2) + 2 \log (x + 2) - \dfrac{1}{x + 2} + C$

8. $10 \log (x + 1) - \log x + \dfrac{2}{x} + \dfrac{3}{2x^2} + C$

Ex. 13—8:

1. $4 \log x + \dfrac{5}{2} \log (x^2 - 3) + \dfrac{2}{\sqrt{3}} \arctan \dfrac{x}{\sqrt{3}} + C$

2. $8 \log (2x + 1) - 2 \log (x^2 + 1) + 5 \arctan x + C$

3. $-6 \log x + \dfrac{5}{2} \log (2x^2 + 5) + \dfrac{3}{\sqrt{10}} \arctan \sqrt{\dfrac{2}{5}} x + C$

4. $3 \log x + 3 \log (1 + x^2) - 5 \arctan x + C$

5. $7 \log (3x - 1) - \dfrac{3}{2} \log (x^2 + 2) + \dfrac{5}{\sqrt{2}} \arctan \dfrac{x}{\sqrt{2}} + C$

6. $\dfrac{3}{2} \log (x^2 - 2) + \dfrac{2}{\sqrt{2}} \arctan \dfrac{x}{\sqrt{2}} + \log (x^2 + 1) + 5 \arctan x + C$

7. $\log (1 + x^2) - 2 \log (1 - 2x) + C$

8. $\dfrac{1}{2} \log (x^2 + 1) + 2 \arctan x + 3 \log (x^2 - 1) + C$

Ex. 14—1:

1. 111
2. 15

3. $\dfrac{64a}{5}$

4. $\dfrac{56a^2}{3}$

5. 4
6. 2

7. $\dfrac{k\pi}{2}$

8. $\dfrac{\log 5}{2}$

9. 24
10. $-5\frac{1}{3}$
11. $4(e^3 - 1)$
12. $\frac{1}{2}$

13. $\dfrac{a^{-4} - a^{-2}}{\log a}$

14. $e - \dfrac{1}{e}$

15. -2
16. $2e^3 - 5$

Ex. 14—2:

1. 112
2. $A = 10\frac{2}{3}; B = 5\frac{1}{3}$

3. 96
4. 2

5. 27
6. $A = 4; B = 12$

7. $77\frac{1}{3}$

8. $a^2(\log b - \log a)$

9. $a^2/6$

10. $21\frac{1}{3}$

Ex. 14—3:

1. 1

3. $\dfrac{\pi}{2a}$

5. e

7. $\dfrac{\pi}{2m}$

2. $\frac{1}{8}$

4. 1

6. No meaning

8. No meaning

Ex. 14—4:

1. 8

2. $\dfrac{a^4}{4}$

3. $\dfrac{\pi}{4}$

4. $\dfrac{1}{2}\log 2$

5. $\dfrac{e^2 + 1}{4}$

6. $\dfrac{1}{a}$

7. $4 - 2\log 2$

8. 1

9. 36

10. $\log_e\left(\dfrac{b}{a}\right)$

11. 2

12. $\dfrac{2}{3}a^3$

Ex. 14—5:

1. $\frac{1}{2}e^x(\sin x - \cos x) + C$

2. $e^x(x - 1) + C$

3. $\log(x - 3) - \dfrac{2}{x - 3} + C$

4. $\log\left(\dfrac{x + 1}{x}\right) - \dfrac{2}{x + 1} + C$

5. $x + \log\left(\dfrac{x - 2}{x - 1}\right) + C$

6. $\log\left(\dfrac{x^2 - 1}{x}\right) + C$

7. $\log\left(\dfrac{x}{1 - x^2}\right) + C$

8. $x + 2 - 4\log(x + 2) - \dfrac{4}{x + 2} + C$

9. $\dfrac{1}{2}\left[\dfrac{x}{x^2 + 1} + \text{arc tan } x\right] + C$

10. $\dfrac{x(2 + x)}{1 + x} - 2\log(1 + x) + C$

11. Divergent

12. Convergent; $|x| > 1$

Ex. 15—1:

1. $\frac{1}{6}a^2$

2. $5\log 5 - 4$

3. $+2$; -2; areas on opposite sides of axis; 4

4. $\dfrac{b^4 - a^4}{4}$

5. $\frac{1}{12}a^4$

6. $\dfrac{\pi a^2}{2} - \dfrac{4}{3}a^2$

Ex. 15—2:

1. $12 \log 3$ **2.** 6 **3.** πab **4.** $3\pi a^2$

5. $\dfrac{\pi a^2}{8}$ **6.** $\dfrac{3\pi a^2}{2}$ **7.** a^2 **8.** $\dfrac{\pi a^2}{4}$

9. πa^2 **10.** $\dfrac{4\pi^3 a^2}{3}$

Ex. 15—3:

1. $\frac{1}{2}[3\sqrt{10} + \log (3 + \sqrt{10})]$
2. $\sqrt{2}\,(e - 1)$
3. $a\sqrt{2} + a \log (1 + \sqrt{2})$
4. $\log \sqrt{3}$
5. $2\pi a$
6. $2\pi a$; same curve as in Ex. 5
7. $a[\pi\sqrt{4\pi^2 + 1} + \frac{1}{2} \log (2\pi + \sqrt{4\pi^2 + 1})$
8. $\sqrt{\rho^2 - a^2}$ **9.** $\log 3 - \frac{1}{2}$ **10.** $\frac{3}{2}\pi a$

Ex. 15—4:

1. $\dfrac{4\pi ab^2}{3}$ **2.** $\dfrac{4\pi a^2 b}{3}$ **3.** 9π

4. $\frac{1}{2}\pi^2$ **5.** $8\pi p^3$ **6.** $2\pi^2 r^2 k$

7. $32\pi\sqrt{5}$ **8.** $16\pi\sqrt{5}$ **9.** 72π

10. $\dfrac{8\pi p^2}{3}\,(3\sqrt{3} - 1)$ **11.** $21\sqrt{2}\,\pi$

Ex. 16—1:

1. $\dfrac{5x^4}{12} + C_1 x + C_2$

2. $\dfrac{x^4}{8} + \dfrac{C_1 x^2}{2} + C_2 x + C_3$

3. $\dfrac{x^4}{24} + \dfrac{C_1 x^3}{6} + \dfrac{C_2 x^2}{2} + C_3 x + C_4$

4. $\dfrac{x^6}{30} + C_1x + C_2$

5. $\dfrac{x^4}{12} + \dfrac{x^3}{3} + C_1x + C_2$

6. $e^x + \dfrac{C_1x^2}{2} + C_2x + C_3$

7. $\cos\theta + \dfrac{C_1\theta^2}{2} + C_2\theta + C_3$

8. 8
9. 112
10. $1\frac{1}{2}$
11. π
12. $Ax^2 + Bx + C = 0$

Ex. 16—2:

1. $3\frac{1}{3}$ 2. 0   3. $8\frac{3}{4}$ 4. $\dfrac{1}{2}\left(a^5 - a^4 + \dfrac{a^3}{3}\right)$

5. 18   6. $\dfrac{1}{2}(k^3 - 1)$ 7. $\dfrac{9}{5}$ 8. $\dfrac{\pi}{4}$

9. $\dfrac{a^3}{3}\left(1 - \dfrac{\sqrt{3}}{2}\right)$ 10. $\dfrac{1}{2}e^2 - \dfrac{1}{2}$ 11. $\dfrac{\pi a^4}{2}$ 12. $\dfrac{23}{24}a^4$

13. 6   14. a^5 15. $\dfrac{a^6}{48}$

Ex. 16—3:

1. $\dfrac{16}{3}$ 2. $\dfrac{64\sqrt{2}}{3}$ 3. $\sqrt{2} - 1$ 4. $e^2 - 1$

5. 3   6. $\dfrac{32}{5}$ 7. 16   8. 25π

9. $\dfrac{3\pi a^2}{4}$ 10. 3π

Ex. 16—4:

1. Trap. rule, 1.812; direct integration, 1.792
2. Trap. rule, 332.0; direct integration, 330.7
3. Simpson's rule, 6554.7; direct integration, 6553.6

4. Simpson's rule, 1.247; direct integration, 1.248
5. Trap. rule, 0.683; direct integration, 0.684
6. Simpson's rule, 9.295; direct integration, 9.279
(Four-place logarithmic and trigonometric tables were used in the above verifications.)

Ex. 16—5:

1. (a) 24   (b) $\dfrac{a^2(2b^2 - a^2)}{4}$ (c) $\dfrac{\pi}{4}$

(d) $-\log\left(\dfrac{\sqrt{2}}{2}\right)$ (e) $\dfrac{1}{a}$ (f) $\frac{1}{4}(e^2 + 1)$

2. (a) $2\pi a$ (b) a

3. $\dfrac{3\pi a^2}{2}$

4. (a) 12;   (b) $\sqrt{2} - 1$

5. (a) $dz = 2ay^3 x\, dx + 3ax^2y^2\, dy$
(b) $dz = yx^{y-1}\, dx + x^y \log x\, dy$

6. (a) $\dfrac{1}{y} - \dfrac{y}{x^2}$;   (b) $-\dfrac{x}{y^2} + \dfrac{1}{x}$

7. (a) $-1 \leqq x \leqq 1$;   (b) $-1 \leqq x \leqq 1$

9. (a) 1;   (b) e^2

10. (a) $\dfrac{1}{3} x^3 \left(\log x - \dfrac{1}{3}\right)$;   (b) $\dfrac{e^x}{2} (\sin x + \cos x)$

Index

Abel, Niels Henrik (1802–29), 16
Absolute term, 25
Absolute value of a constant, 19
Acceleration: components of, 144–45; defined, 83
Algebraic functions: differentiation of, 60–75
Algebraic substitution, 283–85
Alternating series, 234–35
Analytic geometry, 13, 16, 18
Angle: between two intersecting curves, 128–29
Anti-derivative, 252–54
Approximate calculations, 163–66
Approximate integration, 350–60
Arc lengths, 169–72, 306–7, 329–32
Archimedes (287?–212 B.C.), 14
Area: approximate, 350–60; between two curves, 347–48; by double integration, 344–48; by summation, 318–20; of surface of revolution, 334–36; parametric equations, 325–26; polar coordinates, 326–28; rectangular coordinates, 322–24; under curve, 299–300, 303–7
Arithmetic series, 222, 224, 228, 229
Asymptotes, 135

Barrow, Isaac (1630–77), 14–15
Bernouilli family, 15
Binomial expansion, 240–41
Binomial theorem, 64

Cauchy, Augustin Louis (1789–1857), 16
Cavalieri, Francesco Bonaventura (1598–1647), 14
Center of curvature, 182, 185–87
Center of gravity, 367–68
Change of limits, 312–13
Change of variable: and change of limits, 312–13; in integration by substitution, 283–87

Circle of curvature, 181–83
Circles, osculating, 182
Circular motion, 145–48
Companion series, 228
Comparison tests, 228–31
Component accelerations, 144–45
Component velocities, 143–44
Conjugate points, 218
Constants: defined, 19; derivatives of, 60–61; of integration, 254–55
Continuity, 26–27
Convergence: conditions for, 227; interval of, 235–37; tests for, 228, 232–33
Cosine function, derivative of, 113–14
Critical values, 89
Curvature: at a point, 177–79; center of, 182, 185–87; circle of, 181–83; defined, 173–74; in polar coordinates, 176–77; in rectangular coordinates, 174–75; of circle, 175; of hyperbola, 178–79; of lemniscate, 178; radius of, 179–81, 184–85
Curve: derived, 313–15; first integral, 357; fundamental, 357, 358; integral, 315–16, 357, 358; length of, 169–72, 306–7, 329–32; slope of, 126–27; tangent to, 58
Curve tracing, 135–39
Cusps: conditions for, 219; of the first kind, 218; of the second kind, 218
Cycloid, 141

D'Alembert, Jean Le Rond (1717?–1783), 15
Decreasing function, 78–79
Definite integrals, 300–1, 302, 307–13
Democritus (fl. c.400 B.C.), 13
Dependent variable, 19
Derivative: analytical representation of, 51–52; anti-, 252–54; as limiting value, 50; as rate of change, 77; as

ratio, 148–49; as slope, 53–54, 78; as time rate of change, 77; from parametric equations, 139–43; geometric interpretations of, 52; higher, 209; of constants, 60–61; of exponential functions, 106–110; of inverse trigonometric functions, 116–19, 120–21; of logarithmic functions, 99–105, 110–12; of power functions, 64–66; of products, 62–63, 67–69; of quotients, 69–71; of sums, 62; of trigonometric functions, 112–16; partial, 207–9; 214–15; relation to differential and increment, 157; successive, 85–86; total, 215–16

Descartes, René (1596–1650), 13, 15, 16

Differential: defined, 155–56; differentiation with, 161–62; notation, 160–61; of independent variable, 158–59; relation to increment and derivative, 157; successive, 162–63; total, 211

Differential arc lengths, 169–72, 329–32

Differentiation: defined, 54; formulae of, 71–72; general rule, 57; of algebraic functions, 60–75; of exponential functions, 106–10; of implicit functions, 73–75; of logarithmic functions, 99–105, 110–12; of transcendental functions, 120–22; partial, 206–20; process of, 54–57; successive, 121–24; with differentials, 161–62

Dirichlet, Peter Gustav Lejeune (1805–59), 16

Discontinuity, 44–45

Distance, 81–83

Divergence, 227–28, 232–33

Domain of the function, 19

Double integral, 339

Double integration, 344–48

Double points, 218

e, 242

Eudoxus (fl. *c.*360 B.C.), 14

Euler, Leonhard (1707–83), 15

Euler's formulae, 247–49

Euler's series, 247

Evolute, 187–89

Explicit functions, 20–21, 73, 206

Exponential function, 106–10

Exponential series, 241–42

Extent, 135

Falling body, laws of, 84–85

Fermat, Pierre de (1601–65), 14–16

Force, 79, 363–65

Functions: algebraic, 60–75; continuous, 26–27, 44–45; decreasing, 78–79; defined, 19–20; domain of the, 19; explicit, 20–21, 73, 206; exponential, 106–10; implicit, 20–21, 73–75, 122–24, 206, 211–13; increasing, 78–79; inverse, 100–2, 252; inverse trigonometric, 116–19, 120–21; limit of, 35; logarithmic, 99–105, 110–12; multiple-valued, 19; notation, 25–26; power, 64–66; range of, 19; rational, 25; single-valued, 19; transcendental, 99–124; trigonometric, 112–16

Fundamental theorem of integral calculus, 321, 327, 330, 333

Galileo Galilei (1564–1642), 16

Gauss, Karl Friedrich (1777–1855), 16

Geometric interpretation of derivatives, 52

Geometric series, 33, 222, 228

Gödel's theorem, 16

Gravity, center of, 367–68

Gregory's series, 247

Harmonic series, 225

Hyperbola: curvature of, 178–79

Implicit function: defined, 21; differentiation of, 73–75; successive differentiation of, 122–24; total differentiation of, 211–13

Improper integrals, 309–11

Increasing function, 78–79

Increment: defined, 47–48; notation, 48–50; relation to derivative and differential, 157

Independent variable, 19, 158–59

Indeterminate forms, 194–203

Infinite series. *See* Series, infinite

Infinitesimals, 159–60

Inflection points, 133–35

Integral: as anti-derivative, 252–54; defined, 253; definite, 300–1, 302, 307–13; double, 339; elementary, 264–72; improper, 309–11; interpretation of, 312; iterated, 339; multiple, 338–43; standard, 264–72; tables, 371–78; trigonometric, 279–83; triple, 339; use of tables, 288–90

Integrand, 253

Integraph, 357–58

Integration: applications, 360–68; approximate, 350–60; as process of summation, 318–36; between limits, 299–302; by infinite series, 350–52; by parts, 275–79; by substitution, 283–87; constants of, 254–55; defined, 252–54; double, 344–48; iterated, 339; limits of, 301–2, 307–7, 340; multiple, 338–43; of rational fractions, 290–97; of series, 350–52; partial, 341–43; principles of, 255–64; successive, 338–43; triple, 339, 348–49

Intercepts, 135

Interval of convergence, 235–37

Intervals, of integration, combining of, 308–9

Inverse functions, 100–2, 252; trigonometric, 116–19, 120–21

Inverse operations, 252

Involute, 187–89

Isolated point, 219

Iterated integration, 339

Jacobi, Karl Gustav Jakob (1804–51), 16

Kepler, Johannes (1571–1630), 14

Lagrange, Joseph Louis (1736–1813), 15

Lambert, Johann Heinrich (1728–77), 15

Laplace, Pierre Simon de (1749–1827), 15

Laws of falling body, 84–85

Laws of limits, 35–36

Legendre, Adrien Marie (1752?–1833), 15

Leibniz, Gottfried Wilhelm von (1646–1716), 13, 15, 16

Leibniz's formula, 124

Lemniscate: curvature of, 178

Length of arc, 169–72, 306–7

L'Hospital, Guillaume François Antoine de (1661–1704), 15

Limit: change of, 312–13; concept of, 16–17, 32–34; defined, 34, 40–42; infinity as, 36–40; laws of, 35–36; lower, 301–2; of function, 35; upper, 301–2; zero as a, 36–40

Liquid pressure, 365–67

Logarithmic functions: differentiation of, 99–105, 110–12

Logarithms, 102–5

Lower limit, 301–2

Maclaurin's series, 238–40, 243

Maxima and minima: applications, 93–96; defined, 88–89; in curve tracing, 136; rule for finding, 89–93

Mean value theorem, 191–94, 245–46

Moment, 367

Motion: circular, 145–48; rectilinear, 81–85, 143–45; uniform, 82–83; uniformly accelerated, 83–85

Multiple integrals, 338–43

Multiple-valued function, 19

Newton, Isaac (1642–1727), 13, 15, 16

Nodes, 217, 219

Normal, equation of, 130–31

Ordered pairs, 19, 25

Ordinary points, 88, 217

Oscillating series, 224

Osculating circle, 182

Osculation, point of, 218, 220

Parabolic rule, 354–56

Parametric equations: area under curve, 325–26; derivatives of, 139–43; differential arc length, 171; radius of curvature, 184–85

Partial derivatives, 207–9, 214–15

Partial differentials, 214–15

Partial fractions, 291–97

Partial integration, 341–43

Pascal, Blaise (1623–62), 16

Pi (π), value of, 247–49

Planimeter, 358–60

Point: of inflection, 133–35, 136; of osculation, 218–20

Polar coordinates: area under curve in, 326–28; curvature in, 176–77; differential arc length in, 171–72; plane areas by double integration, 345–47

Power, 79

Power functions, 64–66

Power series, 235–37, 238, 246

Pressure, 79; liquid, 365–67

Process of summation, 318–36

Quotients, derivatives of, 69–71

Radius of curvature, 179–81, 184–85

Range of function, 19

Rate of change: average, 29–32; defined, 27–28; derivative as, 77; instantaneous, 29–32, 34, 78–80

Ratio tests, 232–33

Rational fractions: integration of, 290–97

Rational functions, 25

Rectangular coordinates: area under curve in, 322–24; curvature in, 174–75; plane areas by double integration, 344–45

Rectilinear motion: defined, 81; uniform, 82–83; uniformly accelerated, 83–85. *See also* Component accelerations; Component velocities

Related time rates, 149–52

Relation between variables, 19

Roberval, Gilles Personne de (1602–75), 14

Rolle's theorem, 191–94

Sequence, 222

Series, infinite: alternating, 234–35; arithmetic, 222, 224, 228, 229; convergent, 223–24, 227, 229; divergent, 224–25, 227, 230; Euler's, 247; exponential, 241–42; geometric, 33, 222, 228; Gregory's, 247; harmonic, 225; integration by, 350–52; Maclaurin's, 238–40, 243; oscillating, 224; *p*-, 228–29; power, 235–37, 238, 246; Taylor's, 242–46; telescopic, 228, 229

Sigma notation, 225–26

Simpson's rule, 354–56

Sine function: derivative of, 112–13

Single-valued function, 19

Singular points, 217–20

Slope, 28, 53–54, 78; of a curve, 126–27

Solids of revolution: volumes of, 332–34

Subnormal: length of, 131–32

Subtangent: length of, 131–32

Sums: derivatives of, 62

Surface of revolution: area, 334–36

Symmetry, 135

Tangent: equation of, 130–31; function, 114–15; slope of, 58

Taylor's series, 242–46

Telescopic series, 228, 229

Temperature, 79

Time rates, 149–52

Total derivatives, 215–16

Total differentials, 211

Transcendental functions, 99–124

Trapezoidal rule, 352–54, 356

Trigonometric functions, 112–16; inverse, 116–19, 120–21

Trigonometric integrals, 279–83

Trigonometric reduction, 279–83

Trigonometric substitution, 286–87

Triple integration, 339, 348–49

Uniformly accelerated motion, 83–85

Uniform motion, 82–83

Upper limit, 301–2

Values of e^x and e^{-x}: tables, 379

Variable: change of, 283–87, 312–13; defined, 18–19; dependent, 19; derivative of, 61–62, 64–66; independent, 19, 158–59

Velocities, component, 143–44

Velocity, 81–86, 361–63

Volume: by triple integration, 348–49; of solids of revolution, 332–34; subject to change, 79

Wallis, John (1616–1703), 14–15

Work, 79, 363–65

Zeno of Elea (*c.*490–*c.*430 B.C.), 13

Zero as a limit, 36–40